Situational Awareness and Protection Technologies for Low-Carbon Economic Operation of New Power Systems

Situational Awareness and Protection Technologies for Low-Carbon Economic Operation of New Power Systems

Guest Editors

Yi Wang

Yonghui Sun

Yaoqiang Wang

Yanxun Guo

MDPI

Basel • Beijing • Wuhan • Barcelona • Belgrade • Novi Sad • Cluj • Manchester

Guest Editors
Yi Wang
School of Electrical and Information Engineering
Zhengzhou University
Zhengzhou
China

Yonghui Sun
College of Energy and Electrical Engineering
Hohai University
Nanjing
China

Yaoqiang Wang
School of Electrical and Information Engineering
Zhengzhou University
Zhengzhou
China

Yanxun Guo
School of Electrical and Information Engineering
Zhengzhou University
Zhengzhou
China

Editorial Office
MDPI AG
Grosspeteranlage 5
4052 Basel, Switzerland

This is a reprint of the Special Issue, published open access by the journal *Electronics* (ISSN 2079-9292), freely accessible at: https://www.mdpi.com/journal/electronics/special_issues/Low-Carbon_Power_System.

For citation purposes, cite each article independently as indicated on the article page online and as indicated below:

Lastname, A.A.; Lastname, B.B. Article Title. *Journal Name* **Year**, *Volume Number*, Page Range.

ISBN 978-3-7258-6322-8 (Hbk)
ISBN 978-3-7258-6323-5 (PDF)
https://doi.org/10.3390/books978-3-7258-6323-5

Contents

About the Editors

Yi Wang

Yi Wang is an associate professor with Zhengzhou University, Zhengzhou, China. Between 2018 and 2019, he was a Visiting Scholar with the University of Alberta, Edmonton, AB, Canada. He has authored more than 40 papers in refereed international journals. His research interests include theoretical and algorithmic studies in power system estimation, parameter identification, power system dynamics, signal processing, cyber security, optimal planning, and the operation of integrated energy systems.

Yonghui Sun

Yonghui Sun is currently a Professor with the College of Energy and Electrical Engineering, Hohai University, Nanjing, China. He has authored more than 100 papers in refereed international journals. His research interests include the stability analysis and control of power systems, optimal planning and operation of integrated energy systems, optimization algorithms, and data analysis. Dr. Sun was the recipient of the First Award of Jiangsu Provincial Progress in Science and Technology in 2010 as the fourth project member.

Yaoqiang Wang

Yaoqiang Wang is currently a Professor with the School of Electrical and Information Engineering, Zhengzhou University, Zhengzhou, China. He is also serving as the Director of the Henan Provincial Engineering Research Center of Power Electronics and Energy Systems (HERC-PEES). He has authored more than 60 technical papers, including over 50 journal papers, and is the holder of more than 20 patents. His current research interests include power electronics and electrical drives, renewable power generation, new energy power systems, and power system operation and control.

Yanxun Guo

Yanxun Guo is currently an Associate Professor with the School of Electrical and Information Engineering, Zhengzhou University, Zhengzhou, China. He has authored more than 30 technical papers and is the holder of more than 20 patents. His current research interests include DC circuit breakers, the protection and control of high-voltage DC systems, and the integration of renewable energy.

electronics

MDPI

Article

PredXGBR: A Machine Learning Framework for Short-Term Electrical Load Prediction †

Rifat Zabin [1], Khandaker Foysal Haque [2,*] and Ahmed Abdelgawad [3]

1 Department of Computer Science and Engineering, Brac University, Dhaka 1212, Bangladesh; rifat.zabin@g.bracu.ac.bd
2 Institute for the Wireless Internet of Things, Northeastern University, Boston, MA 02115, USA
3 College of Science and Engineering, Central Michigan University, Mount Pleasant, ME 48849, USA; abdel1a@cmich.edu
* Correspondence: haque.k@northeastern.edu

† This paper is an extended version of our paper published in International Conference on Information and Communication Technology for Development (ICICTD 2022).

Abstract: The growing demand for consumer-end electrical load is driving the need for smarter management of power sector utilities. In today's technologically advanced society, efficient energy usage is critical, leaving no room for waste. To prevent both electricity shortage and wastage, electrical load forecasting becomes the most convenient way out. However, the conventional and probabilistic methods are less adaptive to the acute, micro, and unusual changes in the demand trend. With the recent development of artificial intelligence (AI), machine learning (ML) has become the most popular choice due to its higher accuracy based on time-, demand-, and trend-based feature extractions. Thus, we propose an Extreme Gradient Boosting (XGBoost) regression-based model—PredXGBR-1, which employs short-term lag features to predict hourly load demand. The novelty of PredXGBR-1 lies in its focus on short-term lag autocorrelations to enhance adaptability to micro-trends and demand fluctuations. Validation across five datasets, representing electrical load in the eastern and western USA over a 20-year period, shows that PredXGBR-1 outperforms a long-term feature-based XGBoost model, PredXGBR-2, and state-of-the-art recurrent neural network (RNN) and long short-term memory (LSTM) models. Specifically, PredXGBR-1 achieves an mean absolute percentage error (MAPE) between 0.98 and 1.2% and an R^2 value of 0.99, significantly surpassing PredXGBR-2's R^2 of 0.61 and delivering up to 86.8% improvement in MAPE compared to LSTM models. These results confirm the superior performance of PredXGBR-1 in accurately forecasting short-term load demand.

Keywords: electrical load forecasting; load prediction; XGBoost; regression; ML-based load prediction

1. Introduction

Electricity generation that aligns with fluctuating demand has long been a critical challenge for the power sector. Balancing the needs of industrial and domestic consumers while minimizing excess generation to prevent energy waste is an ongoing struggle. The rapid advancement of technology, coupled with the growing emphasis on sustainable energy, has ushered in numerous innovations within the sector. The integration of photovoltaic systems, wind energy, and other renewable sources has facilitated the development of decentralized, stand-alone grid stations [1]. However, despite these advancements, their full potential remains unrealized if system losses are not adequately mitigated. Consequently, accurate demand prediction has emerged as a key focus area for researchers and industry alike.

While the concept of load prediction is not new, having been applied to grid networks for decades, the evolution of predictive methodologies has been significant. Early approaches relied on qualitative and quantitative methods such as curve fitting, decomposition, regression analysis, and exponential smoothing. These traditional models, while effective to an extent, gave way to more complex statistical techniques like auto regression (AR), the auto regressive moving average (ARMA), the auto regressive integrated moving average (ARIMA), and the support vector machine (SVM), all of which introduced intricate, multivariable mathematical models [2–5]. As the central grid expanded, these models became increasingly prone to NP-hard problems, exacerbating the complexity of demand forecasting.

To address this growing complexity, recent research has shifted toward data-driven, ML-based approaches, which offer the potential to significantly reduce system intricacies [6–9]. Over the past decade, ML has ascended to the forefront of predictive analytics, particularly in time series forecasting. By mimicking human learning processes, ML algorithms process vast datasets, extract features, and gain insights, offering unprecedented computational speed, accuracy, and adaptability [10]. These characteristics have made ML indispensable in a variety of practical applications, from image and handwriting recognition [11] to home automation and IoT-based smart systems, such as waste management [12]. In the realm of load forecasting, ML methods, particularly those involving supervised learning, have proven transformative. By utilizing labeled datasets for training, these methods not only streamline the prediction process but also enhance the speed and robustness of the resulting models.

Most contemporary research on electric load forecasting focuses on models like LSTM, the RNN, and the convolutional neural network (CNN), and statistical methods based on the ARIMA and the SVM. While effective, these approaches often overlook short-term, definite time-lag features—dependencies between data points over short, fixed intervals, such as the relationship between electricity demand at one hour and demand from earlier hours. These time-lagged features are crucial for capturing the immediate effects of factors like weather or peak usage that drive sudden load fluctuations. Without them, models struggle with randomness and nonlinearity, leading to less precise, generalized predictions that fail to account for short-term variations. Neglecting these temporal dynamics results in models that capture broad trends but miss critical short-term fluctuations, particularly in environments with rapidly changing demand. This lack of specificity can reduce the model's effectiveness in real-time scenarios, leading to either over- or under-generation of electricity. Incorporating definite time-lag features would enable models to better predict short-term variations, improving accuracy and reliability, and addressing a crucial gap in current forecasting methodologies. This integration could lead to more adaptive, precise load management, ensuring better efficiency in power distribution.

To address these challenges, we introduce `PredXGBR-1`, an XGBoost-based regression model carefully designed to incorporate short-term lag features, which substantially enhance the model's accuracy and robustness in electric load forecasting. Traditional models often overlook the immediate temporal dependencies critical for capturing sudden shifts in electricity demand. In contrast, `PredXGBR-1` leverages key short-term lag features, including the mean and standard deviation of load data over recent 6, 12, and 24 h periods, to monitor and adapt to these rapid demand changes. By focusing on these short-term intervals, `PredXGBR-1` effectively captures the impact of transient factors—such as abrupt weather fluctuations, spontaneous industrial activity spikes, or peak residential usage—providing a detailed understanding of demand volatility that is crucial for precise, real-time forecasting. This nuanced approach allows `PredXGBR-1` not only to deliver high-resolution forecasts but also to maintain resilience against unpredictable demand patterns that might otherwise lead to inefficiencies in power distribution. The model's predictive accuracy and adaptability have been rigorously validated across five diverse datasets, demonstrating that the incorporation of short-term lag features plays an indispensable role in producing precise, responsive forecasts that support dynamic and efficient power management in real-world scenarios.

In addition to the previously outlined attributes of `PredXGBR-1` and `PredXGBR-2`, we emphasize that the primary distinction lies in their feature selection approaches, specifically focusing on different time-lag intervals to capture unique temporal dependencies in the data. `PredXGBR-1` incorporates only short-term lag features, such as the mean and standard deviation of load over recent intervals like the past 6, 12, and 24 h. This design allows it to adapt swiftly to immediate demand fluctuations, which is crucial for short-term load forecasting. Conversely, `PredXGBR-2` is tailored for scenarios requiring longer forecasting horizons by integrating long-term lag features that extend to broader time frames, including the mean and standard deviation of load data over the previous 24 h and 48 h, and weekly intervals. These long-term features enable `PredXGBR-2` to capture seasonal and weekly patterns that influence load demand, offering stability and improved performance for forecasts that rely on recurring patterns over time. This tailored approach for each model version explains the notable variance in their prediction accuracies across different datasets and time periods, as highlighted in our experimental results.

Summary of Contribution:

- We have developed and implemented `PredXGBR-1`, a short-term feature-based XG-Boost model with time-lagged features. `PredXGBR-1` is designed to capture short-term fluctuations in electricity demand by leveraging data from the previous 24 h, and it has been rigorously evaluated and validated across five different datasets. The integration of time-lagged features significantly improved prediction accuracy, addressing a key gap in existing forecasting methods.
- We performed an extensive analysis to explore how different feature sets influence the performance of the model. This comparative study proved our intuition that short-term lag features are essential for enhancing predictive accuracy, especially in rapidly fluctuating demand environments.
- The proposed model demonstrated high accuracy, achieving an MAPE of 0.98–1.2% across all datasets. This result underscores the model's robustness and reliability for short-term load forecasting in diverse scenarios.
- We pledge to share the whole code repository and the dataset with the community to promote reproducibility and advancements in the field of electric load forecasting.

2. Related Works

In recent years, various approaches of short-term load forecasting have been developed, emerging as some of the most effective methods for electric load prediction. ML and deep learning (DL) models have become prominent due to their ability to handle complex data and provide more accurate forecasts, thus facilitating efficient management, economic dispatch, and scheduling of electrical loads [13]. Load forecasting techniques can generally be categorized into three main groups: statistical models, ML-based models, and hybrid models [14].

One of the most widely used techniques is based on artificial neural networks (ANN)s, which have been found to be highly effective for load forecasting. Aly et al. [15] introduced six hybrid models combining ANNs with Wavelet Neural Network (WNNs) and Kalman Filtering (KF), demonstrating improvements in prediction accuracy. Similarly, Singh et al. [16] conducted a regional load forecasting study for the NEPOOL region of ISO New England, utilizing hourly temperature, humidity, and historical load data. However, this study did not consider yearly holiday schedules, which could have an impact on load prediction. A Boosted Neural Network (BooNN), an enhancement of the traditional ANN, was presented by Khwaja et al. [17]. The model reduced forecasting errors by iteratively improving predictions based on the output of previous iterations. Another popular model is LSTM, which has been widely used for accurate load forecasting. Many researchers have proposed both classic and hybrid models involving LSTM. Marino et al. [18] compared conventional LSTM with a Sequence-to-Sequence (S2S) architecture for individual building-level load forecasting, while Ageng et al. [19] designed an hourly load forecasting model for domestic households, combining LSTM with advanced data preparation strategies. This work also considered the segmentation of a day into patterns such as weekends and weekdays, while addressing data quality issues through interpolation and de-noising.

In a comparative study, Ogunjuyigbe et al. [20] evaluated multiple load forecasting models, including Multiple Linear Regression (MLR), the seasonal auto regressive integrated moving average with exogenous variables (SARIMAX), and LSTM. The study highlighted the limitations of univariate approaches, such as the exclusion of holidays, weather conditions, and climate data. Mubashar et al. [21] conducted another comparison, where different classical models were tested with real-time series data. The study confirmed the superiority of LSTM over traditional models such as exponential smoothing and the ARIMA. Bashir et al. [22] proposed a hybrid model combining the Back Propagation Neural Network (BPNN) with Prophet and LSTM. In this model, datasets were trained using the SARIMA and Prophet models, while the residual nonlinear data were trained using LSTM. The outputs were linearly added together and further optimized with the BPNN model. Neeraj and Mathew [23] developed a Singular Spectrum Analysis–Long Short-Term Memory (SSA-LSTM) model, where the dataset was filtered for noise using signal processing techniques. The model's performance was compared to that of traditional ML and DL models, such as Support Vector Regression (SVR), ANNs, and Deep Belief Networks (DBNs), among others. Although no weather or holiday data were considered in the study, this group validated the model using a diverse set of data. In addition, a Discrete Particle Swarm Optimization (DPSO)-LSTM approach was introduced by Yang et al. [24], where DPSO was employed to optimize the selection of features, improving the model's accuracy for weekly load forecasting. However, weekend and working day distinctions were not included in this approach, and no weather forecasting data were incorporated.

In recent years, RNNs have also gained attention in the field of load forecasting. Kong and Dong [25] studied an RNN-based LSTM model to forecast electrical loads, demonstrating that aggregating individual load forecasts provided more reliable results than directly forecasting aggregated loads. The authors also explored the influence of weather forecasting on load prediction accuracy, showing that models incorporating weather data outperformed those that did not. Additionally, a 2D CNN model was examined [26,27], though it struggled to predict full-day load profiles accurately. CNNs have been applied to load forecasting with promising results. Amarasinghe et al. [18] benchmarked a classical CNN against an LSTM-(S2S) model, showing that CNN models could accurately predict peak load demand for a power station. Ibrahim and Rabelo [28] extended this work by applying various CNN structures, such as multivariate CNN, CNN-LSTM, and multi-headed CNNs. Their study found that the multivariate CNN model outperformed LSTM under both noisy and noise-free conditions.

Recent developments in regression models, particularly XGBoost, have contributed to enhanced load forecasting. Wang et al. [27] proposed a model combining linear regression for trend series and XGBoost for fluctuating sub-series, with data decomposed using variational mode decomposition (VMD) and singular value mode decomposition (SVMD). This approach was particularly effective for industrial load forecasting, where environmental factors such as temperature and holiday schedules were considered. Although the model struggled with nonlinearity and uncertainty in industrial loads, the VMD helped stabilize its performance.

In another study, Zheng et al. [29] introduced a hybrid model involving similar day (SD), empirical mode decomposition (EMD), and LSTM. This model used XGBoost to capture similarities between forecast data and historical data, further improving load prediction accuracy. SVMs have also been widely applied in load forecasting. Barman et al. [29] developed a Grasshopper Optimization Algorithm-based SVM (GOA-SVM) to minimize the deviation between forecasted and actual load curves. The GOA-SVM model was validated by comparing it to a genetic algorithm (GA)-SVM model and particle swarm optimization (PSO)-SVM model, demonstrating the effectiveness of the proposed approach using regional climate data. These relevant works are summarized in Table 1.

Table 1. Summary of related literature.

Research	Model	Contribution	Limitation
Aly et al. [15]	ANN, WNN, KF	Six clustering hybrid models combining ANN, WNN, and KF for load forecasting.	Did not account for weekday/weekend patterns.
Singh et al. [16]	Standard ANN	Regional load forecasting using historical temperature, humidity, and load data.	Excludes yearly holiday schedules; limited datasets.
Khwaja et al. [17]	Boosted ANN	Iterative minimization of forecasting error using BooNN, improving prediction accuracy.	No specific limitations mentioned.
Marino et al. [18]	LSTM, S2S	Comparison of LSTM and S2S architectures for building-level forecasting.	Only focused on a single building-level dataset.
Ageng et al. [19]	LSTM, Data Preparation	Addressed data interpolation and de-noising for household load forecasting.	Weather and atmospheric conditions not considered.
Ogunjuyigbe et al. [20]	LSTM, ARIMA	Comparative analysis of LSTM with ARIMA for improved accuracy.	Limited consideration of holidays and weather data.
Mubashar et al. [21]	MLR, ANN, SVR	Use of Gaussian filtering and validation across academic, commercial, and residential datasets.	Did not consider long holidays or special events.
Bashir & Haoyong [22]	Prophet, LSTM	Hybrid Prophet-LSTM model with residual nonlinear data trained by LSTM.	Excluded weekend/weekday patterns; limited dataset validation.
Neeraj & Mathew [23]	SSA-LSTM	Proposed SSA-LSTM model with noise reduction via signal processing.	No weather- or climate-related data; holidays not considered.
Yang et al. [24]	DPSO-LSTM	Combined DPSO algorithm with LSTM for weekly load forecasting.	Did not distinguish weekday patterns or consider weather conditions.
Kong & Dong [25]	RNN, LSTM	Demonstrated improvement in forecast accuracy when using weather data with RNN-based LSTM.	Limited historical data used in evaluation.
Amarasinghe et al. [18]	CNN	Benchmarking classical CNN against LSTM for peak load demand forecasting.	Model was not validated with diverse datasets.
Imani et al. [30]	CNN, SVR	Proposed Nonlinear Relationship Extraction (NRE) using CNN and SVR for load–temperature correlation.	Socio-demographic data and household occupancy not considered.

Table 1. *Cont.*

Research	Model	Contribution	Limitation
Alhussein et al. [31]	CNN-LSTM	Proposed hybrid CNN-LSTM model for feature extraction and sequence learning.	Did not consider socio-demographic data or household occupancy.
Wang et al. [27]	XGBoost, VMD, SVMD	Hybrid XGBoost model combined with trend analysis using VMD for industrial load forecasting.	Model not tested on multiple datasets.
Zheng et al. [29]	LSTM, XGBoost, EMD	Developed a hybrid model combining EMD, LSTM, and XGBoost for similarity-based forecasting.	No major limitations reported.
Barman et al. [32]	GOA-SVM, GA-SVM, PSO-SVM	Proposed Grasshopper Optimization Algorithm-based SVM for minimizing forecasting errors.	Did not incorporate comprehensive regional climate factors.

3. Background and Preliminaries

Traditionally, electric load forecasting, including daily load demand and long-term load prediction, has relied on statistical and probabilistic models. Notable algorithms in this domain include the ARIMA [33] and SVM [34], which have been widely used in earlier load forecasting approaches, as presented in Section 2.

3.1. ARIMA and Time Series Methods

The ARIMA model is one of the most commonly used methods in time series forecasting [33,35–37]. The ARIMA is designed to handle nonstationary time series data by differencing it to make it stationary. It has been extensively applied in fields such as digital signal processing, economic forecasting, and electric load prediction. Extensions of the ARIMA, including the auto regressive integrated moving average with exogenous variables (ARIMAX) [38,39], which incorporates exogenous variables such as weather data and the ARMA [40,41], which assumes stationary data, are also widely used in load forecasting. The working principle of the ARIMA is presented in Figure 1a. The process begins by checking whether the time series is stationary; if not, differencing or power transformation is applied to achieve stationarity. Once the series is stationary, the model identifies key parameters (P, d, q) through the analysis of the autocorrelation and partial autocorrelation functions. The coefficients are then estimated, followed by a diagnostic check to ensure the model fits the data. If the model passes, it is finalized; otherwise, adjustments are made to improve its accuracy. The ARIMA model uses time and load as its primary input parameters, making it particularly effective for time series analysis where seasonality and trends are key features. Among the ARIMA family, the ARIMAX has gained prominence in electric load forecasting due to its ability to integrate weather variables, which are crucial factors influencing load demand. However, the complexity of interconnected grids and the increasing number of variables in real-world scenarios often lead to large, complex mathematical models, making probabilistic models less practical for some applications.

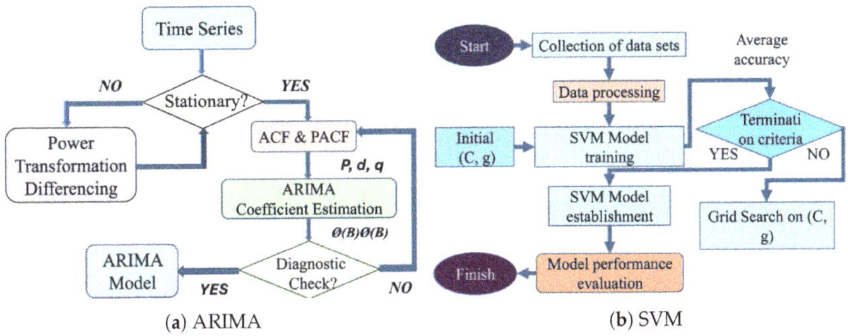

Figure 1. Main steps of ARIMA and SVM.

3.2. SVM

The SVM is another popular approach for load forecasting [42–44]. The SVM is a supervised learning algorithm that excels in both regression and classification tasks. It works by mapping input data to a high-dimensional space, enabling it to classify and predict large volumes of data efficiently. Known for its ability to prevent overfitting, the SVM maximizes the margin between classes, making it particularly effective for large datasets. Kernel methods and large-margin classifiers are central to the SVM's success in handling nonlinear data. Figure 1b depicts the main steps of SVM operation. After collecting and preprocessing the dataset, the SVM initializes hyperparameters (C, g) for the model. The SVM is then trained using the given data, and its performance is evaluated based on criteria such as accuracy. If the model meets the accuracy criteria, it is finalized. Otherwise, a grid search is conducted to optimize the hyperparameters before retraining to improve performance. The SVM has found applications in a wide range of areas, including time series prediction, feature selection, solar and wind energy prediction, lake water level forecasting, and more [45–47]. However, the SVM's scalability can become a limitation when handling the large, interconnected systems typical in modern electric load forecasting.

3.3. DL Approaches

DL, a subset of ML, has emerged as a powerful tool for electric load forecasting. DL models, particularly neural networks, consist of multiple layers that automatically learn features from data. These layers, often referred to as "deep" due to the depth of the network, allow the model to capture complex patterns in the data.

3.3.1. RNN

RNNs are connectionist models designed to process sequential data. Unlike traditional feedforward networks, RNNs maintain internal memory, enabling them to process historical data effectively. This makes them well suited for tasks like electric load forecasting, where past load data are crucial for making predictions [48–50]. Despite their strengths, RNNs were initially difficult to train, limiting their widespread use until recent advancements in training techniques. RNNs are designed to handle sequential data by having connections that loop back within the network. As shown in Figure 2a, the RNN takes an input sequence X and passes it through recurrent layers, where each output O at time step i depends on both the current input x_i and the previous output O_{i-1}, with shared weights w. This recurrent mechanism allows RNNs to capture temporal dependencies in sequences, which makes them suitable for tasks like time series forecasting or natural language processing.

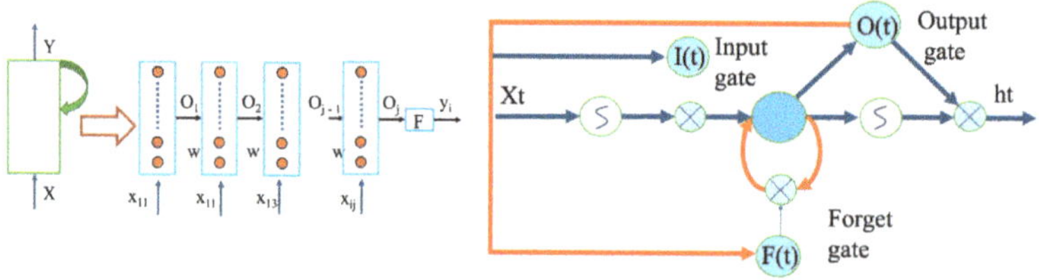

(**a**) RNN (**b**) LSTM

Figure 2. Main steps of RNN and LSTM.

3.3.2. LSTM

LSTM, a variant of RNN, is designed to handle long-term dependencies in data. Its central unit, known as the cell state, acts as memory, retaining relevant information while discarding unnecessary data. This architecture allows LSTM to process historical data more effectively than standard RNNs, making it ideal for applications like handwriting recognition, image processing, and, notably, electric load forecasting [51,52]. Recent developments in LSTM-based algorithms have led to robust models for electric load forecasting [53–55]. LSTM is designed to handle long-term dependencies and mitigate the vanishing gradient problem that can occur in standard RNNs. Figure 2b shows the LSTM's internal architecture, which consists of various gates. The *forget gate* decides what information to discard, the *input gate* regulates the new information added to the cell state, and the *output gate* determines the final output h_t. These gates work together to control the flow of information, making LSTM effective for learning patterns in data over longer time intervals.

3.3.3. Temporal Convolutional Networks (TCN)s

TCNs are convolutional neural network-based architectures specifically designed for sequence modeling tasks, utilizing causal convolutions to capture temporal dependencies in data without the recurrent connections found in RNNs. By stacking layers of 1D convolutions with dilation, TCNs can learn from both short- and long-range dependencies efficiently [56,57]. This makes them suitable for applications like electric load forecasting, where capturing temporal patterns is crucial. Unlike RNNs and LSTM, which rely on sequential data processing, TCNs enable parallel computation, offering lower inference times and improved computational efficiency [58].

3.3.4. Transformer

The Transformer model, introduced by Vaswani et al., revolutionized sequence modeling with its self-attention mechanism, which allows the model to focus on relevant parts of the input sequence without relying on recurrent connections [59]. This attention mechanism enables Transformers to capture dependencies at varying distances in the data, making them highly effective for tasks requiring long-term dependency modeling, such as time series forecasting and natural language processing. Transformers have demonstrated strong predictive performance in electric load forecasting by efficiently handling sequential data with both global and local dependencies [60]. The basic architecture of a Transformer comprises multiple layers of self-attention and feedforward networks, allowing it to process input sequences in parallel, reducing computational complexity and improving scalability.

3.4. XGBoost

XGBoost is a gradient-boosting algorithm known for its high efficiency and predictive accuracy. It uses decision trees as weak predictors and incorporates both linear solvers and tree learning algorithms [61]. The model builds successive trees using residual errors from previous trees, ultimately producing a more optimized prediction. Its application spans various domains, including fingerprint localization, sales forecasting, chronic disease diagnosis, and electric load forecasting [6,62–64]. Due to its ability to handle large datasets with high computational speed, XGBoost has become a preferred model in many forecasting tasks.

While traditional models like the ARIMA and SVM have provided reasonable accuracy in load forecasting, their limitations, particularly with the increasing complexity of interconnected grids, have led to a shift toward ML-based models. Specially, LSTM and XGBoost have been shown to provide superior performance by adapting to complex data patterns and handling nonlinear relationships more effectively.

The superior performance of LSTM and XGBoost in handling nonlinear relationships arises from their unique model architectures and learning mechanisms. Specifically, LSTM networks are designed with memory cells and gating mechanisms that allow them to capture complex, nonlinear dependencies across time steps by retaining relevant information over extended sequences [65,66]. This design enables LSTM models to adapt to intricate patterns in sequential data, making them well suited for electric load forecasting [53,67]. Similarly, XGBoost leverages a gradient boosting framework, which iteratively refines an ensemble of decision trees. Each tree in this ensemble learns from the residual errors of the previous trees, thereby adapting to complex, nonlinear patterns within the data [68]. This boosting technique, combined with efficient computational methods, enables XGBoost to model nonlinear relationships effectively [63]. These theoretical foundations underscore why both LSTM and XGBoost outperform traditional linear methods, particularly in capturing the nonlinear dynamics essential for accurate load forecasting.

4. Proposed Model: `PredXGBR-1`

Short-term electrical load prediction poses significant challenges due to the inherent complexity of the data, which is nonlinear, nonstationary, and often imbalanced. Traditional time series forecasting methods such as the ARIMA and exponential smoothing typically fail to capture the nonlinear dependencies and dynamic patterns present in electrical load data. To address these limitations, we propose `PredXGBR-1`, an XGBoost-based regression model tailored for short-term load forecasting.

The `PredXGBR-1` model utilizes the Extreme Gradient Boosting (XGBoost) framework, a high-performance machine learning algorithm widely recognized for its accuracy, efficiency, and scalability. XGBoost operates on the principle of gradient boosting, where an ensemble of decision trees is sequentially added, with each tree aiming to correct the errors of its predecessors. This iterative process allows the model to continuously refine its predictions, making it exceptionally powerful for capturing complex patterns in time series data, such as the short-term fluctuations in electric load.

A key strength of XGBoost lies in its use of a regularized objective function, which balances predictive accuracy with model complexity. This objective function has two main components: the loss function and the regularization term. The loss function, typically squared error for regression tasks, measures the difference between predicted and actual values, driving the model to improve accuracy. The regularization term, on the other hand, penalizes the complexity of the model, thereby preventing overfitting—a common issue in high-dimensional and noisy datasets like those in electric load forecasting.

The regularization term in XGBoost includes two critical parameters: γ, which controls the number of leaves in each tree, and α, which regulates the magnitude of the weights assigned to these leaves. The parameter γ penalizes the model for adding additional leaves, discouraging overly complex trees, while α ensures that the model does not assign excessively large weights to any leaf. Together, these parameters constrain the model's complexity, ensuring that it generalizes well to new data and remains robust across diverse forecasting scenarios. Additionally, XGBoost's use of shrinkage and subsampling during training further enhances model stability by reducing variance and improving resilience to noise.

The `PredXGBR-1` model benefits from these regularization strategies by achieving a high level of accuracy while maintaining computational efficiency and avoiding overfitting. Specifically tailored for short-term load forecasting, `PredXGBR-1` integrates short-term lag features—such as the mean and standard deviation of load over recent intervals—which makes it responsive to sudden changes in demand. This targeted use of short-term features, combined with XGBoost's regularization framework, enables `PredXGBR-1` to deliver accurate, adaptive predictions in real time, which is essential for efficient energy management in modern power systems.

`PredXGBR-1` leverages short-term lag features to better handle the temporal dependencies and nonlinear relationships within the data while integrating the benefits of gradient-boosting trees to address the challenges of short-term electrical load forecasting.

4.1. Challenges in Short-Term Electrical Load Forecasting

The primary challenges in short-term electrical load forecasting are the following:

- **Nonlinear Relationships:** Electrical load is influenced by various external factors such as weather conditions, time of day, and sudden shifts in demand. These complex and nonlinear relationships are difficult to model using conventional linear methods.
- **Unbalanced Data:** Load datasets are often characterized by periods of stable usage interspersed with sudden spikes or drops in consumption. This imbalance can negatively impact the performance of standard regression models.
- **Temporal Dependencies:** Load at a particular time is dependent on both short-term and long-term historical data, making the selection of features and modeling of temporal dependencies critical.

4.2. How `PredXGBR-1` Addresses These Challenges

To address these challenges, `PredXGBR-1` incorporates several key innovations based on the strengths of the XGBoost algorithm:

- **Tree-Based Regression:** The model employs the classification and regression tree (CART) as a base learner, enabling it to capture complex, nonlinear relationships within the data. The tree-based structure allows the model to perform well in unbalanced datasets by focusing on regions of the data with the highest residuals.
- **Boosting Mechanism:** XGBoost uses boosting to iteratively refine predictions by correcting the residual errors from previous iterations. This iterative process enables `PredXGBR-1` to focus on improving short-term predictions, which are typically more volatile and difficult to forecast.
- **Feature Selection:** The model utilizes short-term lag features—mean and standard deviation of load over the prior 6, 12, and 24 h intervals—which capture the immediate temporal dependencies. This is critical in load forecasting, where short-term variations can greatly impact overall prediction accuracy.

In contrast to conventional XGBoost models, `PredXGBR-1` is tailored specifically for short-term load forecasting through its strategic use of short-term lag features, which allow the model to capture immediate fluctuations and micro-trends in demand. By focusing on these recent temporal dependencies, `PredXGBR-1` can respond quickly to short-term changes—such as peak demand spikes or abrupt shifts driven by weather conditions—offering a more responsive and accurate forecasting capability. This specialized feature selection makes `PredXGBR-1` uniquely suited for short-term predictions, providing it with a nuanced understanding of recent trends that standard XGBoost models, designed without this emphasis on short-term lags, may overlook. As a result, `PredXGBR-1` outperforms traditional approaches in short-term load forecasting, as demonstrated by its consistently lower MAPE and higher R^2 values across datasets.

To address different temporal dependencies, our approach incorporates both short-term and long-term lag features. For the models with long-term features, we consider temporal windows that extend to broader intervals, enabling the model to capture extended seasonal patterns and weekly or monthly trends. Specifically, the long-term lag features include aggregate statistics (mean and standard deviation) of load data over the previous 24 h and 48 h, and weekly intervals. These broader windows provide essential context for models tasked with longer forecasting horizons, allowing them to adapt to recurring cycles in the data.

In contrast, PredXGBR-1 focuses on short-term lag features using intervals from the previous 6, 12, and 24 h. This design choice emphasizes recent data, optimizing the model for accurate short-term forecasting by capturing immediate, short-duration fluctuations.

4.3. Model Structure and Formalization

PredXGBR-1 is a gradient-boosted tree model designed to minimize a convex loss function over time series data. Let N denote the total number of trees in the ensemble, and $f_k(x_i)$ represent the output of the k-th tree for sample x_i. The predicted value $\hat{y}_{x_i}$ for the i-th sample is computed as the sum of the outputs of all trees:

$$\hat{y}_{x_i} = \sum_{k=1}^{N} f_k(x_i), \quad f_k \in \zeta \tag{1}$$

where f_k belongs to the space ζ of regression trees. Unlike decision trees used for classification, regression trees output continuous values, which are better suited for the prediction of time series data. Each tree is built iteratively, with each subsequent tree aiming to correct the errors (residuals) of the previous trees.

4.4. Illustration of the Model Structure

To illustrate the core working mechanism of PredXGBR-1, Figure 3 provides a diagrammatic view of the iterative process. The figure shows how each regression tree in the ensemble progressively adjusts the model's predictions by learning from the residuals of the previous trees. In this framework, the model begins with an initial prediction $\hat{y}_0$, and each subsequent tree $f_k(x;\phi_k)$ adds its output to the prediction based on the residuals from the previous tree.

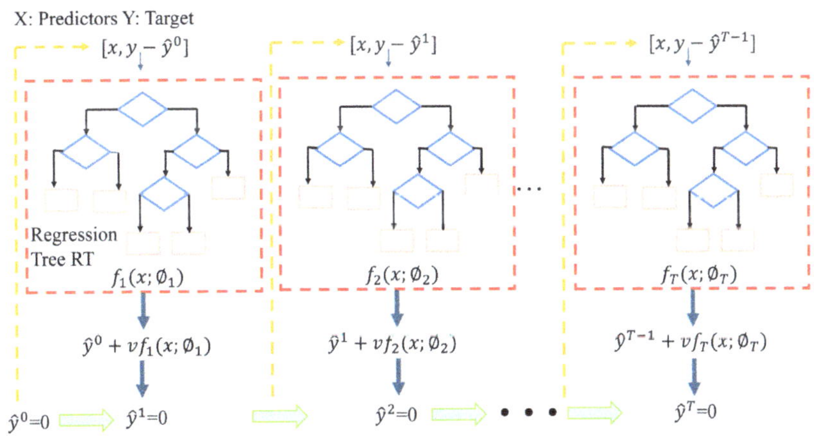

Figure 3. Working principle of the proposed PredXGBR-1 model. The model iteratively refines its prediction by minimizing residuals using successive regression trees. Each new tree improves upon the predictions of its predecessor by learning from the residuals.

In Figure 3, the first tree $f_1(x;\phi_1)$ produces an initial adjustment based on the predictors x. The second tree $f_2(x;\phi_2)$ further refines the prediction $\hat{y}_1$ by addressing the remaining error. This process continues iteratively until the model converges, with the final prediction $\hat{y}_T$ representing the cumulative output of all trees in the ensemble.

4.5. Objective Function

The learning objective is to minimize a regularized loss function that balances predictive accuracy and model complexity. The objective function is defined as

$$\mathcal{L}(\phi) = \sum_{i} l(\hat{y}_i, y_i) + \sum_{k} \Omega(f_k) \quad (2)$$

where $l(\hat{y}_i, y_i)$ is the loss function, typically the squared loss for regression tasks:

$$l(\hat{y}_i, y_i) = \frac{1}{2}(\hat{y}_i - y_i)^2 \quad (3)$$

The term $\Omega(f_k)$ represents the regularization function, which penalizes model complexity to prevent overfitting:

$$\Omega(f_k) = \gamma T + \frac{1}{2}\alpha\|\omega\|^2 \quad (4)$$

Here, T is the number of leaves in the tree, and ω represents the leaf weights. The regularization parameters γ and α help control the depth and weight magnitudes of the trees, respectively, ensuring that the model does not become overly complex and overfit the training data.

4.6. Leaf Weight Optimization

The optimization of the weights assigned to each leaf is critical for improving the predictive power of the model. For each leaf node j, the optimal weight ω_j^* is computed by minimizing the regularized loss function. The weight ω_j^* is given by

$$\omega_j^* = -\frac{\sum_{i \in I_j} g_i}{\sum_{i \in I_j}(h_i + \alpha)} \quad (5)$$

where g_i and h_i are the first and second derivatives of the loss function with respect to the prediction $\hat{y}_i$, known as the gradient and Hessian, respectively. The index I_j denotes the set of instances assigned to leaf j. This equation is derived from a second-order Taylor expansion of the loss function, allowing the model to efficiently minimize the loss while accounting for both the gradient and curvature of the objective function.

4.7. Tree Quality Evaluation

The quality of each regression tree is assessed by evaluating the reduction in the objective function after adding a new tree. The reduction in the residual sum of squares (RSS) is computed as

$$\tilde{\mathcal{L}}_t(q) = -\frac{1}{2}\sum_{j=1}^{T} \frac{\left(\sum_{i \in I_j} g_i\right)^2}{\sum_{i \in I_j}(h_i + \alpha)} + \gamma T \quad (6)$$

This equation represents the improvement in the model's predictions after the addition of each tree, with the regularization term γT acting to penalize overly complex trees.

4.8. Split Candidate Evaluation

The selection of optimal split points is crucial for building effective regression trees. To evaluate the effectiveness of a split, the gain from splitting the data into two subsets, I_L and I_R, is calculated as

$$\mathcal{L}_{\text{split}} = \frac{1}{2}\left[\frac{\left(\sum_{i \in I_L} g_i\right)^2}{\sum_{i \in I_L}(h_i + \alpha)} + \frac{\left(\sum_{i \in I_R} g_i\right)^2}{\sum_{i \in I_R}(h_i + \alpha)} - \frac{\left(\sum_{i \in I} g_i\right)^2}{\sum_{i \in I}(h_i + \alpha)}\right] - \gamma \quad (7)$$

The gain reflects the improvement in the objective function as a result of the split. A higher gain indicates a better split, leading to more accurate predictions and an overall reduction in prediction error.

4.9. Model Generalization and Overfitting Control

A key strength of `PredXGBR-1` is its ability to prevent overfitting through the inclusion of regularization terms in the objective function and the use of early stopping criteria. The parameters γ and α control the complexity of the trees, ensuring that the model does not overfit the training data. Additionally, by monitoring the validation error during training, the model can halt the training process if further iterations do not lead to significant improvements.

`PredXGBR-1` combines the flexibility and power of XGBoost with domain-specific adaptations for short-term load forecasting. The integration of short-term lag features and tree-based regression enables the model to accurately predict short-term fluctuations in electrical load. By employing a robust regularization framework, `PredXGBR-1` ensures that the model generalizes well to unseen data, making it a reliable tool for short-term load forecasting in real-world applications.

5. Datasets and Feature Extraction

To rigorously evaluate the performance of our model, we selected five diverse datasets from centralized grid stations across different regions in the USA. The selection of these datasets was based on their distinct characteristics, such as geographical location, load patterns, and time span, which together provide a comprehensive validation of our model's robustness. Table 2 summarizes each dataset along with its description and time span.

Table 2. Electrical load forecasting datasets for model evaluation.

Dataset	Description	Time Span
PJM—PJM Interconnection LLC [69]	Regional transmission organization (RTO) in the USA, serving Delaware, Illinois, Indiana, Kentucky, Maryland, Michigan, New Jersey, North Carolina, Ohio, etc.	1998–2002
PJME—PJM East [70]	Hourly data from the PJM East region.	2001–2018
PJMW—PJM West [71]	Hourly data from the PJM West region.	2001–2018
AEP—American Electric Power (AEP) [72]	A major investor-owned electric utility in the USA, delivering electricity across 11 states.	2004–2018
Dayton—Dayton Power and Light Company [73]	Serving over 500,000 customers within a 6000-square-mile (16,000 km^2) area in West Central Ohio, around Dayton.	2004–2018

5.1. Data Preprocessing

To ensure the quality and consistency of the data used in our model, we employed a systematic preprocessing approach to address issues such as missing values, timestamp inconsistencies, and temporal variations in load patterns. These preprocessing steps are detailed below, highlighting the methods and rationale behind each step to illustrate their impact on the model's performance:

- **Handling Missing Values:**
  Missing entries in the datasets were primarily due to interruptions in data logging or transmission errors, which can introduce biases and disrupt model training. To address this, we applied two techniques: linear interpolation and forward filling. For extended periods with missing data, linear interpolation was used to generate intermediate values based on surrounding data points, creating smooth transitions and preserving underlying trends. This approach was particularly useful for restoring temporal continuity over multiple time steps. For isolated missing points, forward

filling was employed, where the last available valid value was repeated to fill gaps. This method was beneficial for short, intermittent gaps, ensuring the continuity of time series patterns without distorting the data structure. By addressing missing values, we maintained the temporal integrity of the data, allowing the model to capture continuous patterns in load fluctuations accurately.

- **Organizing Data in Temporal Order:**
  Accurate temporal sequencing is essential for load forecasting models, especially those that rely on time-lagged features to capture dependencies over time. To achieve this, we standardized all timestamps across datasets to a 24 h format using Python's `datetime` module, which enabled us to resolve inconsistencies, such as incorrect AM/PM labels. These inconsistencies, if left unaddressed, could lead to misalignment of hourly data, resulting in misleading trends and poor model performance. After standardizing the timestamps, we sorted records in ascending order by hour, ensuring that each observation followed a natural temporal progression. This careful sequencing allowed our model to accurately interpret time-dependent features and reliably capture the structure of load demand patterns.
- **Removing Duplicate Entries:**
  Duplicate entries in time series data can skew the model's learning process by overemphasizing certain observations, potentially leading to biased predictions. We performed a systematic search for duplicate records within each dataset, focusing on entries with identical timestamps and load values. Once identified, these duplicates were removed to ensure that each data point represented a unique, distinct observation. This step preserved the dataset's integrity, enabling the model to generalize well by learning from an unbiased representation of historical load patterns.
- **Segmentation of Peak and Off-Peak Hours:**
  Load demand often fluctuates significantly between peak and off-peak periods, driven by factors such as residential and industrial activity levels. To capture these fluctuations, we segmented each day's load data into peak and off-peak hours. Specifically, we recorded the maximum load observed during peak hours (typically between 5 p.m. and 9 p.m.) and the minimum load during off-peak hours (usually from midnight to early morning). This segmentation helped the model to distinguish between periods of high and low demand, enhancing its ability to forecast accurately across different times of the day. By providing the model with these segmented values, we enabled it to capture and adapt to the distinct patterns characteristic of peak and off-peak demand, which are critical for short-term load forecasting accuracy.
- **Resolving AM/PM Inconsistencies:**
  Time inconsistencies related to AM/PM formatting were common in some datasets and could interfere with temporal ordering. For instance, an entry incorrectly marked as "PM" instead of "AM" could cause significant deviations in the load pattern analysis, leading to inaccurate predictions. Using the `datetime` module, we converted all timestamps to a uniform 24 h format, thus eliminating ambiguity and ensuring that each record corresponded to the correct time of day. This consistency allowed the model to extract reliable time-dependent features and improved its capability to capture daily load cycles accurately.

Through these comprehensive preprocessing steps, we established a high-quality, consistent foundation for our model's training and evaluation phases. By addressing critical issues such as missing values, duplicate entries, and timestamp inconsistencies, we minimized noise and biases in the data. This systematic preprocessing framework enabled our model, `PredXGBR-1`, to accurately capture short-term fluctuations in load demand, thus enhancing its robustness and predictive power in real-world applications.

5.2. Feature Extraction and Analysis

5.2.1. Seasonal Decomposition

The time series analysis of load consumption has been further explored using *Seasonal Decomposition* to break down the data into its core components: *trend*, *seasonal*, and *residual*. These components provide a deeper insight into the patterns of consumer load behavior over the years. Seasonal decomposition typically employs a moving average method to extract the *trend*, which reflects the long-term direction of the load consumption. The *seasonal* component isolates the recurring patterns that happen at specific times, such as daily, monthly, or yearly cycles, and the *residual* represents the remaining fluctuations once both the trend and seasonal effects have been removed. These residuals help capture any irregularities or anomalies in the data. Figure 4 illustrates the original data along with the *trend*, *periodic*, and *residual* patterns of electrical load consumption for the PJM and Dayton datasets, which are shown here as representative examples due to space limitations, although additional datasets were also analyzed.

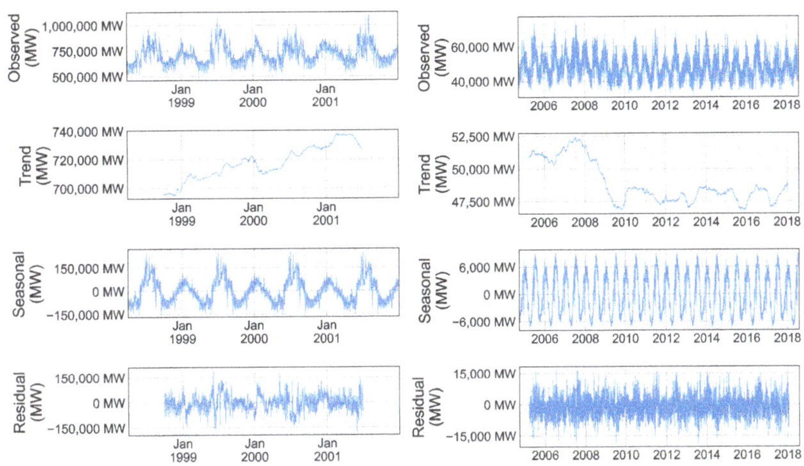

(**a**) PJM (**b**) Dayton

Figure 4. The original data along with the *trend*, *periodic*, and *residual* patterns of electrical load consumption for the PJM and Dayton datasets.

We can observe distinct differences in the load consumption behavior between the *PJM* and *Dayton* datasets. The *PJM dataset* exhibits a steady upward trend in load consumption over the years, reflecting a general increase in energy demand. The seasonal component also shows consistent periodic fluctuations, which are likely influenced by recurring yearly or seasonal cycles. The residuals, while present, do not show significant deviations, indicating relatively stable patterns beyond the trend and seasonality. In contrast, the *Dayton dataset* reveals more variable behavior. The trend shows less consistency, with noticeable shifts, particularly between 2008 and 2010, where a decline in load consumption is evident. Additionally, the seasonal component appears more irregular in amplitude compared to PJM, suggesting more pronounced or unpredictable seasonal effects in the Dayton area. The residuals in the Dayton dataset also display greater variability, highlighting more frequent anomalies or short-term fluctuations that cannot be attributed to trend or seasonality.

Moreover, a gradual upward trend, or gentle acclivity, is observed in the *PJM dataset*, indicating a steady increase in load consumption over time. In contrast, the *AEP dataset* reveals an opposing pattern, where the load consumption trend shows a more declining or stagnant behavior compared to PJM. For the *PJMW dataset*, a sharp rise in load consumption is noticeable between 2005 and 2006, reflecting a sudden increase in energy demand during this period. However, this trend reverses between 2008 and 2010, where a significant decline is evident.

The variation in load consumption trends across these datasets can be attributed to several factors, including changes in human behavior, regional weather conditions, and the broader climate patterns of the respective areas. These factors exert a significant influence on energy usage, highlighting how external conditions and social habits shape the overall trend in energy consumption over time.

5.2.2. Temporal Features of Electric Load Consumption

In this study, three critical features are used to understand the electric load consumption patterns: *hour of the day*, *day of the week*, and *month of the year*. These features offer insights into how electricity demand fluctuates over different temporal dimensions:

- **Hour of the Day:**
  This feature captures how electricity consumption varies throughout the 24 h daily cycle. It provides a granular view of consumption patterns on an hourly basis, which is crucial for identifying peak and off-peak hours. Typically, demand is lower during late night and early morning hours (11 p.m. to 7 a.m.), when most residential, commercial, and industrial activities are minimal. Conversely, consumption often peaks during the morning and early evening, when individuals and businesses are most active. This feature allows for a detailed examination of daily demand cycles and helps in load forecasting and grid management.
- **Day of the Week:**
  The *day of the week* feature distinguishes between weekdays and weekends, capturing the variation in electricity demand that occurs based on the socio-economic activities typically scheduled during the week. Weekdays (Monday through Friday) usually show higher and more stable demand patterns due to the regular operation of industries, offices, and commercial establishments. Weekends, on the other hand, may exhibit a drop in demand, particularly in commercial and industrial sectors, though residential consumption may fluctuate depending on lifestyle habits.
- **Month of the Year:**
  This feature reflects the seasonal variation in electric load consumption over the twelve months, providing insight into how different times of the year impact electricity demand. Seasonal changes drive consumption patterns, with summer months (e.g., July, August) generally showing higher demand due to increased use of cooling systems, while winter months (e.g., December, January) may reflect higher consumption due to heating needs. Transitional seasons, such as fall and spring, tend to exhibit lower and more stable consumption levels compared to the extremes of summer and winter. Analyzing monthly data helps understand the impact of climatic conditions on load demand, allowing for better planning and resource allocation.

This detailed examination of the *hour of the day*, *day of the week*, and *month of the year* features allows for a comprehensive understanding of how electricity consumption fluctuates on multiple temporal scales, which is critical for improving load forecasting models and enhancing grid efficiency. Due to space limitations, Figures 5 and 6 present heatmaps of the PJM and Dayton datasets as representative examples of temporal feature analysis, although other datasets were also considered.

The analysis of load consumption patterns reveals consistent trends across both the PJM and Dayton datasets, particularly in terms of temporal features. **Hourly load variations** are similar between the two regions, with off-peak hours (11:00 p.m. to 7:00 a.m.) showing the lowest demand. This period corresponds to reduced activity in residential, commercial, and industrial sectors. Peak consumption occurs between 11:00 a.m. and 7:00 p.m., driven by daytime activities. These hourly patterns are consistent over multiple years, indicating stable consumption behaviors across both datasets.

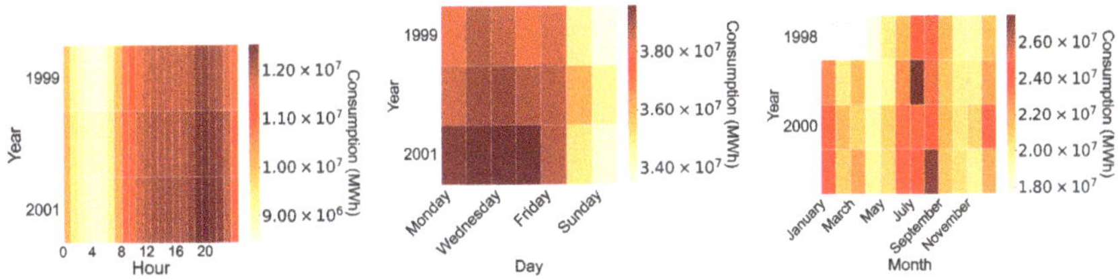

(**a**) Hour of day (**b**) Day of week (**c**) Month of year

Figure 5. Heatmaps of different temporal features of PJM dataset.

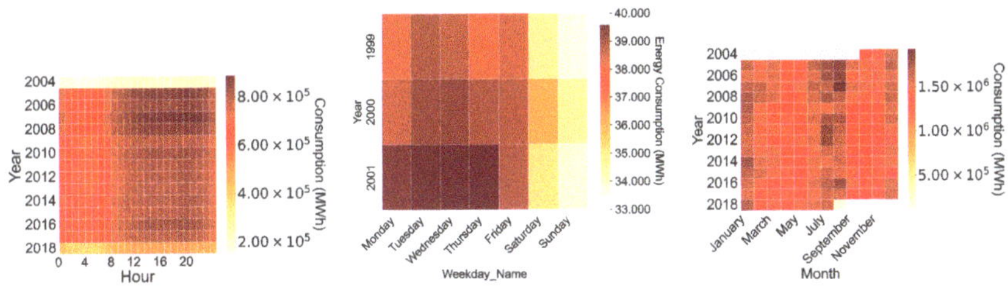

(**a**) Hour of day (**b**) Day of week (**c**) Month of year

Figure 6. Heatmaps of different temporal features of Dayton dataset.

In terms of **weekly consumption patterns**, although not explicitly visualized in the data, it is likely that both datasets follow typical trends, with higher electricity demand observed during weekdays due to increased industrial and commercial activity. Weekends are expected to show relatively lower demand, predominantly reflecting residential consumption, aligning with common load distribution patterns.

Regarding **seasonal variations**, both datasets exhibit notable changes in consumption across the months of the year. Peak demand is observed during the summer months (July and August), driven primarily by cooling needs. Conversely, the fall months (September and October) show a reduction in load due to milder weather conditions. While the overall trends remain consistent, the PJM dataset exhibits more pronounced seasonal fluctuations compared to Dayton, potentially due to regional differences in population density and economic activity.

6. Performance Evaluation

6.1. Evaluation Metrics

To rigorously evaluate the performance of `PredXGBR-1` in short-term load forecasting, we employed the MAPE and coefficient of determination (R^2) as our primary metrics. These metrics were chosen for their interpretability and relevance in assessing predictive accuracy and goodness-of-fit, which are essential for reliable electric load forecasting.

The MAPE provides a measure of prediction accuracy in terms of percentage, making it a useful metric for understanding relative forecasting errors. For each dataset, the MAPE is calculated as follows:

$$\text{MAPE} = \frac{1}{n}\sum_{i=1}^{n}\left|\frac{y_i - \hat{y}_i}{y_i}\right| \times 100 \tag{8}$$

where y_i is the actual load demand for the i-th instance, $\hat{y}_i$ is the predicted load demand for the i-th instance, and n is the total number of instances in the dataset. This metric captures the average percentage deviation between the predicted and actual values, allowing us to compare model performance across datasets with varying load profiles. Lower MAPE values indicate higher accuracy in the model's ability to forecast short-term fluctuations in load demand.

The coefficient of determination (R^2) provides an indication of how well the model's predictions match the actual load demand values. It is defined as

$$R^2 = 1 - \frac{\sum_{i=1}^{n}(y_i - \hat{y}_i)^2}{\sum_{i=1}^{n}(y_i - \bar{y})^2} \quad (9)$$

where y_i is the actual load demand, $\hat{y}_i$ is the predicted load demand, and $\bar{y}$ is the mean of the actual load demand values. The R^2 metric captures the proportion of variance in the actual load that is predictable from the model's outputs, with values closer to 1 indicating a better fit.

For each dataset, MAPE and R^2 values were computed to assess the model's performance across different temporal patterns, providing a comprehensive evaluation of the accuracy and reliability of `PredXGBR-1`. This evaluation approach, applied across five datasets representing diverse load profiles, allowed us to validate the robustness and generalizability of our model.

6.2. Optimal Parameter Selection

To achieve optimal performance for the `PredXGBR-1` model in short-term load forecasting, we carried out a systematic hyperparameter tuning process, focusing on adjusting critical parameters that influence model accuracy and efficiency. This process involved a grid search approach combined with cross-validation, allowing us to test a range of values for each parameter and identify the best configuration.

Key parameters tuned for `PredXGBR-1` included the learning rate, maximum depth of trees, number of estimators, and regularization terms. The learning rate was varied between 0.01 and 0.3 to balance convergence speed and stability. We tested the maximum depth of trees between 3 and 10 to control model complexity and mitigate the risk of overfitting, particularly for datasets with diverse patterns. For the number of estimators, we explored values from 50 to 300, enhancing model robustness while considering computational efficiency. Regularization parameters, such as gamma (γ) and alpha (α), were adjusted to further prevent overfitting by controlling tree complexity and the magnitude of leaf weights.

Each parameter configuration was evaluated using five-fold cross-validation, providing a rigorous assessment of model performance across different data subsets. This cross-validation approach ensured that the final parameter set selected offered the best balance between accuracy and generalizability, thereby reducing the risk of overfitting.

6.3. `PredXGBR` Performance Across Different Datasets

We conducted a comprehensive performance evaluation of the proposed short-term lag-based XGBoost model, `PredXGBR-1`, alongside baselines including SVM, TCN, RNN, LSTM, and Transformer models. This evaluation, conducted across five datasets (PJM [69], PJME [70], PJMW [71], AEP [72], and Dayton [73]), utilized MAPE and R^2 metrics to assess the accuracy of short-term (Model1) and long-term (Model2) lag-based predictions. The visualization in Figure 7 further supports these insights.

The results demonstrate that `PredXGBR-1` consistently achieves the lowest MAPE and highest R^2 values across all datasets, indicating its superior performance in short-term forecasting. For instance, in the PJM dataset, `PredXGBR-1` attains an MAPE of 1.07%, significantly outperforming both the SVM and TCN, which have MAPE values of 5.13% and 19.46%, respectively. This represents an approximate 79% improvement over the SVM, highlighting `PredXGBR-1`'s ability to capture short-term micro-trends effectively.

Additionally, `PredXGBR-1` excels in R^2 performance, achieving values close to 0.99 in datasets like PJM and AEP. In contrast, TCN and Transformer models, while competitive, are less responsive to short-term demand shifts due to their inherent design, which leans toward long-term data patterns. This contrast underscores `PredXGBR-1`'s strength in reacting to immediate fluctuations, an essential characteristic for short-term electric load forecasting.

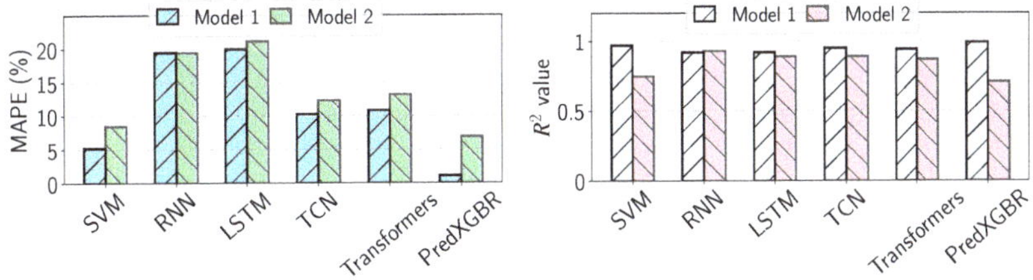

(**a**) PJM dataset (1998–2002)

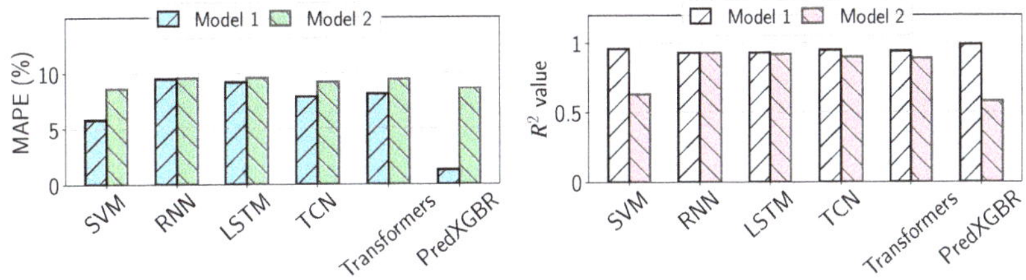

(**b**) PJME dataset (2001–2018)

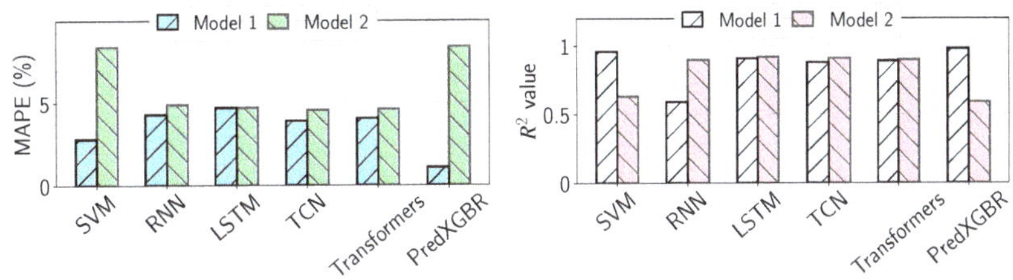

(**c**) PJMW dataset (2001–2018)

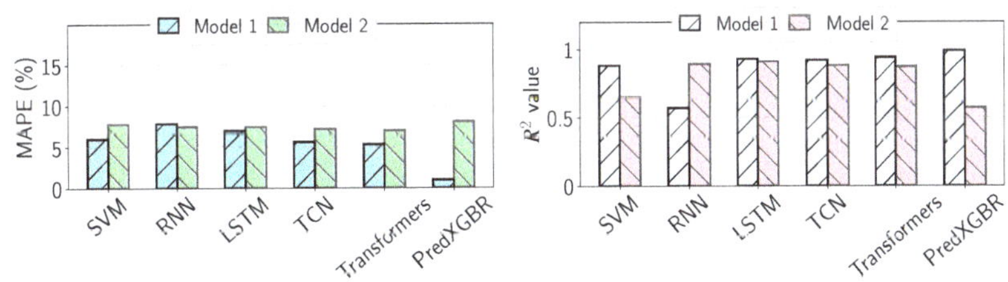

(**d**) AEP dataset (2004–2018)

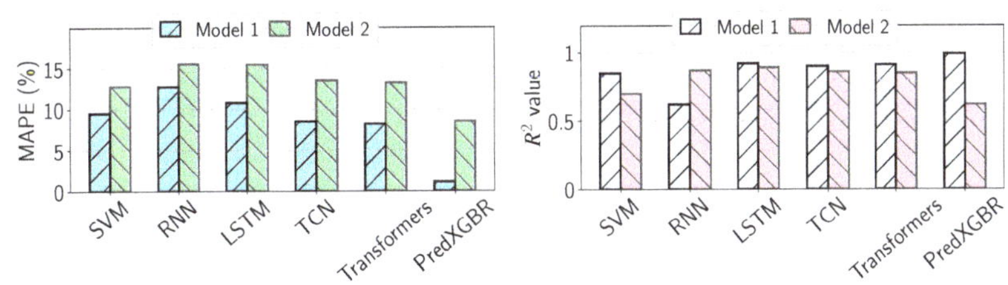

(**e**) Dayton dataset (2004–2018)

Figure 7. Comparative analysis of the MAPE and R^2 value of the proposed approach: `PredXGBR-1`.

The TCN and Transformer models, though effective, do not surpass `PredXGBR-1` in capturing rapid, short-term changes in demand, as seen in more volatile datasets like PJMW. While TCN's sequence-processing architecture offers advantages over RNNs and LSTM, it remains less accurate than `PredXGBR-1` in settings requiring high sensitivity to micro-trends. Similarly, Transformer models excel in complex, long-sequence patterns but are comparatively less effective in addressing immediate changes in short-term load demand.

The observed differences in model performance across datasets, especially for the RNN model, can be attributed to the unique temporal characteristics and demand patterns within each dataset. RNNs, which rely on sequential dependencies to capture temporal patterns, may encounter difficulties in generalizing effectively across datasets with varying seasonality, load volatility, and trend shifts. For example, datasets like PJM and Dayton exhibit more irregular and volatile patterns, posing challenges for RNN models due to their sensitivity to sequential dependencies and susceptibility to issues like vanishing gradients. These factors can result in inconsistent performance across different datasets. In contrast, `PredXGBR-1`, with its XGBoost-based structure, demonstrates greater resilience to these variations. By leveraging short-term lag features and a gradient-boosting framework, `PredXGBR-1` effectively captures nonlinear relationships and adapts to different demand profiles, achieving more stable and accurate predictions across diverse datasets. This adaptability underscores `PredXGBR-1`'s robustness in handling a range of temporal characteristics, setting it apart from traditional sequential models like the RNN.

These findings, summarized in Table 3, confirm that `PredXGBR-1`'s design, centered on short-term lag features, not only yields precise and stable forecasts but also enhances performances across various datasets. This makes `PredXGBR-1` a robust and responsive solution for real-world applications where adaptability and accuracy are paramount.

Table 3. Detailed results for different models and datasets with MAPE and R^2 values for Model1 (short-term lag) and Model2 (long-term lag).

Model	Dataset	Model1 MAPE	Model2 MAPE	Model1 R^2	Model2 R^2
SVM	PJM	5.13	6.87	0.96	0.71
	PJME	5.80	8.59	0.96	0.63
	PJMW	2.80	8.42	0.96	0.63
	AEP	6.23	8.08	0.94	0.57
	Dayton	7.36	8.49	0.93	0.62
RNN	PJM	19.46	19.44	0.92	0.93
	PJME	9.49	9.58	0.93	0.93
	PJMW	4.28	4.87	0.59	0.90
	AEP	7.86	7.49	0.57	0.89
	Dayton	12.74	15.54	0.62	0.87
LSTM	PJM	19.96	21.12	0.92	0.89
	PJME	9.21	9.57	0.93	0.92
	PJMW	4.70	4.71	0.91	0.92
	AEP	7.00	7.46	0.93	0.91
	Dayton	10.80	15.46	0.92	0.89
TCN	PJM	19.46	19.44	0.92	0.93
	PJME	7.85	9.20	0.95	0.90
	PJMW	3.90	4.55	0.88	0.91
	AEP	7.86	7.49	0.57	0.89
	Dayton	12.74	15.54	0.62	0.87

Table 3. *Cont.*

Model	Dataset	Model1 MAPE	Model2 MAPE	Model1 R^2	Model2 R^2
Transformer	PJM	19.96	21.12	0.92	0.89
	PJME	8.10	9.45	0.94	0.89
	PJMW	4.05	4.60	0.89	0.90
	AEP	7.00	7.46	0.93	0.91
	Dayton	10.80	15.46	0.92	0.89
PredXGBR	PJM	**1.07**	6.87	**0.99**	0.71
	PJME	**1.28**	8.59	**0.99**	0.58
	PJMW	**1.07**	8.42	**0.98**	0.59
	AEP	**0.98**	8.08	**0.99**	0.57
	Dayton	**1.12**	8.49	**0.99**	0.62

6.4. `PredXGBR` Generalization Performance

The generalization performance of the proposed `PredXGBR-1` model, as shown in Figure 8, underscores its effectiveness in maintaining accuracy when trained on one dataset and tested on others. Across all scenarios, `PredXGBR-1` achieves lower MAPE and higher R^2 values compared to the baseline models SVM-1 and TCN-1, indicating a clear advantage in adapting to unseen data.

In detail, when trained on the PJM dataset, `PredXGBR-1` demonstrates substantial improvements in MAPE across all test datasets, consistently achieving values below 5%, whereas SVM-1 and TCN-1 reach MAPE values between 7% and 10% for most cases. This pattern persists when trained on other datasets such as PJME, PJMW, and AEP, where `PredXGBR-1` consistently records a lower MAPE, often by a margin of 3–5% compared to SVM-1 and TCN-1. For instance, when trained on the AEP dataset, `PredXGBR-1` achieves an MAPE of around 1% on other datasets, while SVM-1 and TCN-1 exhibit higher MAPE values, underscoring `PredXGBR-1`'s robustness in preserving forecast accuracy across different test sets.

In terms of R^2 values, `PredXGBR-1` maintains values close to or above 0.95 across most test scenarios, highlighting its strong correlation with actual data. In contrast, SVM-1 and TCN-1 show more variability, with R^2 values occasionally dropping below 0.9, particularly in cross-dataset scenarios. This discrepancy reinforces that `PredXGBR-1` captures and generalizes short-term load variations better than the baseline models.

Overall, Figure 8 demonstrates that `PredXGBR-1` not only achieves a lower MAPE but also maintains high R^2 values across diverse datasets, emphasizing its superior generalization capability. This performance makes `PredXGBR-1` a more reliable choice for applications requiring stable short-term forecasting across regions with varying load characteristics.

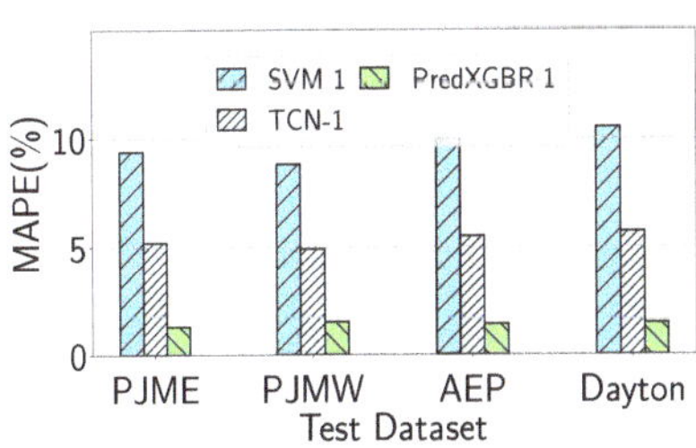

(**a**) MAPE—trained with PJM dataset

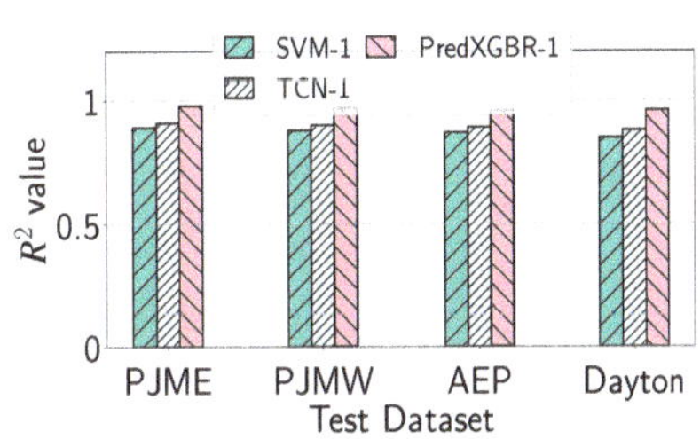

(**b**) R^2 value—trained with PJM dataset

Figure 8. *Cont.*

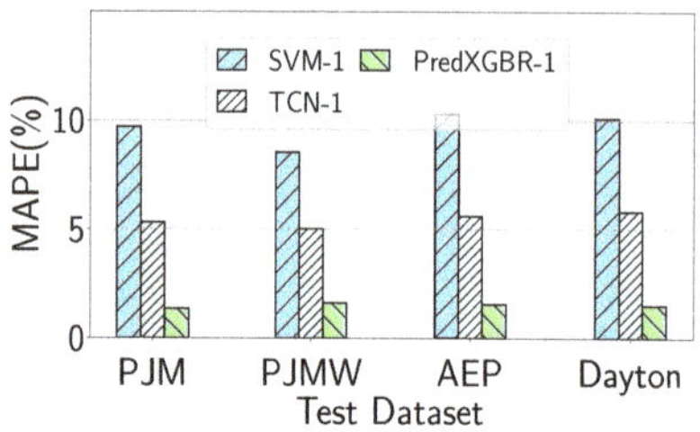

(**c**) MAPE—trained with PJME dataset

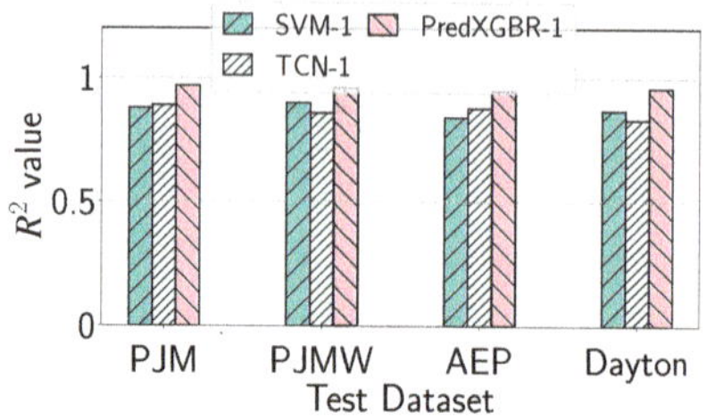

(**d**) R^2 value—trained with PJME dataset

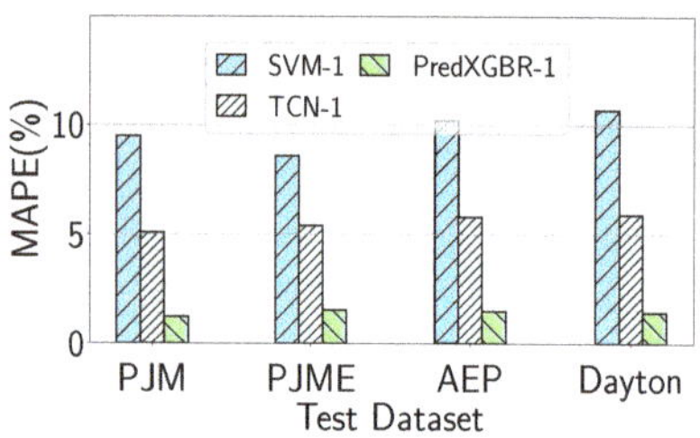

(**e**) MAPE—trained with PJMW dataset

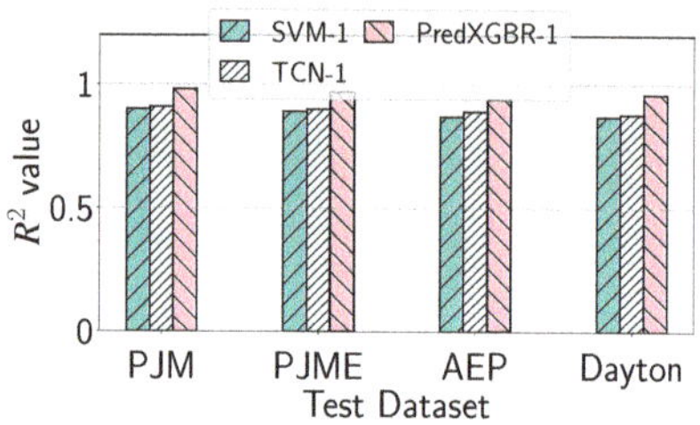

(**f**) R^2 value—trained with PJMW dataset

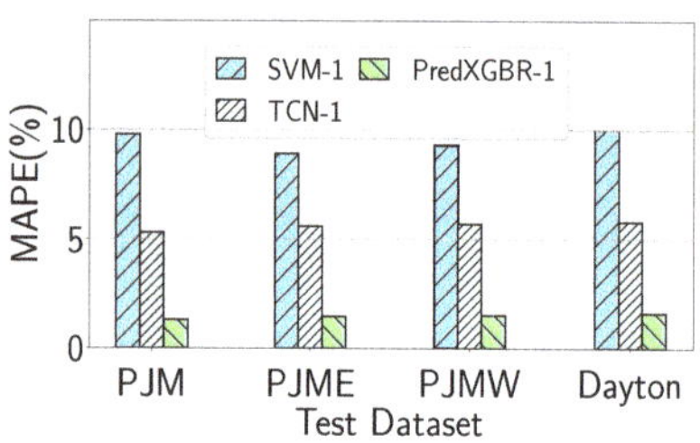

(**g**) MAPE—trained with AEP dataset

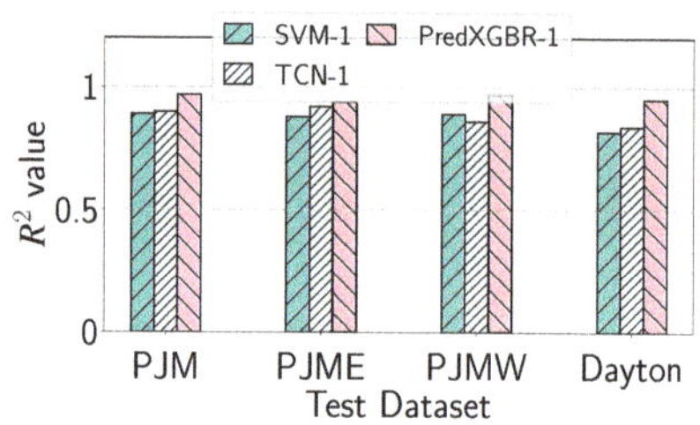

(**h**) R^2 value—trained with AEP dataset

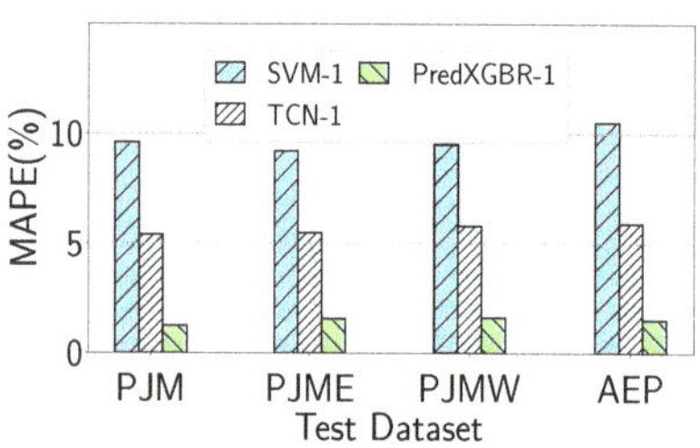

(**i**) MAPE—trained with DAYTON dataset

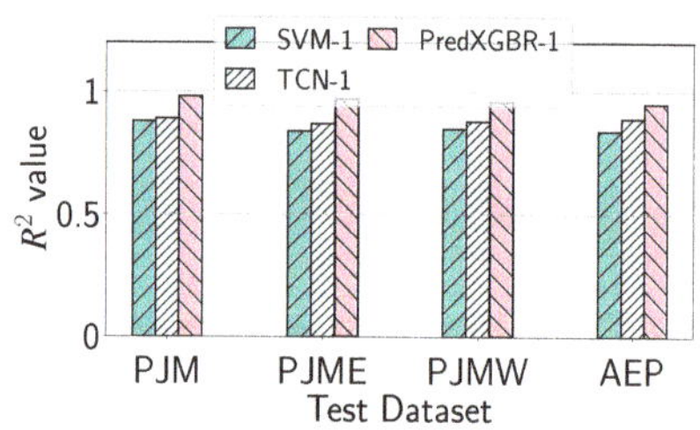

(**j**) R^2 value–trained with DAYTON dataset

Figure 8. Analysis of the generalization performance of `PredXGBR-1` when compared with two of the best-performing models—SVM and TCN. Models are trained with one dataset and tested with others.

6.5. Computational Complexity and Inference Time

In this section, we analyze the computational complexity and inference time of the evaluated models, as shown in Figure 9. The computational complexity, represented in giga floating point operations per second (GFLOPS), estimates the processing power required by each model, while inference time (in milliseconds) reflects the time taken for each model to generate predictions. These inference times were measured on a Linux-based workstation equipped with a 12th Gen Intel(R) Core(TM) i7-12700K CPU, an NVIDIA RTX A4000 GPU, and 64 GB of RAM.

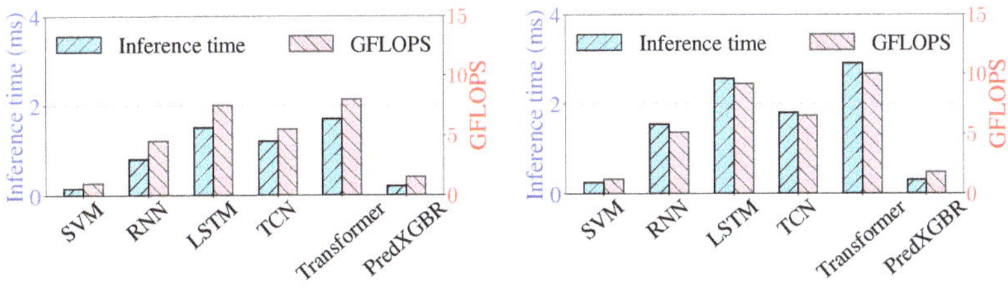

(**a**) Model1 (based on short-term lag feature) (**b**) Model2 (based on long-term lag feature)

Figure 9. Comparative analysis of the computational complexity (FLOPS) and inference time of `PredXGBR`-1 (Model1).

The results indicate that models with long-term lag features, such as Transformer-2 and LSTM-2, demand higher computational resources and exhibit longer inference times compared to their short-term counterparts. For instance, Transformer-2 has the highest computational burden at 10 GFLOPS and an inference time of 2.9 ms, followed closely by LSTM-2 at 9.18 GFLOPS and 2.56 ms. This elevated demand stems from the need to capture extended temporal patterns (e.g., weekly and monthly trends), which increases input dimensionality and processing requirements.

Conversely, short-term lag models, particularly `PredXGBR`-1, are significantly more efficient. `PredXGBR`-1, with a computational burden of only 1.5 GFLOPS and an inference time of 0.2 ms, is approximately 93% faster in inference time and 85% less computationally demanding than Transformer-2. This efficiency makes `PredXGBR`-1 well suited for real-time applications, where low latency is crucial. Among neural network models, TCN-1 and SVM-1 also show competitive performance, with TCN-1 requiring 5.5 GFLOPS and an inference time of 1.2 ms, and SVM-1 being the most efficient at only 1.0 GFLOPS and an inference time of 0.15 ms, underscoring their potential for low-resource environments.

Overall, while long-term lag models provide a more comprehensive temporal context, they incur higher computational costs. Short-term lag models, particularly `PredXGBR`-1, achieve a balance between computational efficiency and predictive accuracy, making them ideal for scenarios requiring rapid responses and low-latency predictions.

7. Conclusions

In this paper, we introduced a short-term feature-based XGBoost model, `PredXGBR`-1, designed to address the limitations of traditional and probabilistic methods in electric load forecasting. By leveraging time-lagged features from the previous 24 h, our model captures short-term fluctuations in demand with remarkable accuracy. We rigorously evaluated `PredXGBR`-1 across five different datasets representing twenty years of electrical load data from various regions in the USA. Our comparative analysis between long-term and short-term feature-based models highlighted the significance of focusing on short-term lag features. While traditional models like the RNN and LSTM demonstrated moderate accuracy, `PredXGBR`-1 consistently outperformed them with an average MAPE of 0.98–1.2% and an R^2 score close to 1, signifying near-perfect predictive performance. This level of accuracy makes `PredXGBR`-1 a highly reliable model for real-time, short-term load forecasting, especially in environments with rapidly fluctuating demand. The findings presented in this paper underscore the robustness and adaptability of `PredXGBR`-1, paving the way for more efficient and precise forecasting in the electric power sector.

Author Contributions: Conceptualization, methodology: R.Z. and K.F.H.; Algorithm formation, formal analysis, resources, and data processing, R.Z.; Validation, original draft, and review preparation, R.Z., K.F.H. and A.A. All authors have read and agreed to the published version of the manuscript.

Funding: This research received no external funding.

Data Availability Statement: The datasets and algorithms supporting the reported results can be found at https://github.com/rifatzabin/PredXGBR accessed on 8 November 2024.

Conflicts of Interest: The authors declare no conflict of interest.

References

1. Saqib, N.; Haque, K.F.; Zabin, R.; Preonto, S.N. Analysis of grid integrated PV system as home RES with net metering scheme. In Proceedings of the 2019 International Conference on Robotics, Electrical and Signal Processing Techniques (ICREST), Dhaka, Bangladesh, 10–12 January 2019; pp. 395–399.
2. Rodrigues, F.; Cardeira, C.; Calado, J.M.; Melicio, R. Short-term load forecasting of electricity demand for the residential sector based on modelling techniques: A systematic review. *Energies* **2023**, *16*, 4098. [CrossRef]
3. Habbak, H.; Mahmoud, M.; Metwally, K.; Fouda, M.M.; Ibrahem, M.I. Load forecasting techniques and their applications in smart grids. *Energies* **2023**, *16*, 1480. [CrossRef]
4. Akhtar, S.; Shahzad, S.; Zaheer, A.; Ullah, H.S.; Kilic, H.; Gono, R.; Jasiński, M.; Leonowicz, Z. Short-term load forecasting models: A review of challenges, progress, and the road ahead. *Energies* **2023**, *16*, 4060. [CrossRef]
5. Eren, Y.; Küçükdemiral, İ. A comprehensive review on deep learning approaches for short-term load forecasting. *Renew. Sustain. Energy Rev.* **2024**, *189*, 114031. [CrossRef]
6. Cordeiro-Costas, M.; Villanueva, D.; Eguía-Oller, P.; Martínez-Comesaña, M.; Ramos, S. Load forecasting with machine learning and deep learning methods. *Appl. Sci.* **2023**, *13*, 7933. [CrossRef]
7. Avendano, I.A.C.; Javan, F.D.; Najafi, B.; Moazami, A.; Rinaldi, F. Assessing the impact of employing machine learning-based baseline load prediction pipelines with sliding-window training scheme on offered flexibility estimation for different building categories. *Energy Build.* **2023**, *294*, 113217. [CrossRef]
8. Zhang, D.; Wang, S.; Liang, Y.; Du, Z. A novel combined model for probabilistic load forecasting based on deep learning and improved optimizer. *Energy* **2023**, *264*, 126172. [CrossRef]
9. Aduama, P.; Zhang, Z.; Al-Sumaiti, A.S. Multi-feature data fusion-based load forecasting of electric vehicle charging stations using a deep learning model. *Energies* **2023**, *16*, 1309. [CrossRef]
10. Zhang, L.; Wen, J.; Li, Y.; Chen, J.; Ye, Y.; Fu, Y.; Livingood, W. A Review of Machine Learning in Building Load Prediction. *Appl. Energy* **2021**, *285*, 116452. [CrossRef]
11. Saqib, N.; Haque, K.F.; Yanambaka, V.P.; Abdelgawad, A. Convolutional-neural-network-based handwritten character recognition: an approach with massive multisource data. *Algorithms* **2022**, *15*, 129. [CrossRef]
12. Haque, K.F.; Zabin, R.; Yelamarthi, K.; Yanambaka, P.; Abdelgawad, A. An IoT based efficient waste collection system with smart bins. In Proceedings of the 2020 IEEE 6th World Forum on Internet of Things (WF-IoT), New Orleans, LA, USA, 2–16 June 2020; pp. 1–5.
13. Chen, K.; Chen, K.; Wang, Q.; He, Z.; Hu, J.; He, J. Short-Term Load Forecasting with Deep Residual Networks. *IEEE Trans. Smart Grid* **2018**, *10*, 3943–3952. [CrossRef]
14. Hammad, M.A.; Jereb, B.; Rosi, B.; Dragan, D. Methods and Mdels for Electric Load Forecasting: A Comprehensive Review. *Logist. Sustain. Transp.* **2020**, *11*, 51–76. [CrossRef]
15. Aly, H.H. A Proposed Intelligent Short-Term Load Forecasting Hybrid Models of ANN, WNN and KF based on Clustering Techniques for Smart Grid. *Electr. Power Syst. Res.* **2020**, *182*, 106191. [CrossRef]
16. Singh, S.; Hussain, S.; Bazaz, M.A. Short Term Load Forecasting using Artificial Neural Network. In Proceedings of the 2017 Fourth International Conference on Image Information Processing (ICIIP), Shimla, India, 21–23 December 2017; pp. 1–5.
17. Khwaja, A.; Zhang, X.; Anpalagan, A.; Venkatesh, B. Boosted Neural Networks for Improved Short-Term Electric Load Forecasting. *Electr. Power Syst. Res.* **2017**, *143*, 431–437. [CrossRef]
18. Amarasinghe, K.; Marino, D.L.; Manic, M. Deep Neural Networks for Energy Load Forecasting. In Proceedings of the 2017 IEEE 26th International Symposium on Industrial Electronics (ISIE), Edinburgh, UK, 19–21 June 2017; pp. 1483–1488.
19. Ageng, D.; Huang, C.Y.; Cheng, R.G. A Short-Term Household Load Forecasting Framework using LSTM and Data Preparation. *IEEE Access* **2021**, *9*, 167911–167919. [CrossRef]
20. Ogunjuyigbe, A.S.; Ayodele, T.R.; Lasarus, C.P.; Yusuff, A.A.; Mosetlhe, T.C. Comparative Analysis of Short-Term Load Forecasting Methods. In Proceedings of the 2021 IEEE AFRICON, Arusha, Tanzania, 13–15 September 2021; pp. 1–6.
21. Mubashar, R.; Awan, M.J.; Ahsan, M.; Yasin, A.; Singh, V.P. Efficient residential load forecasting using deep learning approach. *Int. J. Comput. Appl. Technol.* **2022**, *68*, 205–214. [CrossRef]
22. Bashir, T.; Haoyong, C.; Tahir, M.F.; Liqiang, Z. Short term electricity load forecasting using hybrid prophet-LSTM model optimized by BPNN. *Energy Rep.* **2022**, *8*, 1678–1686. [CrossRef]
23. Neeraj, N.; Mathew, J.; Agarwal, M.; Behera, R.K. Long short-term memory-singular spectrum analysis-based model for electric load forecasting. *Electr. Eng.* **2021**, *103*, 1067–1082. [CrossRef]
24. Yang, J.; Zhang, X.; Bao, Y. Short-term load forecasting of central China based on DPSO-LSTM. In Proceedings of the 2021 IEEE 4th International Electrical and Energy Conference (CIEEC), Wuhan, China, 28–30 May 2021; pp. 1–5.

25. Nespoli, A.; Ogliari, E.; Pretto, S.; Gavazzeni, M.; Vigani, S.; Paccanelli, F. Electrical load forecast by means of lstm: The impact of data quality. *Forecasting* **2021**, *3*, 91–101. [CrossRef]
26. Orr, M. Short-Term Electrical Load Forecasting for Irish Supermarkets with Weather Forecast Data. Ph.D. Thesis, National College of Ireland, Dublin, Ireland, 2021.
27. Wang, Y.; Sun, S.; Chen, X.; Zeng, X.; Kong, Y.; Chen, J.; Guo, Y.; Wang, T. Short-term load forecasting of industrial customers based on SVMD and XGBoost. *Int. J. Electr. Power Energy Syst.* **2021**, *129*, 106830. [CrossRef]
28. Ibrahim, B.; Rabelo, L. A deep learning approach for peak load forecasting: A case study on panama. *Energies* **2021**, *14*, 3039. [CrossRef]
29. Zheng, H.; Yuan, J.; Chen, L. Short-term load forecasting using EMD-LSTM neural networks with a Xgboost algorithm for feature importance evaluation. *Energies* **2017**, *10*, 1168. [CrossRef]
30. Imani, M. Electrical load-temperature CNN for residential load forecasting. *Energy* **2021**, *227*, 120480. [CrossRef]
31. Alhussein, M.; Aurangzeb, K.; Haider, S.I. Hybrid CNN-LSTM model for short-term individual household load forecasting. *IEEE Access* **2020**, *8*, 180544–180557. [CrossRef]
32. Barman, M.; Choudhury, N.D.; Sutradhar, S. A regional hybrid GOA-SVM model based on similar day approach for short-term load forecasting in Assam, India. *Energy* **2018**, *145*, 710–720. [CrossRef]
33. Chodakowska, E.; Nazarko, J.; Nazarko, Ł. Arima models in electrical load forecasting and their robustness to noise. *Energies* **2021**, *14*, 7952. [CrossRef]
34. Mohandes, M. Support vector machines for short-term electrical load forecasting. *Int. J. Energy Res.* **2002**, *26*, 335–345. [CrossRef]
35. Kontopoulou, V.I.; Panagopoulos, A.D.; Kakkos, I.; Matsopoulos, G.K. A review of ARIMA vs. machine learning approaches for time series forecasting in data driven networks. *Future Internet* **2023**, *15*, 255. [CrossRef]
36. Sharma, R.R.; Kumar, M.; Maheshwari, S.; Ray, K.P. EVDHM-ARIMA-based time series forecasting model and its application for COVID-19 cases. *IEEE Trans. Instrum. Meas.* **2020**, *70*, 6502210. [CrossRef]
37. Sirisha, U.M.; Belavagi, M.C.; Attigeri, G. Profit prediction using ARIMA, SARIMA and LSTM models in time series forecasting: A comparison. *IEEE Access* **2022**, *10*, 124715–124727. [CrossRef]
38. Atabay, F.V.; Pagkalinawan, R.M.; Pajarillo, S.D.; Villanueva, A.R.; Taylar, J.V. Multivariate time series forecasting using arimax, sarimax, and rnn-based deep learning models on electricity consumption. In Proceedings of the 2022 3rd International Informatics and Software Engineering Conference (IISEC), Ankara, Turkey, 15–16 December 2022; pp. 1–6.
39. Ahn, E.; Hur, J. A short-term forecasting of wind power outputs using the enhanced wavelet transform and arimax techniques. *Renew. Energy* **2023**, *212*, 394–402. [CrossRef]
40. He, K.; Yang, Q.; Ji, L.; Pan, J.; Zou, Y. Financial time series forecasting with the deep learning ensemble model. *Mathematics* **2023**, *11*, 1054. [CrossRef]
41. Song, Z.; Yang, L. Statistical inference for ARMA time series with moving average trend. *J. Nonparametr. Stat.* **2022**, *34*, 357–376. [CrossRef]
42. Ahmad, W.; Ayub, N.; Ali, T.; Irfan, M.; Awais, M.; Shiraz, M.; Glowacz, A. Towards short term electricity load forecasting using improved support vector machine and extreme learning machine. *Energies* **2020**, *13*, 2907. [CrossRef]
43. Aisyah, S.; Simaremare, A.A.; Adytia, D.; Aditya, I.A.; Alamsyah, A. Exploratory weather data analysis for electricity load forecasting using SVM and GRNN, case study in Bali, Indonesia. *Energies* **2022**, *15*, 3566. [CrossRef]
44. Emhamed, A.A.; Shrivastava, J. Electrical load distribution forecasting utilizing support vector model (SVM). *Mater. Today Proc.* **2021**, *47*, 41–46. [CrossRef]
45. Pant, M.; Kumar, S. Fuzzy time series forecasting based on hesitant fuzzy sets, particle swarm optimization and support vector machine-based hybrid method. *Granul. Comput.* **2022**, *7*, 861–879. [CrossRef]
46. Gao, W.; Li, Z.; Chen, Q.; Jiang, W.; Feng, Y. Modelling and prediction of GNSS time series using GBDT, LSTM and SVM machine learning approaches. *J. Geod.* **2022**, *96*, 71. [CrossRef]
47. Ramadevi, B.; Bingi, K. Chaotic time series forecasting approaches using machine learning techniques: A review. *Symmetry* **2022**, *14*, 955. [CrossRef]
48. Aseeri, A.O. Effective RNN-based forecasting methodology design for improving short-term power load forecasts: Application to large-scale power-grid time series. *J. Comput. Sci.* **2023**, *68*, 101984. [CrossRef]
49. Zhao, C.; Ye, J.; Zhu, Z.; Huang, Y. FLRNN-FGA: Fractional-Order Lipschitz Recurrent Neural Network with Frequency-Domain Gated Attention Mechanism for Time Series Forecasting. *Fractal Fract.* **2024**, *8*, 433. [CrossRef]
50. Zhang, X.; Zhong, C.; Zhang, J.; Wang, T.; Ng, W.W. Robust recurrent neural networks for time series forecasting. *Neurocomputing* **2023**, *526*, 143–157. [CrossRef]
51. Misgar, M.M.; Mushtaq, F.; Khurana, S.S.; Kumar, M. Recognition of offline handwritten Urdu characters using RNN and LSTM models. *Multimed. Tools Appl.* **2023**, *82*, 2053–2076. [CrossRef]
52. Masood, F.; Khan, W.U.; Ullah, K.; Khan, A.; Alghamedy, F.H.; Aljuaid, H. A hybrid CNN-LSTM random forest model for dysgraphia classification from hand-written characters with uniform/normal distribution. *Appl. Sci.* **2023**, *13*, 4275. [CrossRef]
53. Abumohsen, M.; Owda, A.Y.; Owda, M. Electrical load forecasting using LSTM, GRU, and RNN algorithms. *Energies* **2023**, *16*, 2283. [CrossRef]
54. Zaboli, A.; Tuyet-Doan, V.N.; Kim, Y.H.; Hong, J.; Su, W. An lstm-sae-based behind-the-meter load forecasting method. *IEEE Access* **2023**, *11*, 49378–49392. [CrossRef]

55. Pavlatos, C.; Makris, E.; Fotis, G.; Vita, V.; Mladenov, V. Enhancing Electrical Load Prediction Using a Bidirectional LSTM Neural Network. *Electronics* **2023**, *12*, 4652. [CrossRef]
56. Bai, S.; Kolter, J.Z.; Koltun, V. An empirical evaluation of generic convolutional and recurrent networks for sequence modeling. *arXiv* **2018**, arXiv:1803.01271.
57. Wu, Z.; Pan, S.; Long, G.; Jiang, J.; Zhang, C. Graph wavenet for deep spatial-temporal graph modeling. *arXiv* **2019**, arXiv:1906.00121.
58. Oreshkin, B.N.; Carpov, D.; Chapados, N.; Bengio, Y. N-BEATS: Neural basis expansion analysis for interpretable time series forecasting. *arXiv* **2019**, arXiv:1905.10437.
59. Vaswani, A. Attention Is All You Need. *Advances in Neural Information Processing Systems*. 2017. Available online: https://user.phil.hhu.de/~cwurm/wp-content/uploads/2020/01/7181-attention-is-all-you-need.pdf (accessed on 20 June 2022).
60. Zhou, H.; Zhang, S.; Peng, J.; Zhang, S.; Li, J.; Xiong, H.; Zhang, W. Informer: Beyond efficient transformer for long sequence time-series forecasting. In Proceedings of the AAAI Conference on Artificial Intelligence, Virtually, 2–9 February 2021; Volume 35, pp. 11106–11115.
61. Chen, T.; Guestrin, C. Xgboost: A scalable tree boosting system. In Proceedings of the 22nd ACM Sigkdd International Conference on Knowledge Discovery and Data Mining, San Francisco, CA, USA, 13–17 August 2016; pp. 785–794.
62. Oukhouya, H.; El Himdi, K. Comparing machine learning methods—svr, xgboost, lstm, and mlp—for forecasting the moroccan stock market. In Proceedings of the Computer Sciences & Mathematics Forum, Online, 1–5 May 2023; MDPI: Basel, Switzerland, 2023; Volume 7, p. 39.
63. Zhang, T.; Zhang, X.; Liu, Y.; Chow, Y.H.; Iu, H.H.; Fernando, T. Long-term energy and peak power demand forecasting based on sequential-XGBoost. *IEEE Trans. Power Syst.* **2023**, *39*, 3088–3104. [CrossRef]
64. Zhang, L.; Jánošík, D. Enhanced short-term load forecasting with hybrid machine learning models: CatBoost and XGBoost approaches. *Expert Syst. Appl.* **2024**, *241*, 122686. [CrossRef]
65. Kumar, B.; Yadav, N. A novel hybrid model combining βSARMA and LSTM for time series forecasting. *Appl. Soft Comput.* **2023**, *134*, 110019. [CrossRef]
66. Kumar, I.; Tripathi, B.K.; Singh, A. Attention-based LSTM network-assisted time series forecasting models for petroleum production. *Eng. Appl. Artif. Intell.* **2023**, *123*, 106440. [CrossRef]
67. Mounir, N.; Ouadi, H.; Jrhilifa, I. Short-term electric load forecasting using an EMD-BI-LSTM approach for smart grid energy management system. *Energy Build.* **2023**, *288*, 113022. [CrossRef]
68. Mishra, P.; Al Khatib, A.M.G.; Yadav, S.; Ray, S.; Lama, A.; Kumari, B.; Sharma, D.; Yadav, R. Modeling and forecasting rainfall patterns in India: A time series analysis with XGBoost algorithm. *Environ. Earth Sci.* **2024**, *83*, 163. [CrossRef]
69. PJM Interconnection LLC. *Regional Transmission Organization (RTO) in the USA, Serving Delaware, Illinois, Indiana, Kentucky, Maryland, Michigan, New Jersey, North Carolina, Ohio, etc. (1998–2002)*; PJM Interconnection LLC: Norristown, PA, USA, 2024.
70. PJM Interconnection. *PJM East Hourly Data from the PJM East Region (2001–2018)*; PJM Interconnection: Norristown, PA, USA, 2018.
71. PJM Interconnection. *PJM West Hourly Data from the PJM West Region (2001–2018)*; PJM Interconnection: Norristown, PA, USA, 2018.
72. American Electric Power (AEP). *A Major Investor-Owned Electric Utility in the USA, Delivering Electricity Across 11 States (2004–2018)*; American Electric Power (AEP): Columbus, OH, USA, 2018.
73. Dayton Power and Light Company. *Serving Over 500,000 Customers Within a 6,000-Square-Mile Area in West Central Ohio, Around Dayton (2004–2018)*; Dayton Power and Light Company: Dayton, OH, USA, 2018.

electronics

MDPI

Article

Dynamic Aggregation Method for Load Aggregators Considering Users' Deviation Electricity

Linxi Li [1], Xun Dou [1,*], Hanyu Yang [1], Yadie Fu [1], Jiancheng Yu [2], Xianxu Huo [3] and Chao Pang [2]

[1] College of Electrical Engineering and Control Science, Nanjing Tech University, Nanjing 211816, China; monica@njtech.edu.cn (L.L.); hanyu93@njtech.edu.cn (H.Y.); 17766079304@163.com (Y.F.)
[2] State Grid Tianjin Electric Power Company Electric Power Research Institute, Tianjin 300384, China; jiancheng.yu@tj.sgcc.com.cn (J.Y.); chao.pang@tj.sgcc.com.cn (C.P.)
[3] State Grid Tianjin Electric Power Company, Tianjin 300010, China; xianxu.huo@tj.sgcc.com.cn
* Correspondence: dxnjut@njtech.edu.cn

Abstract: Constructing a new energy-based power system is not only an important direction for the transformation and upgrade of China's power system, but also a key to achieving peak carbon and carbon neutrality. How to fully utilize situation awareness technology to adapt to diverse and differentiated scenarios has become a crucial breakthrough point for ensuring the reliable, safe, high-quality, low-carbon, and economical operation of the new power system. Starting from the distribution network demand resources, this paper proposes a dynamic aggregation method for load aggregators considering the user deviation quantity, to deal with the current situation that the adjustable load-side resource points are multi-faceted and wide, and the operating subjects are complex and difficult to participate directly in the grid dispatch. First, considering there is subjectivity in the electricity behavior of users under the jurisdiction of the load aggregator, a deviation amount may be generated during the actual aggregation process, which reduces the profit of the load aggregator. Therefore, a load aggregator-level user deviation dynamic volume forecasting method based on the Markov chain is proposed, which is used to predict the deviation quantity of users during the dispatch cycle and achieve a dynamic status estimate on the load side of the new power system. On this basis, a dynamic aggregation model for load aggregators based on the deviation volume was constructed with the objective of maximizing the revenue of load aggregators. The examples, by comparing the aggregation results of users under three scenarios, show the proposed method can effectively guarantee the income of load aggregators, verify the effectiveness of the proposed dynamic aggregation strategy, and provide technical support for the operation situation awareness of the load side of the new power system.

Keywords: load aggregators; dynamic aggregation; Markov chain; load forecasting; state estimation

1. Introduction

Constructing a new power system under the premise of ensuring energy and power supply has become an important measure to implement the "dual-carbon" goal. With the development of informationization and intelligence of the new power system, situational awareness of the new power system with a high proportion of new energy can help power dispatchers to obtain real-time operation information in a timely manner, so as to quickly discover potential problems. This is an important way to achieve observable and controllable operation of the power system, as well as to improve the safety level of power system operation and ensure the popularization of renewable energy [1].

The safety and stability of new electricity systems containing a significant proportion of renewable energy face great challenges [2]. On the one hand, the power supply side experiences leap in the development of clean energy represented by wind and solar power, and the share of new elements such as controlled loads and electric vehicles on the demand side gradually increases, bringing the potential for rapid response and aggregation to the power grid, while also introducing more uncertainty factors into the system [3]. On the other hand, due to the grid connection of large wind and solar power projects, the proportion of conventional supporting power sources such as coal power is decreasing, the system tends to be electronified, the control modes are diversified, and the safety and stability analysis of the system becomes more complex [4].The rapid development from various sides brings great difficulty in real-time balance of power systems, the issue of power supply shortage becomes more severe. And, it will be difficult to maintain the balance of supply and demand in the power system only by the dispatchable resources on the power generation side [4]. In this context, load aggregators have emerged as a new independent subject. They can optimize the electricity usage methods of residential consumers in many flexible ways, while aggregating user participation in grid management, thereby substantially improving the flexibility and reliability of grid regulation [5].

The current increase in the penetration rate of renewable energy poses challenges to the integration of new energy sources. Balancing the supply and demand of electricity and making accurate predictions have become key research directions in modern power systems. Reference [6] proposes the use of Markov prediction methods for forecasting medium- to long-term electricity loads. Reference [7] suggests the Gray–Fuzzy–Markov Chain Method for three-stage short-term load forecasting, providing a basis for grid planning. Reference [8] introduces a short-term load forecasting model based on Markov chains, designed for continuous training during operation and capable of pre-training, making it universally applicable. However, there is limited research considering the subjective nature of user electricity consumption habits, as well as factors like temporality, randomness, and responsiveness. Utilizing load forecasting techniques to predict deviations in user electricity consumption is an area that is less explored.

However, there are certain challenges in dynamically aggregating users through load aggregators to enable them to participate in grid dispatch at the current stage. One of the main challenges is that the types of adjustable loads are diverse, geographically dispersed, and have small individual adjustment capabilities, which make it difficult to efficiently call upon users.

To address this limitation, domestic and foreign scholars have conducted corresponding research on adjustable load aggregation. Reference [9] introduces active aggregation objects and application scenarios, and then constructs an aggregation model for electric vehicle users aggregated by load aggregators in their jurisdiction, fully tapping into the regulation ability of electric vehicles to participate in ancillary services. Reference [10] proposes a general composite load model aggregation method that comprehensively considers the topology of load networks and conducts detailed equivalent error analysis of load model aggregation to illustrate the possible error factors and their impact mechanisms on equivalent accuracy. Specifically, a large number of researchers focus on the aggregation strategies of single loads of electric vehicles, air conditioners, energy storage, and electric heating in typical scenarios. Reference [11] takes the load of a park as the research object, proposes a proactive aggregation method for park loads, and constructs an aggregation optimization model based on historical data. Reference [12] conducts a comprehensive review of modeling strategies for air conditioning load aggregation and provides research directions for future air conditioning load aggregation technologies. Reference [13] takes electric vehicles as the research object, adopts a heterogeneity parameter equivalent aggregation method, constructs an aggregation model for electric vehicles, and effectively extracts the overall power regulation characteristics of electric vehicles. Reference [14] takes air conditioning load as the research object, models and aggregates air conditioning load in the region through load aggregators, proposes a variable frequency air conditioning total scheduling optimization strategy considering incentive compensation measures, and introduces temperature rise compensation factors for sensitivity analysis to demonstrate the effectiveness of load aggregation technologies for different electricity consumption scenarios of air conditioning loads in the region. Reference [15] takes temperature control load as the research object, constructs an aggregation model based on the physical characteristics of temperature control load, and introduces sparrow search algorithm for model solving. Reference [16] uses load aggregators to aggregate air conditioning loads in the region, and proposes a central air conditioning aggregation strategy from three aspects: physical characteristics of central air conditioning, economic benefits of load aggregators, and impact on user comfort. The example results show the proposed model can be used to smooth out the output of distributed power sources. Reference [17] takes air conditioning load as the research object, considers user decision uncertainty, establishes a dynamic response model based on the risk-averse multi-arm mechanism theory, and proposes a massive air conditioning load aggregation strategy. Reference [18] has developed a two-layer optimization model for an active distribution system, considering the participation of two different load aggregators, to achieve economic profit optimization. However, the above aggregation methods mostly focus on single-type loads, lack aggregation strategies for different types of loads, and only consider the adjustable potential theoretically possessed by users during aggregation, without considering the actual electricity consumption deviation situations of users, which results in load aggregators bearing a portion of the default costs. At the same time, research on aggregation methods from the user's perspective is still in its infancy, and there is a lack of economic analysis of aggregation benefits.

In response to this, the paper proposes a dynamic aggregation method for load aggregators that takes into account the default electricity consumption of users. Unlike traditional research on load aggregation, this paper integrates various types of resources for aggregation and considers deviations in user electricity consumption in aggregation strategies. The dynamic optimization is performed at an hourly time scale based on a Markov chain prediction method. The specific contributions of the paper are outlined as follows:

1. proposes a dynamic electricity consumption deviation prediction method for load aggregators at the user level based on Markov chains, which is used to predict the electricity consumption deviations of users during the dispatch cycle.
2. Constructs a dynamic load aggregation model for load aggregators based on electricity consumption deviations: the optimization objective is to maximize the expected revenue of load aggregators composed of declared electricity consumption revenue and default costs, considering constraints such as user load reduction boundary, power balance, declared capacity constraints of load aggregators, and capacity deviation.
3. Through simulation studies, the effectiveness of the proposed dynamic aggregation strategy is verified, and it is proved it can improve the declared accuracy and expected revenue of load aggregators in capacity declaration.

In Section 2, the dynamic load aggregation architecture for load aggregators that takes into account electricity consumption deviations of users is outlined. In Section 3, a dynamic electricity consumption deviation prediction method is proposed, which is used to predict the electricity consumption deviations of users during the dispatch cycle. In Section 4, a dynamic load aggregation model for load aggregators based on electricity consumption deviations is proposed. The simulation system's test results are presented in Section 5. Finally, in Section 6, this paper is concluded, and future work suggestions are proposed.

2. Dynamic Aggregation for Load Aggregators Considering Users' Deviation Electricity

Load aggregators can aggregate users' adjustable loads through price compensation [19], thereby profiting from the price difference between the selling price to the power grid and the compensation price to the participating users [20]. In the scenario described in this paper, the load aggregator aggregates users in its jurisdiction who are willing to participate in power grid dispatching, taking into account the actual deviation electricity volume of the participating users, and reports the optimal electricity volume to the power grid dispatching center. At the same time, the load aggregator will also provide corresponding financial compensation to the participating users. The aggregation framework of the load aggregator described in this paper is illustrated in Figure 1.

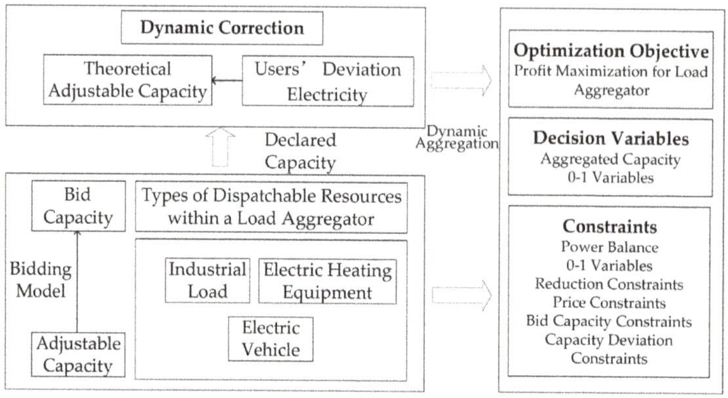

Figure 1. Aggregation framework of the load aggregator.

First, use a demand-side adjustable resource potential calculation method based on data processing to obtain the theoretical adjustment capability of users; second, based on the proposed deviation electricity volume dynamic prediction method, obtain the actual deviation electricity volume of users; finally, with the goal of maximizing the expected profit of the load aggregator, considering power balance constraints, price constraints, capacity deviation constraints, etc., dynamically adjust the declared response capacity of the load aggregator.

3. Dynamic Prediction Model for Deviation Electricity Volume

3.1. User Participation Volume Assessment

Most current studies calculate the adjustable potential of users based on theoretical values, but in reality, the subjective nature of user electricity consumption can lead to the generation of deviation electricity. For the deviation electricity, the power grid will charge corresponding default fees. If the power dispatch center is reported based on the theoretical adjustable potential, it results in high default costs, therefore, it is necessary to establish the relationship between the actual reduction number of residential users and the level of price compensation, as shown in Figure 2. At the same time, the actual load reduction of users at different economic incentive levels given in reference [21] is specifically expressed as follows:

$$p_i^{real}(t) = \gamma_i(j)Q_i^{max}(t) \tag{1}$$

where, $p_i^{real}(t)$ is the actual load reduction of the i-th user in the previous day in period t,$\gamma_i(j)$ is the load reduction rate of the i-th user under economic incentive j, i.e., the ratio of the user's actual load reduction and the theoretical load adjustable potential, and $Q_i^{max}(t)$ is the maximum value of the theoretical load adjustable potential of the i-th user at time t.

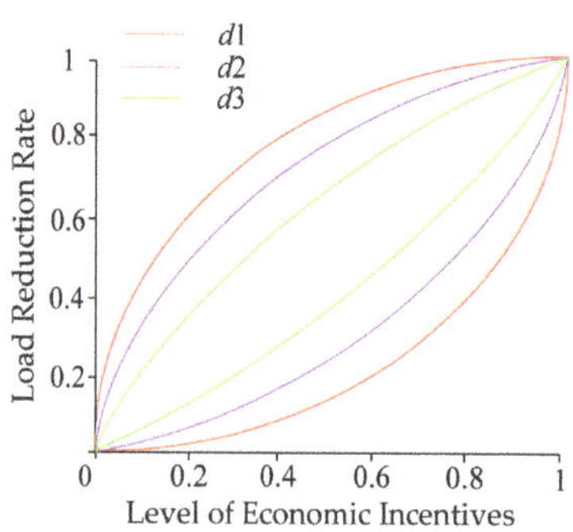

Figure 2. Load response degree.

In the Figure 2, $d1$, $d2$, and $d3$, respectively, represent the upper and lower limits of reduction for users with high, medium, and low levels of variability. As can be seen from the above figure, when the level of price compensation is low, the actual degree of reduction by users is also low. As the level of compensation continues to increase, the response level of users also increases accordingly. However, when the level of price compensation reaches a certain level, the actual load reduction by users approaches the limit of their response capacity.

Based on the actual load reduction amount of the user, the deviation electricity volume that the load aggregator needs to bear for the user can be calculated, as shown in the following formula:

$$w_i(t) = \sum_{i}^{s} Q_i^{max}(t) - \sum_{i}^{s} p_i^{real}(t) \tag{2}$$

where, 's' represents the number of users within the jurisdiction of the load aggregator. It denotes the deviation electricity volume of the load aggregator in period 't' at the initial stage.

3.2. Forecasting of User Deviation Electricity Volume at the Load Aggregator Level

Forecasting user deviation electricity volume plays a crucial role in ensuring the profits of load aggregators. Due to the subjectivity of user electricity usage habits, their response levels also exhibit temporal and random characteristics. How to accurately predict user deviation electricity volume is a question that needs to be researched.

This paper proposes a rolling forecasting mechanism for user deviation electricity volume to update the declared volume at the next moment. This allows for the dynamic aggregation of user volumes at the next moment. The framework of the dynamic prediction method for deviation electricity volume is shown in Figure 3. According to Section 3.1, the value of deviation electricity volume 24 h before moment t1 is set, and the possible deviation electricity volume at this moment is predicted using the Markov chain prediction model. Then, by rolling at a certain step length, the possible deviation electricity volume that users may generate at the moment is obtained. The declaration volume of the load aggregator is dynamically corrected, thereby reducing the default cost incurred by the load aggregator due to user default when declaring the volume. Through this method, the deviation electricity volume can be updated for several future periods.

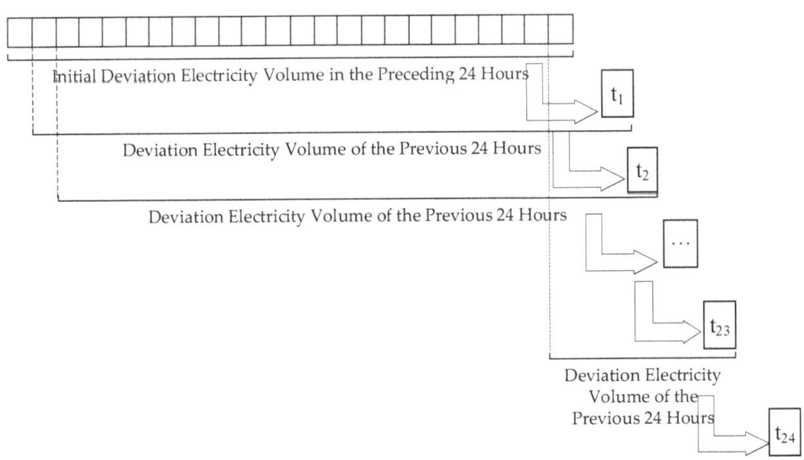

Figure 3. Dynamic forecasting structure for deviation electricity volume.

The steps of model prediction are as follows:

1. grouping of historical data by mean-mean-square deviation grouping method.

Set the input sequence as $x_1, x_2, \cdots x_n$, and find the mean $\overline{x}$ and variance σ. The mean-standard deviation grading method is used to categorize the user response in each period into five states: low response period, low response period, flat response period, high response period, and high response period (equivalent to determining the state space of a Markov chain) as shown in Table 1. Thus, the sequence is divided into the following five groups:

$$(-\infty, \overline{x} - 1.1\sigma), [\overline{x} - 1.1\sigma, \overline{x} - 0.5\sigma), [\overline{x} - 0.5\sigma, \overline{x} + 0.5\sigma), [\overline{x} + 0.5\sigma, \overline{x} + 1.1\sigma), [\overline{x} + 1.1\sigma, +\infty)$$

Table 1. Grouping standards and levels.

State	Grouping Criteria
1	$x < \overline{x} - 1.1\sigma$
2	$\overline{x} - 1.1\sigma \leq x < \overline{x} - 0.5\sigma$
3	$\overline{x} - 0.5\sigma \leq x < \overline{x} + 0.5\sigma$
4	$\overline{x} + 0.5\sigma \leq x < \overline{x} + 1.1\sigma$
5	$x \geq \overline{x} + 1.1\sigma$

2. Identify the state corresponding to the index value in the deviation electricity sequence.

According to the mean and variance in the first step, the deviation power sequence is classified into corresponding intervals based on the grouping criteria. Each interval corresponds to a state that represents a different level of power deviation.

3. Calculate the autocorrelation coefficients and weights of each order.

As previously mentioned, the electricity usage habits of users are subjective. The deviation in power consumption is a random variable, the strength of the relationship between the deviations in power consumption can be represented using autocorrelation coefficients of various orders, the calculation formula is as follows:

$$r_k = \frac{\sum\limits_{i=1}^{n-k} (x_i - \overline{x})(x_{i+k} - \overline{x})}{\sum\limits_{i=1}^{n} (x_i - \overline{x})^2} \tag{3}$$

where, n is the length of the default power series, k = 1,2. . .5, and x_i is the indicator value. The specification of the above equation can be obtained:

$$W_k \triangleq |r_k| / \sum_{k \in E} |r_k| \tag{4}$$

where, $\sum\limits_{k \in E} W_k = 1$ and $W_k \geq 0$, W_k are the magnitude of the weighting of the initial moment of prediction, characterizing the weight of each state in the prediction process in relation to the predicted value. The larger W_k is, the greater the weighting of its corresponding transfer probability.

4. Find the transition matrix.

Based on the state sequence of the deviation in power, the transition probability matrix is calculated as follows:

$$p_{ij} = \frac{f_{ij}}{\sum\limits_{j=1}^{m} f_{ij}}, (p_{ij}) i, j \in (1,2,3,4,5) \tag{5}$$

where, m is the number of states, $f_{ij}(i, j \in 1,2,3,4,5)$ is the transfer frequency matrix, and f_{ij} is the frequency at which state i passes through the transfer and thus arrives at state j.

5. According to the aforementioned steps, we obtain the transition probability matrices at all levels.

Use the predicted indicator values of the previous 5 h as the initial state. Then, calculate the prediction probability for the same state, as shown below:

$$p_i = \sum_{k=1}^{m} W_k P_i^{(k)} \tag{6}$$

where, $P_i^{(k)}$ is the state probability of the indicator value, k is the step size, and k = {1,2,3,4,5}, $max\{p_i\}$ are the predicted states of the indicator value at that time.

6. Calculate the weight set.

The aforementioned steps can predict the range of default electricity, but it is impossible to know the specific predicted value, therefore, it is considered to introduce level feature values to address this issue.

Based on the probabilities of the five states at the predicted moment as found in the previous step, each state corresponds to a weight respectively, then the set of weights it constitutes is denoted as $W = \{w_1, w_2, w_3, w_4, w_5\}$, and the calculation formula is shown below:

$$W_i = \frac{(P_i)^{\eta}}{\sum\limits_{k=1}^{5} (P_k)^{\eta}} \tag{7}$$

where, η is the coefficient of action for maximum probability.

7. Calculation of level eigenvalues.

The specific calculation formula is shown below:

$$H = \sum_{i=1}^{5} i \cdot W_i \tag{8}$$

8. Calculation of deviation in power prediction values.

Based on the above $P_i(i = 1,2,3,4,5)$, the state i with the largest probability among them is selected, and the final predicted value $W(t)$ is $\frac{T_i H}{i+0.5}$ when $H > i$, and the final predicted value $W(t)$ is $\frac{B_i H}{i-0.5}$ when $H < i$. Where T_i and B_i denote the upper and lower bounds of the desired state interval respectively [21].

4. Dynamic Aggregation Model for Load Aggregators Based on Deviation Power

To effectively ensure the profits of load aggregators, this chapter considers the case of user power deviation, and proposes a rolling correction method for the declared capacity of load aggregators, thus constructing a dynamic aggregation model for load aggregators, and updating the optimal aggregation amount for load aggregators.

4.1. Objective Function

This chapter sets the scene with load aggregators aggregating adjustable power from users. It constructs an aggregation optimization model where load aggregators take into account the user's adjustment capabilities. The final profit of the designed load aggregator consists of the declared power profit C_1 by the load aggregator and the default cost C_2. With the aim of maximizing the profit of the load aggregator, the following objective function is established.

$$\begin{aligned} maxC &= C_1 - C_2 \\ &= \sum_{t=1}^{T} \left[p(t)\lambda^s(t) - \sum_{i}^{s} Q_i(t)u_i(t)\lambda^b(t) \right] - \left[p(t) - \sum_{i=1}^{s} Q_i(t)u_i(t) \right] \lambda^v(t) \end{aligned} \tag{9}$$

$$p(t) = \sum_{i=1}^{s} (Q_i^{max}(t) - W_i(t)) \tag{10}$$

where, T is the aggregation period, which is 24 h in this article. $P(t)$ is the declared capacity of the load aggregator in period t, i.e., the declared capacity of the load aggregator after rolling forecast of defaulted electricity, $\lambda^s(t)$ is the unit price of electricity sold to the grid by the load aggregator in period t, $\lambda^b(t)$ is the compensation price of the load aggregator to the customer in period t, $\lambda^v(t)$ is the penalty price of the defaulted electricity of the load aggregator in period t, $Q_i(t)$ is the curtailment amount of the i-th customer in period t, $u_i(t)$ is the participation or non-participation of the i-th customer among load aggregators in period t. is a 0–1 variable, when indicates participation in aggregation in that period, and when indicates participation in aggregation in that period. aggregation, $u_i(t)$ is a 0–1 variable, when $u_t(t) = 1$, it means participating in aggregation in that period, when $u_i(t) = 0$, it means not participating in aggregation in that period. $Q_i^{max}(t)$ is the maximum regulation potential of the i-th user in period t, and $W_i(t)$ is the rolling forecast of defaulted electricity of the i-th user in period t.

4.2. Constraint Condition

1. Constraint on the upper and lower limits of load reduction for a single user.

For each user, to ensure the basic operation of user load, the reduction capacity when participating in aggregation should not exceed the theoretical adjustable capacity value. Meanwhile, for users whose actual reduction is less than the lower limit, their participation in the aggregation should be less than the actual reduction. It can be specifically described as:

$$Q_i^{min}(t) \le Q_i(t) \le Q_i^{max}(t) \tag{11}$$

$$0 \le Q_i(t) \le Q_i^{max}(t) - W_i(t) \tag{12}$$

where, $Q_i^{min}(t)$ is the lower limit of the regulation potential of the i-th user at time t.

2. Constraints on the upper and lower limits of the total load reduction for users.

The total reduction of all users within the load aggregator in the same period should be less than the capacity declared to the grid during that period. The specific constraint is described as:

$$\sum_{i=1}^{h} Q_i(t) \le p(t) \tag{13}$$

where, h is the number of users participating in the aggregation.

3. User compensation price constraint.

The compensation price applied by the user when participating in the aggregation should be less than the maximum compensation price. The specific constraint description is as follows:

$$0 < \lambda^b(t) \le \lambda^b_{max}(t) \tag{14}$$

4. Power Balance Constraint.

Power balance constraints ensure the stable operation of the power system; therefore, it is necessary to establish power balance constraints for load aggregators and the user side. The power balance constraint for the declared capacity of load aggregators is specifically described as:

$$p_i(t) = \sum_{i}^{h} Q_i(t) + W_i(t) \tag{15}$$

5. Load aggregator capacity declaration constraint.

The load aggregator should declare a capacity to the power grid at time t that is less than the sum of the theoretical adjustable capacity of each user at time t. The specific constraint is described as follows:

$$0 \leq p_i(t) \leq \sum_{i=1}^{h} Q_i^{max}(t) \tag{16}$$

6. Load aggregator price declaration constraint.

The sale price of electricity by the load aggregator should be lower than the average market sale price of electricity, with the specific constraint described as follows:

$$\lambda^s(t) \leq \frac{1}{T}\sum_{t=1}^{T} \lambda_t^{\text{maket}} \tag{17}$$

where, λ_t^{maket} is the average value of the market electricity price respectively.

7. Capacity deviation constraint.

To ensure the safe operation of the power grid, the actual response capacity of the load aggregator must be considered when participating in power grid scheduling, therefore, it is necessary to ensure the scheduling capacity of the power grid is within a certain deviation, so as not to affect the safe operation of the power grid. The specific constraint description is as follows:

$$\sum_{t=1}^{T} \frac{\alpha_t}{T} \geq \varsigma \tag{18}$$

$$\alpha_t = \frac{\sum_{i}^{h} Q_i(t)}{p(t)} \tag{19}$$

where, ς is the permissible deviation capacity.

The solution process is shown in Figure 4.

As illustrated in Figure 4, the resolution process begins with the determination of the initial power deviation for each user, using the Markov forecasting algorithm to dynamically predict the power deviation for each user over the next 24 periods, and adjusting the declared capacity of the load aggregator at time t based on the forecasted power deviation. Subsequently, considering the objective function and associated constraints, the dynamic aggregation model of the load aggregator is optimized, optimizing the reduction quantity of users selected by the load aggregator for aggregation at time t. Finally, the reduction quantities of the users participating in aggregation over the 24 periods are output.

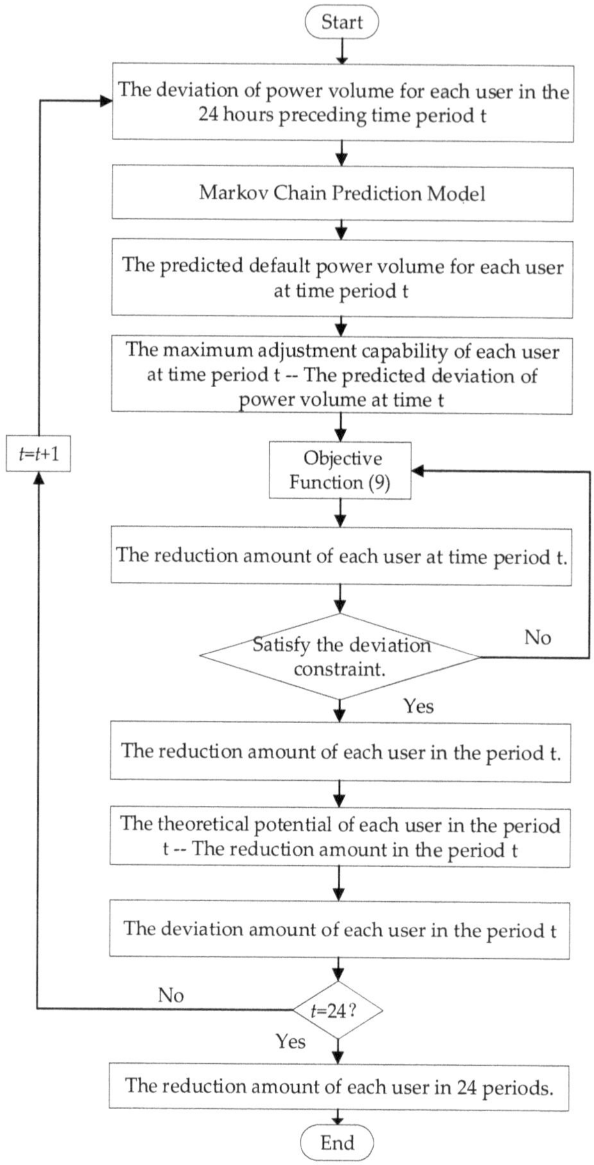

Figure 4. Dynamic aggregation solution process of load aggregator.

5. Example Analysis

5.1. Basic Case Data

This chapter analyses the optimal aggregation scheme and the declared capacity of the load aggregator to aggregate the adjustable loads in the region when it participates in grid scheduling, based on the user regulation potential calculated. The load aggregator is set to aggregate five types of users in the region. To ensure the security and stability of the grid, the capacity deviation of the grid dispatch center δ is set at 10%, and the dispatch is considered successful if and only if the dispatch deviation between the actual aggregated quantity and the declared quantity of the load aggregator is within 10%.

1. Price parameter configuration.

Different load aggregators, with varying scales of aggregated resources, also have different amounts of power deviation. It would be unfair if the market applies the same default price across the board. Therefore, different default prices will be applied to the power deviations of different load aggregators. This paper, referencing [22], sets a graded penalty price. Based on the power deviation situation of the load aggregators, penalty levels are divided according to their declared capacities. Different penalty prices are applied to different penalty levels, with specific penalty prices shown in Table 2, and the time of use electricity price of the power grid is shown in Figure 5.

Table 2. Division of penalty levels for load aggregators.

Penalty Level	Deviation Electricity Volume (kW·h)	Penalty Price CNY/(kW·h)
Level 1	$\leq 0.2\ P(t)$	0.1
Level 2	$(0.2 \sim 0.4)\ P(t)$	0.3
Level 3	$\geq 0.4\ P(t)$	Real-time electricity price

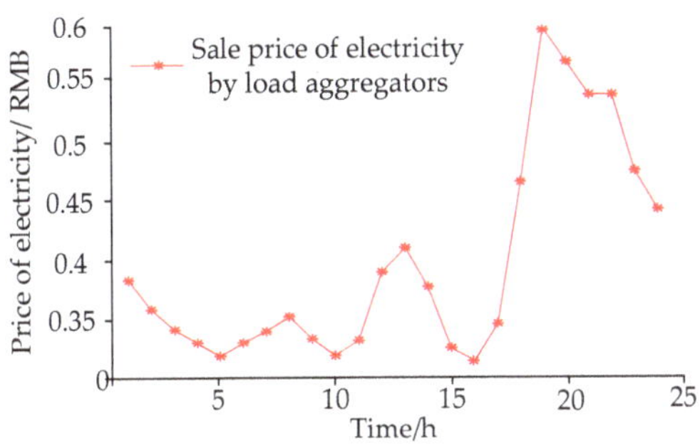

Figure 5. Grid electricity sale price.

To maximize the profits of the load aggregators, the price of electricity they sell to the grid increases during peak periods at noon and in the evening. For example, at 19:00 in the evening, the price of electricity from the load aggregators reaches its highest. However, considering the cost of purchasing electricity from the grid should not be too high, the price of electricity sold by the load aggregators is relatively lower during other normal periods and off-peak periods.

2. Scenario Configuration.

To verify the effectiveness of the proposed dynamic aggregation method, the following three scenarios are set up for comparative analysis.

S1: the load aggregator does not consider user deviation and declares capacity to the grid according to the maximum adjustment ability.

S2: the load aggregator considers users' deviation but does not dynamically adjust the declared capacity according to the deviation.

S3: the load aggregator considers user deviation and dynamically adjusts the declared capacity according to the deviation.

5.2. Analysis of Dynamic Aggregation Effectiveness

5.2.1. Setting and Prediction of Deviation in Electricity Volume

The compensation prices offered by the load aggregator to the users are 0.22 CNY/(kW·h), 0.22 CNY/(kW·h), 0.20 CNY/(kW·h), 0.23 CNY/(kW·h), and 0.21 CNY/(kW·h), respectively. According to Section 3.1, taking into account the impact of user response uncertainty, it is assumed when the economic incentive level $j = 0.9$, the load reduction rate under the condition of user heterogeneity d1 is extracted, determining the actual reduction amount of the user. At the same time, considering the influence of grid electricity prices on users, it is therefore set there is a difference in the reduction rate of users during peak and off-peak periods, as shown in Figure 6, thus obtaining the deviation of electricity for each user.

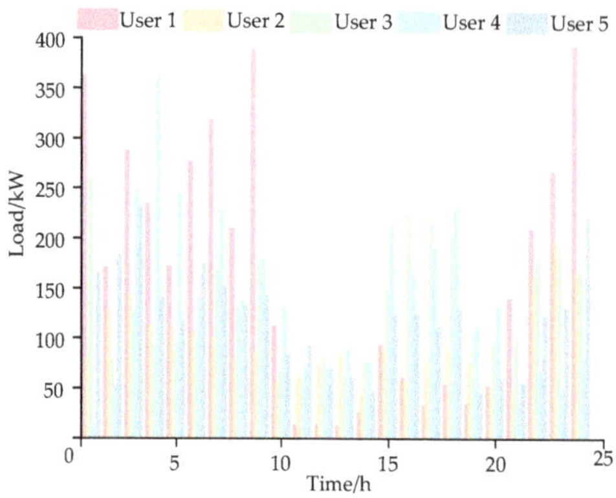

Figure 6. Deviation electricity volume setting.

For the initial deviation in electricity set by the user for the preceding 24 h, the deviation in electricity for the aggregation day is dynamically predicted based on the method proposed in Section 3.2. The characterized deviation in electricity is shown in Figure 7.

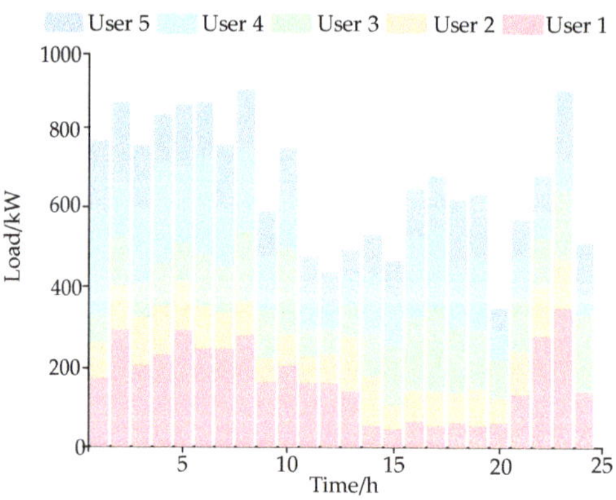

Figure 7. Deviation electricity volume forecast results.

The load aggregator, based on the results of dynamic deviation electricity forecasts, adjusts the declared capacity, and dynamically aggregates users.

5.2.2. Comparative Analysis of Load Aggregator's Declared Amount

The load aggregator declares to the grid the dispatchable capacity that can participate in regulation. The load aggregator declares the capacity based on the maximum adjustable ability of the user when the user's deviation electricity is not considered. When the load aggregator takes into account the deviation electricity, it sets up a comparison between not predicting the user's deviation electricity and making rolling predictions. The resulting declared capacity of the load aggregator is obtained, as shown specifically in Figure 8.

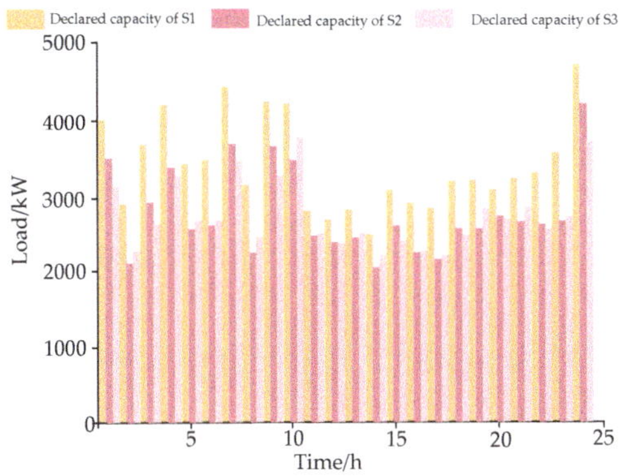

Figure 8. The load aggregator of each scenario declares the capacity.

The graph above illustrates the declared capacity of the load aggregator under different scenarios, providing a relatively intuitive depiction of the declared capacity in three scenarios. As the graph shows, after considering the user's deviation in electricity usage and making adjustments to the declared amount, the load aggregator's declared amount decreases in varying degrees within 24 periods. Overall, considering the subjectivity in users' electricity usage habits, the total declared capacity in Scenario S3 is approximately 14.947 MW·h less than that in Scenario S1 within a day. Meanwhile, for users, as the electricity price during peak hours is higher than during off-peak hours, and the user's deviation amount is greater during off-peak hours, the reduction in the load aggregator's declared amount during peak hours is less than that during off-peak hours.

Through the above analysis, after adjusting the declared capacity of the load aggregator, it can reduce the default cost caused by the users' own reasons, thereby protecting the profits of the load aggregator and enhancing its competitiveness in participating in grid dispatching.

5.2.3. Analysis of Dynamic Aggregation Plan of Load Aggregators

To further analyze how the load aggregator in this area selects users for dynamic aggregation, we examine the aggregation situation under scenario S3, as shown in Figure 9. Figure 9 represents the total load aggregation plan of the load aggregator in this area over 24 periods, considering maximizing its own profit.

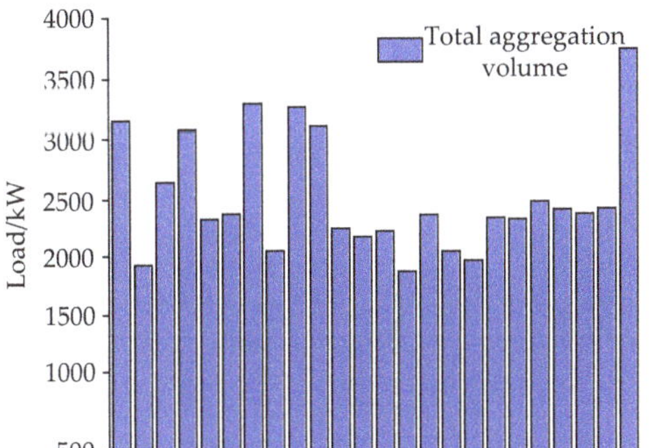

Figure 9. Aggregation volume of the load aggregator at each period.

As the graph indicates, the maximum aggregation volume of the load aggregator occurs at 24:00, reaching up to 3739.559 kW·h. Moreover, the load aggregator's aggregation volume can generally reach 2000 kW·h within the 24 h period. Therefore, by aggregating user-adjusted electricity through the load aggregator, the load in this area can be significantly reduced, thereby effectively improving the load curve of the power grid. Overall, the load aggregator's aggregation capacity is roughly the same as its declared volume, with only a small amount of deviation. The difference is minimal between 11:00–13:00 and 18:00–23:00, which is due to the limitations of user adjustment capabilities. The load aggregator's declared capacity, corrected after rolling forecasts, is close during this period. Although different users will have different adjustment capabilities, the values achieved after aggregation by the load aggregator will be similar.

Meanwhile, this paper presents the specific reduction amounts for each user within the region across 24 periods, as shown in Figure 10.

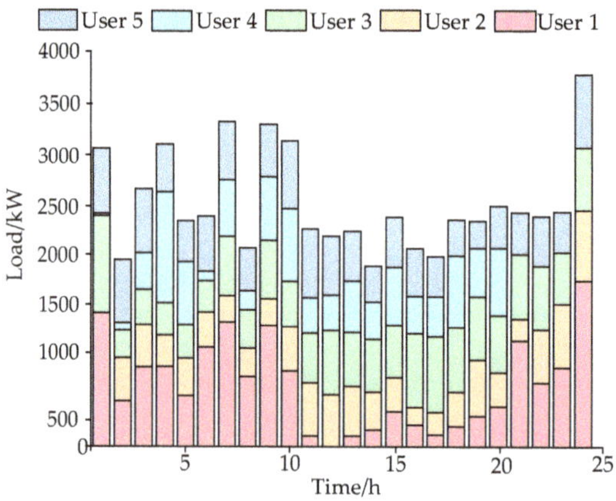

Figure 10. Reduction of each user.

As the graph illustrates, due to the varying adjustment capabilities of different users, the volume of aggregation and whether the load aggregator chooses to include a user in the aggregation can vary. Regarding the load aggregator's selection of users for aggregation, user 1 did not participate in the aggregation at 12:00, user 2 did not partake at 1:00, user 4 participated throughout the period from 21:00 to 24:00, and both user 3 and user 5 were included in the aggregation. This is because the compensation prices for user 3 and user 5 are lower than those of other users, thus, the load aggregator prioritizes aggregating these types of users within their adjustment capability range. It's worth noting among the five users, user 1's aggregation volume is higher than the other four users, with a reduction volume approximately 59%, 12%, 58%, and 20% higher, respectively. This is due to the different adjustment capabilities of each user, to some extent, the reduction volume is directly proportional to the user's adjustment capability. Taking user 1 as an example, the reduction volume reached 1242 kW·h at 7:00, the aggregation volumes of other users were significantly smaller, and their adjustment capabilities were far less than user 1 during this period.

Through the above analysis, the load aggregator, considering the rolling forecast of deviation power, can reasonably allocate and utilize the reduction amounts of users, thereby achieving dynamic aggregation of users, and further improving the economic efficiency of the load aggregator.

5.3. Economic Benefit Analysis of Load Aggregators

The profit of the load aggregator is related to parameters such as the electricity price for the grid, the actual reduction amount of the user, and the economic compensation paid to the user for participating in the aggregation. The profit of the load aggregator is the difference between the revenue from electricity sales obtained from the grid and the compensation amount paid to the user and the default amount to the grid. The effectiveness of the model was verified by comparing the aggregated users when the declared capacity was dynamically corrected and when it was not corrected. The profits, revenues, and costs of each load aggregator are shown in Table 3.

Table 3. Comparison of revenue of load aggregators in different scenarios.

Scenario	Profit/CNY (USD)	Revenue/CNY (USD)	Cost/CNY (USD)
Scenario 1	11,286.51 CNY (1580.74 USD)	30,899.58 CNY (4327.67 USD)	19,643.07 CNY (2757.13 USD)
Scenario 2	11,156.11 CNY (1562.48 USD)	24,992.96 CNY (3500.41 USD)	13,836.85 CNY (1937.93 USD)
Scenario 3	11,430.46 CNY (1600.90 USD)	25,204.28 CNY (3530.01 USD)	13,773.82 CNY (1929.10 USD)

As shown in Table 3, the load aggregator can obtain a revenue of 30,899.58 CNY (4327.67 USD, 1 USD = 7.14) by declaring to the power grid based on the maximum adjustment capacity. However, since Scenario S1 did not consider the existence of user bias, the load aggregator needs to bear a substantial default cost after paying compensation to users, resulting in a final profit of 11,286.51 CNY(1580.74 USD). This is not conducive to the load aggregator's participation in the power grid to enhance its trading competitiveness. Relatively speaking, Scenario S2 considered the deviation of electricity, which would reduce the cost. Although there is a slight decrease in profit compared to Scenario S1, due to the reduction of default costs, it can improve its competitiveness to a certain extent. Scenario S3's profit is 11,430.46 CNY(1600.90 USD), an increase of 1.27% compared to S1 and 2.46% compared to S2. Also, due to the rolling prediction of the deviation of electricity, it performs better than Scenario S2 in terms of both revenue from the power grid and payments to users, as well as default costs. From this, it can be seen the method proposed in this paper can enhance the load aggregator's own income.

Meanwhile, the aggregator achieves compensation and aggregation of the residential load. The profits of the aggregator at each period are shown in Figure 11.

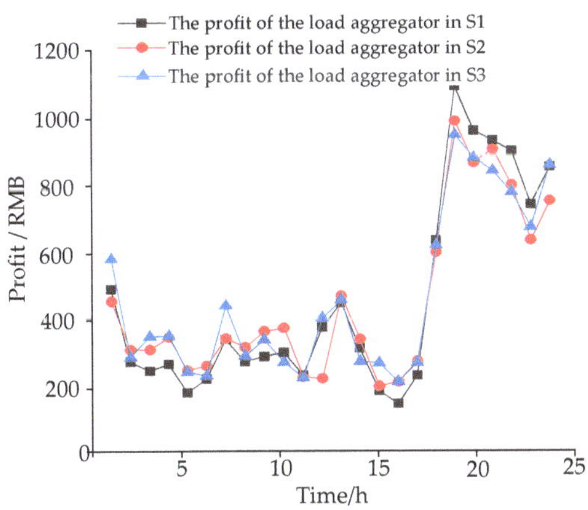

Figure 11. Profit of load aggregator in each scenario.

Figure 11 illustrates the profit variations of the load aggregator when selecting user aggregation and selling electricity to the grid under three scenarios across 24 periods. As can be seen from the figure, the level of profit variation for the load aggregator essentially aligns with the trend of electricity selling price to the grid. It is observable when the selling price of electricity is higher, a larger aggregation capacity can bring more profit to the load aggregator. Between 18:00 and 24:00, because the selling price of electricity by the load aggregator is higher, the profit that the load aggregator can obtain is greater than other periods. Notice the aggregation profit of scene 1 from 18:00–23:00 is higher than the other two scenes, this is because this paper sets the deviation penalty price as a step-type penalty, when the response deviation is greater than 0.4 the penalty price is the real-time price of electricity, and by the aforementioned time period belongs to the valley time period, the declared amount of scene 1 decreases to a larger extent, and the price of electricity is relatively low, so its aggregation of a higher return and a lower cost of the penalty and the obtains a higher profit. However, such a capacity declaration strategy is not conducive to the long-term development of the aggregator and is not conducive to the real-time scheduling of the grid, so this paper does not recommend this form of responsive power declaration.

Through the above analysis, under the guidance of the grid electricity selling price, the load aggregator integrates the users when participating in grid scheduling and obtains the optimal user aggregation plan through dynamic aggregation. This can optimize the way users use electricity, reduce the pressure on the power grid, and enable both users and load aggregators to benefit.

6. Conclusions

Addressing the issue of default power volume due to the subjectivity in user electricity usage, this chapter makes a rolling prediction of the user's default power volume. A method for correcting the deviation of the declared capacity by the load aggregator is proposed, constructing a dynamic aggregation model for the load aggregator. Through the numerical examples in this chapter, the main conclusions are drawn.

1. A dynamic aggregation model for the load aggregator has been established, based on the potential for user adjustment and considering the default electricity volume of users. This can provide a reference for the load aggregator in the optimization selection of users during the aggregation process. It can ensure the overall aggregation volume of the load aggregator can reach 2000 kW·h in 24 periods, and the maximum can reach 3739.559 kW·h.
2. By making a rolling forecast of the default electricity volume and adjusting the declared capacity of the load aggregator, it can effectively safeguard the interests of the load aggregator. Both excessively high and low declared capacities will affect the profits of the load aggregator. Analysis indicates the dynamic aggregation method proposed in this paper can effectively increase the profits of the load aggregator by approximately 1.2%.
3. Case analysis was conducted with users from a specific region, and the dynamic aggregation model for load aggregators was solved, which validated the effectiveness of the proposed dynamic aggregation model and method.

This study considers the situation where the user's subjectivity in electricity usage leads to default electricity volumes. In practice, the user's adjustable capacity is also affected by factors such as environment and policy. In subsequent research, it is necessary to take into account other influencing factors to improve the calculation model of user load adjustment potential. This allows load aggregators to aggregate user participation in grid scheduling more accurately. The model and case analysis proposed in this paper indicate, considering deviations in user electricity consumption, load aggregators have significant profit potential through optimizing aggregation strategies. As the power market continues to mature, load aggregators can actively participate in the market, serving as key players. Future commercial models for load aggregators in the power market can be designed based on this theoretical approach. On one hand, this can provide economically sound decision-making for load aggregators, and on the other hand, it can enhance the flexibility and reliability of the power grid operations.

Author Contributions: Conceptualization, L.L.; methodology, L.L.; data curation, X.D.; writing—original draft, Y.F., J.Y. and H.Y.; writing—review and editing, C.P. and Y.F.; visualization, X.H.; supervision, X.H. and C.P.; project administration, L.L., X.D. and H.Y.; funding acquisition, J.Y. and C.P. All authors have read and agreed to the published version of the manuscript.

Funding: This work was supported by State Grid Corporation of China headquarters science and technology project (5108-202218280A-2-244-XG).

Data Availability Statement: Data is contained within the article.

Conflicts of Interest: Author Xianxu Huo was employed by the company State Grid Tianjin Electric Power Company, and author Jiancheng Yu, Chao Pang were employed by the companyState Grid Tianjin Electric Power Company Electric Power Research Institute. The remaining authors declare that the research was conducted in the absence of any commercial or financial relationships that could be construed as a potential conflict of interest.

References

1. An, Q.; Li, G.; Wang, J.; Song, J.; Ge, L.; Mujeeb, A.; Xu, Y.; Hu, J. Exploring carbon equilibrium in integrated electricity-hydrogen system. *IEEE Trans. Netw. Sci. Eng.* **2023**, *early access*. [CrossRef]
2. Wu, Z.; Wang, J.; Zhou, M.; Xia, Q.; Tan, C.W.; Li, G. Incentivizing Frequency Provision of Power-to-Hydrogen Toward Grid Resiliency Enhancement. *IEEE Trans. Ind. Inform.* **2023**, *19*, 9370–9381. [CrossRef]
3. Seok Ko, K.; Han, S.; Keun Sung, D. A New Mileage Payment for EV Aggregators with Varying Delays in Frequency Regulation Service. *IEEE Trans. Smart Grid* **2018**, *9*, 2616–2624. [CrossRef]
4. Zhigang, Z.; Chongqing, K. Challenges and Prospects for Constructing the New-type Power System Towards a Carbon Neutrality Future. *Proc. CSEE* **2022**, *42*, 2806–2818. [CrossRef]
5. Gul, M.; Khan, O.; El-Saadany, E.F.; Youssef, A.; Shaaban, M.F. Cyber Security of Market-Based Congestion Management Methods in Power Distribution Systems. *IEEE Trans. Ind. Inform.* **2021**, *17*, 8142–8153. [CrossRef]
6. Haoxuan, F. Load Forecasting Research of Markov Chain based on Data Modeling. *J. Phys. Conf. Ser.* **2023**, *2470*, 012001. [CrossRef]
7. Munkhammar, J.; van der Meer, D.; Widén, J. Very short term load forecasting of residential electricity consumption using the Markov-chain mixture distribution (MCM) model. *Appl. Energy* **2021**, *282*, 116180. [CrossRef]
8. Fruh, H.; Groß, D.; Rudion, K. Short Term Load Forecasting for Individual Consumers based on Markov Chains. In Proceedings of the 2019 Modern Electric Power Systems (MEPS), Wroclaw, Poland, 9–12 September 2019. [CrossRef]
9. Wang, L.; Dong, Y.; Liu, N.; Liang, X.; Yu, J.; Dou, X. A Novel Modeling Method for Multi-Regional Flexible Load Aggregation based on Monte Carlo Method. In Proceedings of the 2021 IEEE 11th Annual International Conference on CYBER Technology in Automation, Control, and Intelligent Systems (CYBER), Jiaxing, China, 27–31 July 2021; pp. 632–637. [CrossRef]
10. Vaish, J.; Datta, S.S. Short-term Load Forecasting using Bootstrap Aggregation based Ensemble Method. In Proceedings of the 2021 7th International Conference on Electrical Energy Systems (ICEES), Chennai, India, 11–13 February 2021; pp. 245–249. [CrossRef]
11. Zhang, W.; Lian, J.; Chang, C.Y.; Kalsi, K. Aggregated modeling and control of air conditioning loads for demand response. In Proceedings of the 2014 IEEE PES General Meeting | Conference & Exposition, National Harbor, MD, USA, 27–31 July 2014. [CrossRef]
12. He, W.; Hongfeng, C.; Yan, L.; Sumei, L. A Review of Air Conditioning Load Aggregation in Distribution Networks. *Front. Energy Res.* **2022**, *10*, 890899. [CrossRef]
13. Congying, W.; Jian, X.; Siyang, L.; Yuanzhang, S. Aggregation and Scheduling Models for Electric Vehicles in Distribution Networks Considering Power Fluctuations and Load Rebound. *IEEE Trans. Sustain. Energy* **2020**, *11*, 2755–2764. [CrossRef]
14. Li, Q.; Zhao, Y.; Yang, Y.; Zhang, L.; Ju, C. Demand-Response-Oriented Load Aggregation Scheduling Optimization Strategy for Inverter Air Conditioner. *Energies* **2023**, *16*, 337. [CrossRef]
15. Zhang, W.; Shigang, L.; Jie, T.; Yongli, B. Joint Planning of Renewable Energy and Storage Considering Thermostatically Controlled Loads Aggregation Regulation. *Energy Storage Sci. Technol.* **2023**, *12*, 1901–1912. [CrossRef]
16. Gao, C.; Zhang, L.; Yang, X. Research on Load Aggregation of Central Air Conditioning and Its Participation in the Operation of Power System. *Proc. CSEE* **2017**, *37*, 3184–3191+3373. [CrossRef]
17. Hu, J.; Zheng, T.; Jin, Y.; Chen, K.; Xu, J. An Aggregation Strategy of Air Conditioning Loads Considering Uncertainty of Customer Behavior and Frequency Regulation Demand. *Power Syst. Technol.* **2022**, *46*, 3534–3542. [CrossRef]
18. Evangelopoulos, V.A.; Kontopoulos, T.P.; Georgilakis, P.S. Heterogeneous aggregators competing in a local flexibility market for active distribution system management: A bi-level programming approach. *Int. J. Electr. Power Energy Syst.* **2022**, *136*, 107639. [CrossRef]
19. Xiong, X.; Xue, C.; Yang, L.; Bing, Z.; Jia, T.; Haitang, L. Multi-stakeholder Demand Response Strategy Based on Stackelberg Game. *Electr. Meas. Instrum.* **2021**, *58*, 81–88. [CrossRef]

20. Yongxiu, H.E.; Qian, C.; Yunzhi, F.; Chengran, F.U.; Yan, C.; Yuexia, P.A.N.G. Typical Foreign Ancillary Service Market Products and Enlightenment to China. *Power Syst. Technol.* **2018**, *42*, 2915–2922. [CrossRef]
21. Miao, Z.; Xu, L. Predicting Annual Precipitation Using the Weighted Markov Chain Solved by the Improved FCM Algorithm. *J. Irrig. Drain.* **2017**, *36*, 114–121. [CrossRef]
22. Yanping, S. *Research on Dynamic Aggregation Model and Scheduling Mechanism of Demand Side Resources for Urban Power Grid*; North China Electric Power University: Beijing, China, 2019.

electronics

MDPI

Article

A Current Differential Protection Scheme for Distribution Networks with Inverter-Interfaced Distributed Generators Considering Delay Behaviors of Sequence Component Extractors

Gang Wang [1], Min Huang [1], Hao Bai [2], Jie Li [1,*], Ruotian Yao [2], Haoming Wang [1] and Chengxin Li [3]

[1] School of Electric Power Engineering, South China University of Technology, Guangzhou 510641, China; wangg@scut.edu.cn (G.W.); ephuangm@mail.scut.edu.cn (M.H.); 202211082774@mail.scut.edu.cn (H.W.)
[2] Electric Power Research Institute, China Southern Power Grid, Guangzhou 510663, China
[3] Guangzhou Electric Power Design Institute, China Southern Power Grid, Guangzhou 510630, China
* Correspondence: jli94@scut.edu.cn; Tel.: +86-130-1600-5946

Abstract: The high-level proliferation of inverter-interfaced distributed generators (IIDGs) in modern distribution networks (DNs) has changed system topologies and fault current signatures, which compromises the protective relays in DNs. Investigating IIDG fault behaviors-based protection scheme will benefit the grid's safety and stability. This paper proposes a novel current differential protection (CDP) scheme that considers the delay behaviors of positive- and negative-sequence component extractors for IIDGs in DNs. A frequency-domain analytical model of the fault current for a grid-connected IIDG with the PQ control strategy and a low-voltage ride-through (LVRT) capability is investigated. The dynamic behavior of the IIDGs considering the sequence-component extractor based on the Pade approximation is presented, where the T/4 delay extractor of the IIDGs causes a two-stage behavior in the fault transient process. It is found that a 5 ms error between the measured and actual values after the fault will affect the transient characteristics of the IIDGs. The transient current generated by the IIDGs during grid faults contains a large number of low-order harmonic components within the range of 0–200 Hz, which is significantly different to the current provided by the power grid. Therefore, the proposed CDP scheme uses protective relays at both terminals to obtain the required transient electric quantity using the Prony method. By constructing the frequency-characteristics ratio (FCR) and the exchanging FCR between two terminal relays, the developed protection criteria are implemented. The accuracy of the fault analysis method, whose maximum computational error is below 0.1%, and the feasibility of the proposed protection scheme are demonstrated by using a 10 kV DN in a PSCAD/EMTDC simulation, which can be applied to various fault conditions and traditional DNs without IIDGs.

Keywords: distribution network; fault analysis; protective relays; current differential protection; inverter-interfaced distributed generators

1. Introduction

1.1. Background and Motivation

Distribution networks (DNs) are the essential link between transmission systems and electricity users. With the increase in electricity demands and the development of power grids, it is essential that modern DNs can accurately and reliably isolate the faulty sections. An applicable protection relay ensures the secure and stable operation of modern DNs, which can prevent damage to the equipment and minimize power outages over a widespread area.

Renewable clean energy plays a crucial role in alleviating the energy crisis and improving the environment. An increasing variety of distributed generations (DGs) are being incorporated into modern medium-voltage DNs. The high-level proliferation of DGs, especially inverter-interfaced distributed generators (IIDGs), enables the DNs to be green and sustainable [1,2]. However, grid-connected IIDGs that are significantly different to the traditional synchronous generators [3–5] dramatically exacerbate the complexity of the grid operation and fault levels [6,7], which change the topologies and fault characteristics of the DNs. The large-scale deployment of IIDGs undoubtedly challenges the selectivity and sensitivity of the existing protective devices of DNs. In addition, the fault analysis and feature extraction of DNs are the theoretical basis for the construction of the protection methodology. Each change in the DN would require a change in fault analysis and relay protection. Hence, exploring the protection scheme according to the fault analysis of IIDG behaviors to enhance the DN's safety and stability is highly demanded.

1.2. State of the Art

The current characteristics have been utilized to construct overcurrent protection (OCP) schemes in recent decades [8]. An OCP approach based on negative- and zero-sequence currents is provided for multi-phase faults that occur in traditional DNs without IIDGs [9]. The authors of [10] present an adaptive OCP method by using local information and the Thevenin equivalent circuit of DNs, without extra infrastructure. The OCP techniques can automatically adjust the protection settings to meet the relay-tripping requirements based on different current characteristics and the effects of the IIDGs in DNs [11,12]. The authors of [13] studied an IIDG mitigation strategy to eliminate the protection blindness of the OCP of DNs without the installation of new relays. An adaptive OCP criterion was investigated through the MMS service for DNs with the PQ-IIDG [14]. Furthermore, the IIDGs will aggravate the protection coordination of overcurrent relays in DNs [15–17]. There is a challenge in determining the direction of fault currents for the overcurrent relays that generally employ nondirectional types of DNs [18]. The directional elements also fail to run properly under the influence of the inverter control of IIDGs [19]. Therefore, the above OCP schemes or other improvements may not be suitable for DNs with the high-level proliferation of IIDGs.

Current differential protection (CDP) seems to be the preferred choice for modern DNs in the presence of IIDGs due to its sufficient selectivity, higher sensitivity and various applications for communication [20–22]. A positive-sequence component-based CDP scheme is suggested by using data self-synchronization and communication techniques for DNs combined with IIDGs [23]. A time-graded CDP method is proposed to achieve the coordination effectively, with lateral protection devices in the post-fault DN topology [24]. CDP approaches with improved fault data self-synchronization are proposed in radial DNs with IIDGs [25,26]. A virtual multi-terminal CDP methodology is implemented via the simultaneous exchange of remote and local currents [27]. The authors of [28] provide a differential protection criterion based on impedance differences during normal operation and the internal/external faults of DNs. The sensitivity of the traditional CDP can be improved by changing the restraint area of the protection criteria [29]; however, the dual-end data synchronization issue is ignored. The authors of [30,31] propose a positive-sequence current-based CDP criterion for low-cost communication. Several amplitude-based CDP methods are presented for the DNs to improve the flexibility of IIDG integration [32–34]. Moreover, abrupt changes in the electrical quantities during faults involve a wealth of transient information. A differential protection scheme is presented based on a high-frequency impedance model for the IIDG-contained DN [35]. Combined with phasor-based protection, the transient current polarity is selected for fault identification to eliminate the coordination delays of relays in DNs with IIDGs [36], which are incapable of adapting to the negative influence of the inverter control of IIDGs.

It is worth noting that IIDGs commonly use T/4 delay extractors to split positive- and negative-sequence components, which leads to delay behaviors and measurement errors during the fault transient stage [11,27,37]. Positive- and negative-sequence components of the point of common coupling (PCC) voltage and current measured by the control system are inconsistent with instantaneous components of IIDGs with a delay dynamic. Measurement errors caused by T/4 delay behaviors will affect the fault current characteristics of IIDGs and grids. The fault transient current of DNs with IIDGs is a nonlinear, time-varying, wide-frequency signal, involving a large number of harmonic components. However, the above protection methods do not consider the dynamic characteristics of T/4 delay behaviors. The impact of measurement errors in the positive- and negative-sequence components on the IIDG transient current is also neglected.

1.3. Scope and Main Contributions

This article addresses the above problems regarding the fault transient analysis and protection principle of DNs with IIDGs to guarantee the secure and reliable operation of power systems. The objective of this article consists of developing the frequency-domain analytical model for calculating the fault current of a grid-connected PQ-IIDG and investigating the delay behaviors of IIDGs considering the T/4 sequence-component extractor. Furthermore, by fully understanding the frequency characteristics of a 5 ms two-stage behavior of IIDGs, a novel CDP scheme for DNs with IIDGs considering T/4 delay behaviors is proposed, which can enhance the sensitivity and selectivity of the protection relay.

This paper presents an accurate fault analysis and calculation model for IIDGs compared to existing methods [38–40], and at the same time, makes the protection robust and reliable. This method allows for the installation of the IIDG in any location within the protected area and bus, which is applicable to different fault types, fault locations and fault resistances, with good generality. Furthermore, the protection principle can be performed via the existing communication channel at both terminals, without extra channels to collect the local information of each IIDG. Additionally, protective relays can be operated correctly for a traditional DN without IIDGs.

1.4. Structure of the Paper

The remainder of this paper is organized as follows. Section 2 describes the dynamic process of the T/4 delay behaviors and its effect on the fault transient current of IIDGs. The amplitude-frequency characteristics of the fault transient current of the IIDG are also obtained. Section 3 proposes a CDP scheme based on the frequency-characteristics ratio (FCR). Section 4 tests the effectiveness of the proposed protection scheme in the PSCAD/EMTDC simulation platform. The discussion section is presented in Section 5. Finally, Section 6 summarizes this paper.

2. Fault Transient Analysis of IIDGs Considering T/4 Delay Behaviors

2.1. Fault Model

The typical control configuration of a grid-connected IIDG is shown in Figure 1. The closed-loop control system mainly consists of the measurement, outer-loop, inner-loop and modulation, where u_{abc} is the three-phase voltage at the point of common coupling (PCC); i_{abc} is the IIDG output current; L and R are the equivalent inductance and resistance, respectively; $\dot{U}$ is the vector of PCC voltage, $\dot{I}$ is the vector of IIDG current output in the dq coordinate system; the superscripts d and q are the d-axis and q-axis in the rotating reference frame; the subscripts m and ref denote the measured and reference values; and the subscripts (1) and (2) represent the positive- and negative-sequence components.

During the measurement process shown in Figure 1, three-phase voltages u_{abc} and currents i_{abc} are split using a T/4 delay extractor to obtain relevant positive- and negative-sequence components. To enable decoupled control, the extracted positive- and negative-sequence voltages and currents are mapped to the dq coordinate system, respectively. The outer-loop consists of the PQ controller and a low-voltage ride-through (LVRT) control strategy, which generates reference currents $\dot{I}_{ref(1)}^{dq}$ based on the voltage detected at PCCs. The inner-loop controls positive- and negative-sequence currents using the proportional-integral (PI) controller to track their reference values. Based on the pulse width-modulated

(PWM) modulation, the reference voltages obtained from the inner-loop are converted into triggering pulses to control the power electronic devices.

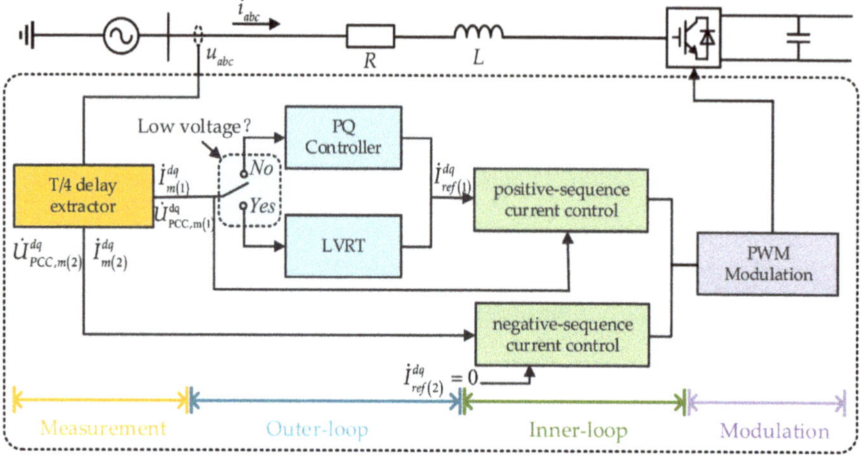

Figure 1. Closed-loop control system of a grid-connected IIDG.

Considering the dynamic characteristics of the T/4 delay sequence component extractor from the measurement shown in Figure 1, the relationship between the measured and actual values of the PCC voltage ($\dot{U}^{dq}_{PCC,m}$ and $\dot{U}^{dq}_{PCC}$) is as follows:

$$\begin{cases} \dot{U}^{dq}_{PCC,m(1)} = H_1(s+j\omega)\dot{U}^{dq}_{PCC(1)} + e^{-j2\omega t}H_1(s-j\omega)\dot{U}^{dq}_{PCC(2)} \\ \dot{U}^{dq}_{PCC,m(2)} = e^{j2\omega t}H_2(s+j\omega)\dot{U}^{dq}_{PCC(1)} + H_2(s-j\omega)\dot{U}^{dq}_{PCC(2)} \end{cases} \tag{1}$$

where s is the Laplace operator; ω is the angular frequency; and H_1 and H_2 are the transfer functions of the positive- and negative-sequence components that consider the T/4 delay behaviors, respectively, which can be given as:

$$\begin{cases} H_1 = \frac{1}{2}\left(1 + je^{-sT/4}\right) \\ H_2 = \frac{1}{2}\left(1 - je^{-sT/4}\right) \end{cases} \tag{2}$$

From (1) and (2), the voltage measurement errors of the PCC $\Delta\dot{U}_{PCC}$ caused by the T/4 delay behavior can be expressed as:

$$\begin{cases} \Delta\dot{U}_{PCC(1)} = [H_1(s+j\omega) - 1]\dot{U}^{dq}_{PCC(1)} + e^{-j2\omega t}H_2(s-j\omega)\dot{U}^{dq}_{PCC(2)} \\ \Delta\dot{U}_{PCC(2)} = [H_2(s-j\omega) - 21]\dot{U}^{dq}_{PCC(2)} + e^{j2\omega t}H_2(s+j\omega)\dot{U}^{dq}_{PCC(1)} \end{cases} \tag{3}$$

From (1)–(3) and Figure 1, the positive- and negative-sequence components can be accurately separated and measured using the sequence component extractor with T/4 delay behavior in the steady state of the IIDGs (i.e., $\dot{U}^{dq}_{PCC,m(1)} = \dot{U}^{dq}_{PCC(1)}$, $\dot{U}^{dq}_{PCC,m(2)} = \dot{U}^{dq}_{PCC(2)}$). However, since the outer loop and inner loop use positive- and negative-sequence components of measured voltages and currents as feedback inputs, a measurement error $\Delta\dot{U}_{PCC}$ that is affected by couplings between the positive- and negative-sequence components in the transient process will profoundly affect the fault current characteristics of IIDGs.

Combined with the T/4 delay extractor, LVRT strategy, inner-loop control of positive- and negative-sequence currents and modulation shown in Figure 1, a transfer function diagram of fault currents of IIDGs considering the T/4 delay behaviors is shown in Figure 2, where C represents the transformation among different coordinate systems; the subscript 2r, 3 s and 2 s are the rotating coordinate system (dq-axis), natural coordinate system (abc-axis) and stationary coordinate system (αβ-axis), respectively; $G_{PI} = K_P + K_I/s$ is the transfer function of the PI controller; and K_P and K_I are the proportional and integral gains of the PI controller, respectively. In Figure 2, the measurement transfer function involves coordinate transformations and separations of positive- and negative-sequence currents using the T/4 delay extractor. The reference voltages calculated by difference currents via the PI controller in the current inner-loop transfer function are fed into the primary circuits, which can obtain the actual positive- and negative-currents of the IIDGs.

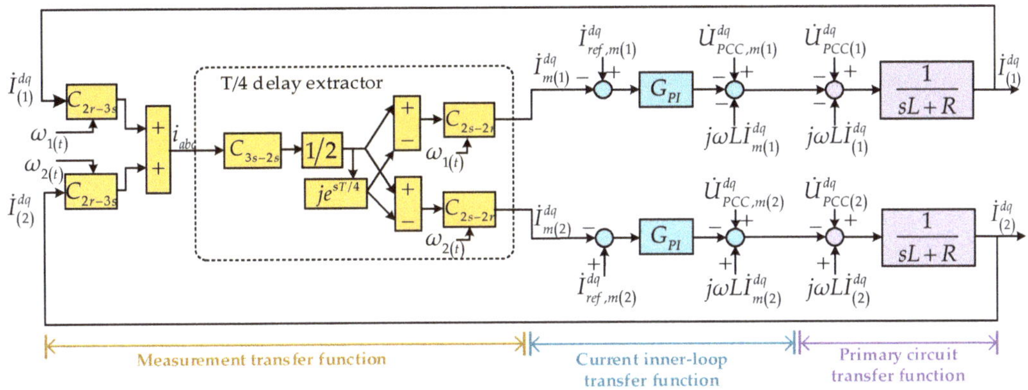

Figure 2. Transfer function diagram of fault currents of IIDGs considering the T/4 delay behaviors.

From Figure 2, it can be seen that the transfer function of IIDG fault currents is composed of the positive- and negative-sequence current extractor, the current inner-loop and the primary topology of the grid-connected converter. It is worth noting that the positive- and negative-sequence currents are independent at the inner loop and primary topology, while there is coupling between the positive- and negative-sequence components in the measurement with the T/4 delay extractor. Therefore, the fault current of a grid-connected IIDG considering the T/4 delay behaviors is:

$$\begin{cases} \dot{I}_{(1)}^{dq} = \frac{1}{sL+R}\left[\left(\dot{I}_{ref,m(1)}^{dq} - \dot{I}_{m(1)}^{dq}\right)G_{PI} + \dot{U}_{PCC,(1)}^{dq} - \dot{U}_{PCC,m(1)}^{dq} + j\omega L\dot{I}_{m(1)}^{dq} - j\omega L\dot{I}_{(1)}^{dq}\right] \\ \dot{I}_{(2)}^{dq} = \frac{1}{sL+R}\left[\left(\dot{I}_{ref,m(2)}^{dq} - \dot{I}_{m(2)}^{dq}\right)G_{PI} + \dot{U}_{PCC,(2)}^{dq} - \dot{U}_{PCC,m(2)}^{dq} - j\omega L\dot{I}_{m(2)}^{dq} + j\omega L\dot{I}_{(2)}^{dq}\right] \end{cases} \quad (4)$$

2.2. Fault Characteristics

According to (1)–(4) and the transfer function shown in Figure 2, the irrational functions H_1 and H_2 that contain the delay $e^{-sT/4}$, cannot be directly used in the fault analysis. In this regard, the delay in (2), based on the Pade approximation method [41], can be expressed as:

$$e^{-sT_d} = \frac{\sum_{i=0}^{n} (-1)^n \frac{(T_d s)^i}{i!2^i}}{\sum_{i=0}^{n} \frac{(T_d s)^i}{i!2^i}} \quad (5)$$

where delay duration T_d is 5 ms, and n is the number of approximate orders, which is equal to 30 in this paper.

Based on (1)–(5), the unit step response of the measured positive- and negative-sequence components of PCC voltages $\dot{U}_{PCC,m}^{dq}$, the actual value with T/4 delay behaviors $\dot{U}_{PCC,T}$, and the actual value ignoring the T/4 delay behaviors $\dot{U}_{PCC,NT}$, are shown in

Figure 3, where the measured positive-sequence voltage in the q-axis $\dot{U}^{q}_{PCC(1)}$ that is not affected by that in the d-axis voltage $\dot{U}^{d}_{PCC(1)}$ remains 0.

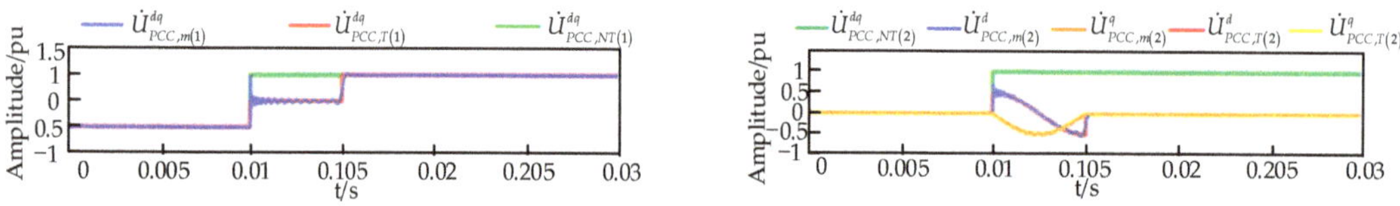

(**a**) Positive-sequence measurement error. (**b**) Negative-sequence measurement error.

Figure 3. Unit step responses for positive-and negative-sequence components of PCC voltages.

It is clear from Figure 3a,b that $\Delta\dot{U}^{d}_{PCC(1)}$ and $\Delta\dot{U}^{dq}_{PCC(2)}$ are 0 during the IIDG steady state, that is, $\dot{U}^{d}_{PCC,m}$ can track $\dot{U}^{d}_{PCC}$ in real-time. However, there are measurement errors, $\Delta\dot{U}^{d}_{PCC(1)}$ and $\Delta\dot{U}^{dq}_{PCC(2)}$, in the transient process. The delay behavior in which $\dot{U}^{d}_{PCC,m(1)}$ increases to $\dot{U}^{d}_{PCC(1)}$ after the unit step responses will remain at 5 ms for the positive-sequence measurement. The second-order harmonic components dominate the transient coupling between the positive- and negative-sequence components caused by the T/4 delay extractor. $\dot{U}^{dq}_{PCC,m(2)}$ also has a 5 ms delay behavior with a transient error whose magnitude is 0.5 $\dot{U}^{dq}_{PCC,m(2)}$ for the negative-sequence measurement.

Moreover, the reference negative-sequence current $\dot{I}^{dq}_{ref(2)}$ in the current inner loop shown in Figure 2 is zero to eliminate its negative impact. The reference positive-sequence current $\dot{I}^{dq}_{ref(1)}$ is generated by $\dot{U}^{dq}_{PCC,m(1)}$ based on the LVRT strategy. $\Delta\dot{U}^{dq}_{PCC}$ will affect the $\dot{I}^{dq}_{ref(1)}$, which causes the two-stage behavior in the fault transient process. It is clear that the reference current and measurement errors of voltages and currents affect the fault characteristics of IIDGs. For a better understanding of the above effects on the fault transient currents of IIDGs, based on (1), (2), (4) and Figure 2, a transfer function diagram of reference currents to actual currents can be obtained (i.e., the process from the current inner loop transfer function to the primary circuit transfer function in Figure 2), as shown in Figure 4.

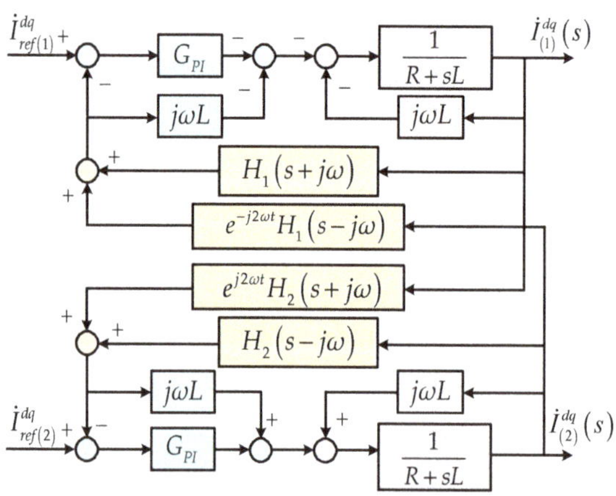

Figure 4. Transfer function diagram of reference positive- and negative-sequence current values to those actual currents.

From Figure 4, the actual positive- and negative-sequence currents affected by the reference current are expressed as:

$$\begin{cases} \dot{I}^{dq}_{(1)}(s) = & \frac{\dot{I}^{dq}_{ref(1)}(s)G_{PI}}{N_1(s)}\{G_{PI}H_1(s-j\omega)+G_{RL}-j\omega L[1-H_2(s-j\omega)]\}+ \\ & \frac{\dot{I}^{dq}_{ref(2)}(s)G_{PI}}{N_1(s)}[(G_{PI}+j\omega L)e^{-j2\omega_1 t}H_1(s-j\omega)] \\ \dot{I}^{dq}_{(2)}(s) = & \frac{\dot{I}^{dq}_{ref(2)}(s)G_{PI}}{N_2(s)}\{G_{RL}(s)+G_{PI}(s)H_1(s+j\omega)+j\omega L[1-H_1(s+j\omega)]\}+ \\ & \frac{\dot{I}^{dq}_{ref(1)}(s)G_{PI}}{N_1(s)}[G_{PI}-j\omega L]e^{j2\omega_1 t}H_2(s+j\omega) \end{cases} \tag{6}$$

subject to:

$$\begin{cases} G_{RL}=sL+R \\ N_1(s)= & [G_{RL}+G_{PI}H_1\,(s+j\omega)]^2+\omega^2L^2[1-H_1\,(s+j\omega)]^2+ \\ & [j\omega L-G_{PI}(s+j2\omega)](j\omega L-G_{PI})H_1\,(s+j\omega)H_2\,(s+j\omega) \\ N_2(s)= & [G_{RL}+G_{PI}H_2\,(s-j\omega)]^2+\omega^2L^2[1-H_2\,(s-j\omega)]^2+ \\ & [j\omega L-G_{PI}(s-j2\omega)](j\omega L-G_{PI})H_2\,(s-j\omega)H_1\,(s-j\omega) \end{cases} \tag{7}$$

As is shown in (6) and (7), the transfer functions of actual currents consist of a post-fault steady-state component and a transient-state decay component. The amplitude-frequency and phase-frequency transfer function characteristics of the IIDG output current in (6) are shown in Figure 5a. In Figure 5a, the amplitude-frequency and phase-frequency characteristics of the current inner-loop transfer function show pulsating waveforms with a period of 200 Hz when the reference current value is perturbed by the T/4 delay behaviors. Response characteristics with and without the T/4 delay behaviors are almost the same for high-order harmonic components above 500 Hz, which indicates that the T/4 delay behaviors do not significantly affect medium-order and high-order harmonic components. However, their fault response characteristics are quite different in the interval of 0–200 Hz. The low-order harmonic components of the IIDG fault current are higher due to the delay behaviors of the sequence-component extractor, which will have a greater impact on the transient characteristics of the IIDG fault current.

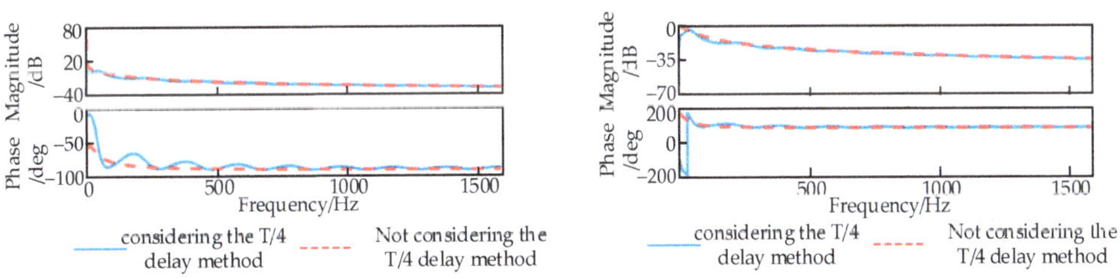

(**a**) Effect of current reference (**b**) Effect of voltage measurement errors

Figure 5. The amplitude-frequency and phase-frequency characteristic curves of the transfer function of the output current of the IIDG.

In addition, the voltage measurement error $\Delta\dot{U}^{dq}_{PCC}$ will affect the feedforward of the grid-side voltage in the current inner loop shown in Figure 5a. The difference between measured and actual voltages is large within the 5 ms after the fault occurs, where the voltage feedforward in the inner loop cannot match the actual value. It will inevitably cause an abrupt change for each sequence component current of IIDGs, which is not an active response of the control system to the fault condition. The more severe the fault, the greater the change in the voltages and currents will be. The effect of the current measurement error is similar to the voltage measurement error.

Similarly, the amplitude-frequency and phase-frequency characteristic of the transfer function of the output current of the IIDG under the influence of the voltage measurement error are shown in Figure 5b. As can be seen from Figure 5b, the transfer function of the current inner loop of IIDGs affected by the voltage measurement error still exhibits a pulsating waveform with a period of 200 Hz when the delay behaviors of the sequence component extractor are considered. By comparing Figure 5a,b, it can be seen that the voltage measurement error has a small effect on the amplitude-frequency and phase-frequency characteristics.

It can be concluded that the T/4 delay sequence component extractor will contribute to the two-stage behavior in the fault transient process of IIDGs. In stage I, the T/4 delay behaviors cause measurement errors between measured and actual voltage/current (i.e., $\dot{U}_{PCC,m}^{dq}$ and $\dot{I}_{m}^{dq}$) in the inner-loop controller. The transient components of PCC voltages and currents are affected by reference currents determined by the LVRT strategy, the voltage measurement error, and the current measurement error, which will remain 5 ms. In stage II, the voltage and current measurement errors are zero and the measured values can track the actual values in real time. The transient components of PCC voltages and currents are only affected by the changed reference current and gradually decrease to 0, that is, the IIDG is a fault steady-state process and unaffected by the T/4 delay behavior.

3. Proposed Protection Scheme

When a fault occurs in DNs with IIDGs, the fault current supplied by IIDGs contains a large number of low-order harmonic components, where the second-order harmonic is a dominant component, while the power-frequency component is the dominant component in low-frequency components for the short-circuit current contributed by the system source. Hence, low-order harmonic components caused by T/4 delay behaviors of IIDGs can be used in the protection scheme proposed in this paper.

3.1. Protection Principle

The CDP principle is commonly used in modern DNs due to its high sensitivity and selectivity. Here, a novel CDP principle based on low-order harmonic components is proposed in this paper. The current variation is adopted as the start-up criterion. Once protection relays are picked up, protective relays at two terminals acquire the required transient electric quantity based on the Prony method. The frequency-characteristic ratio (FCR) is constructed to adaptively adjust the restraint coefficient. Finally, an FCR-based protection criterion is implemented to overcome issues of false or failed tripping of relays in DNs with IIDGs.

To avoid the unbalanced current during normal operations, the proposed start-up criterion based on the current variation is as follows:

$$\Delta I_{st} > I_{st,set} \tag{8}$$

where $I_{st,set}$ is the pickup threshold of the phase current.

The ratio of the sum of the amplitude of power-frequency components to the sum of the amplitudes of low-order harmonic components for transient currents within a frequency band is defined as the FCR, which is given as:

$$\eta = \frac{\sum\limits_{49 \leq f_i \leq 51} A_i}{\sum\limits_{20 \leq f_i \leq 220} A_i} \tag{9}$$

where A_i is the amplitude of transient currents current at a frequency f_i. Considering the effect of decaying components of transient currents, the frequency band of low-order harmonic components is selected to be in the interval of 20~220 Hz, which includes second-order harmonic components, third-order harmonic components and fourth-order harmonic components. The frequency band of power-frequency components is selected to be in the interval of 49~51 Hz.

To further improve the sensitivity and reliability of protection relays, a frequency factor F_η based on the FCRs in (9) is defined as follows:

$$F_\eta = \frac{\min(\eta_m, \eta_n)}{\max(\eta_m, \eta_n)} \tag{10}$$

where m and n are two-terminal relays. The frequency factor F_η can be adaptively decreased when there is a large difference in FCRs between two-terminal relays, otherwise it will remain the same when there is a small difference in FCRs.

Based on (10), the action criterion of the CDP is as follows:

$$\left| \left| \dot{I}_m \right| - \left| \dot{I}_n \right| \right| - F_\eta K \left(\left| \dot{I}_m \right| + \left| \dot{I}_n \right| \right) \geq I_{set} \tag{11}$$

where $\left| \dot{I}_m \right|$ and $\left| \dot{I}_n \right|$ are the amplitude of currents at two-terminal relays m and n of the feeder, respectively; K is the restraint coefficient; and I_{set} is the relay setting of the CDP. Based on (10) and (11), the larger the FCR η for both terminal relays, the smaller the F_η will be. The sensitivity of relays can be improved adaptively for a fault feeder. The FCRs of both terminal relays are closed for a non-fault feeder, which ensures protection selectivity and reliability. Furthermore, the protection scheme is also applicable to the traditional DN without an IIDG, due to insignificant differences in FCRs.

3.2. Prony Method

To extract the low-order harmonic components of the fault current, the nth-order Prony method is employed to fit the required electrical quantity of relays in DNs with IIDGs. The current i at a time interval ΔT is sampled uniformly and discretely for a certain period with a total number of samples M. The discrete function is:

$$\begin{cases} \hat{i}(m) = \sum\limits_{x=1}^{n} c_x y_x^m, m = 0, 1, \ldots, M-1 \\ c_x = A_x e^{j\varphi_x} \\ y_x^m = e^{(\beta_x + j2\pi f_x)\Delta T} \end{cases} \tag{12}$$

where $\hat{i}(m)$ is the fitted approximation of $i(m)$; A_x, φ_x, f_x and β_x are the amplitude, initial phase, damping coefficient and frequency of the x-th complex exponential function, respectively. The solution process is as follows:

1. According to (12), the sample function can be defined as:

$$r(i, j) = \sum\limits_{m=n_e}^{M-1} \hat{i}(m-j)\hat{i}^*(m-i) \quad (i, j = 0, 1, \ldots, n) \tag{13}$$

where $\hat{i}^*$ is the conjugate of $\hat{i}$.

2. Construct the sample matrix $\mathbf{Q}$:

$$\mathbf{Q} = \begin{bmatrix} q(1,0) & q(1,1) & \cdots & q(1,n_e) \\ q(2,0) & q(1,1) & \cdots & q(1,n_e) \\ \vdots & \vdots & & \vdots \\ q(n_e,0) & q(n_e,1) & \cdots & q(n_e,n_e) \end{bmatrix} \tag{14}$$

where n_e is the order after the expansion, which is $M/2$.

3. Using the singular value decomposition and least square estimation to determine the effective rank and all the parameters of the characteristic equation $a_x (x = 1, 2, \ldots, n)$, substitute a_x into (15) to solve $y_x (i = 1, 2, \ldots, n)$:

$$\sum_{x=0}^{n} a_x y^{n-i} = 0 \tag{15}$$

4. Obtain the linear fit values in (16) based on the parameter a_x.

$$\hat{i}(m) = -\sum_{x=1}^{n} a_x \hat{i}(m-x) \tag{16}$$

5. Calculate the parameters $c_x(i = 1, 2, \ldots, n)$ using (17)

$$\begin{bmatrix} 1 & 1 & \cdots & 1 \\ y_1 & y_2 & \cdots & y_n \\ \vdots & \vdots & & \vdots \\ y_1^{M-1} & y_2^{M-1} & \cdots & y_n^{M-1} \end{bmatrix} \begin{bmatrix} c_1 \\ c_2 \\ \vdots \\ c_n \end{bmatrix} = \begin{bmatrix} \hat{i}(0) \\ \hat{i}(1) \\ \vdots \\ \hat{i}(M-1) \end{bmatrix} \tag{17}$$

6. A_x, φ_x, f_x and β_x can be yielded using the relationship between the parameters shown in (18).

$$\begin{cases} A_x = |c_x| \\ \varphi_x = \arctan\left[\frac{\mathrm{Im}(c_x)}{\mathrm{Re}(c_x)}\right] \\ f_x = \arctan\left[\frac{\mathrm{Im}(y_x)}{\mathrm{Re}(y_x)}\right] / 2\pi\Delta T \\ \beta_x = \ln|y_x| / \Delta T \end{cases} \tag{18}$$

Thus, the overall flowchart of the proposed CDP scheme for DNs with IIDGs considering delay behaviors of the sequence-component extractor is illustrated in Figure 6. Algorithm 1 shows the pseudocode of the proposed CDP scheme.

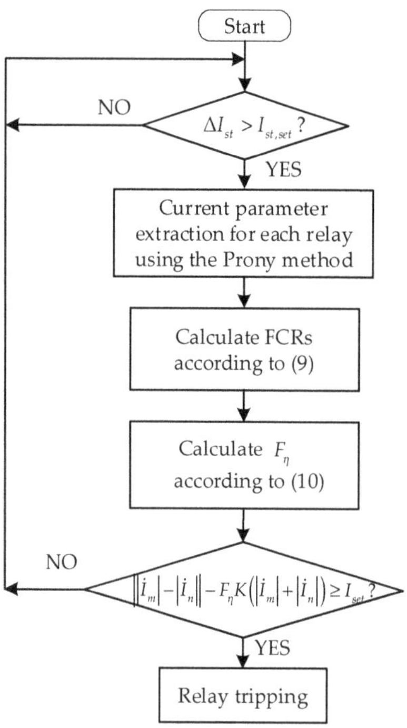

Figure 6. Flowchart of the proposed CDP scheme.

Algorithm 1: Pseudocode of the proposed CDP scheme

```
Load current data from PSCAD (Im, In)
Obtain magnitude and phase information of currents by the Prony algorithm (result_A, result_B)
result_A= [Ax_m, fx_m, phix_m, betax_m]
result_B= [Ax_n, fx_n, phix_n, betax_n]
n=length(result_A)
F1_m=0
F2_m=0
for i=1: n−1
    if (result_A(i,2)>=49 & result_A(i,2) <=51)
        F1_m= F1_m+result_A(i,1)
    end
    if (result_A(i,2)>=20 && result_A(i,2) <=220
        F2_m= F2_2+result_A(i,1);
    end
end
η_A=F1_m/F2_m;
F1_n=0
F2_n=0
for i=1: n−1
    if (result_B(i,2)>=49 & result_B(i,2)<=51)
        F1_n= F1_n+result_B(i,1);
    end
    if (result_A(i,2)>=20 && result_B(i,2)<=220
        F2_n= F2_n+result_B(i,1);
    end
end
η_B=F1_n/F2_n
F_η=min(η_A,η_B)/max(η_A,η_B)
if abs(abs(Im)-abs(In))-K*F_η*(abs(Im)+abs(In))>=I_set
disp('YES')
end
```

4. Case Study

In this section, a typical radial DN with IIDGs is built using the PSCAD/EMTDC simulation platform as shown in Figure 7, where two feeders (L1 and L2) are modeled with lengths of 9 km and 7 km. The rest of the parameters in the DN are shown in Table 1.

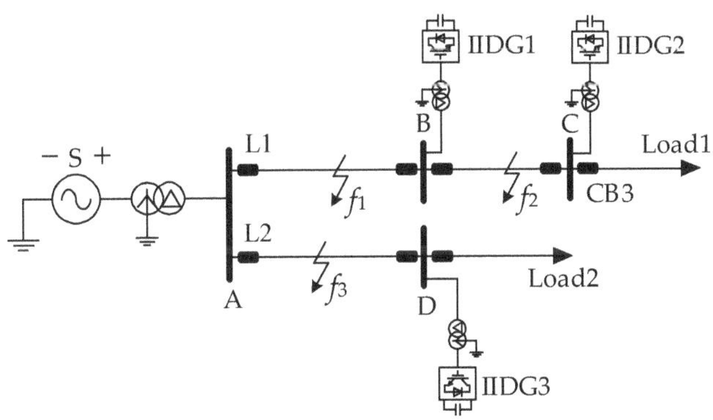

Figure 7. Typical radial DNs with IIDGs.

Table 1. Parameters of DNs.

Parameter	Value
System equivalent impedance	0.0001 + j0.628 (Ω)
Main transformer capacity	50 (MVA)
Main transformer ratio	110 kV/10.5 kV
Load capacity	10 (MVA)
Load power factor	0.9
Line impedance	0.06 + j0.089 (Ω/km)
Length of line AB	3 (km)
Length of line BC	4 (km)
Length of line AD	5 (km)
IIDG1, IIDG2 and IIDG3 capacity	3, 1.5 and 4 (MVA)
Fault time	0.2 (s)

4.1. Verification of Fault Transient Analysis Method

4.1.1. Fault Characteristics

To verify the correctness of the effects of measurement errors generated by T/4 delay behaviors of the sequence component extractor on IIDG transient currents, Figure 7 shows a three-phase symmetrical fault occurring at f_1 of the DN, where the PCC voltage drops to 60%. The three-phase PCC voltage u_{abc}, actual PCC voltage $\dot{U}_{PCC}^{dq}$ and the PCC voltage measurement $\dot{U}_{PCC,m}^{dq}$ are shown in Figure 8.

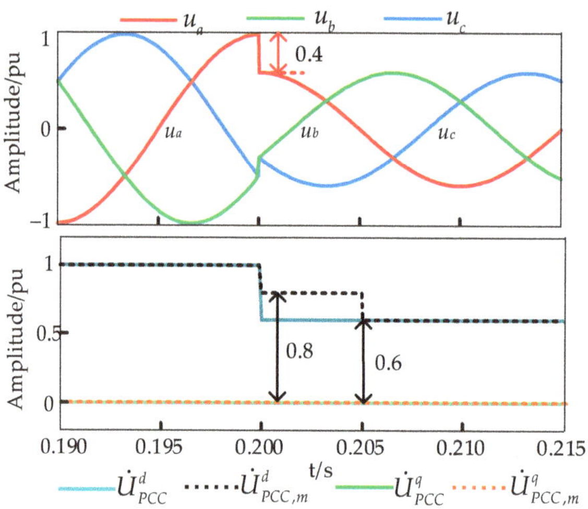

Figure 8. Fault waveforms for PCC voltages.

From Figure 8, three-phase voltages u_{abc} remain symmetrical after the fault. PCC voltages in the q-axis are zero because the d-axis is oriented to the PCC voltages. The amplitude of the actual PCC voltage $\dot{U}_{PCC(1)}^{dq}$ falls from 1.0 pu to 0.6 pu, while the measured PCC voltage $\dot{U}_{PCC,m(1)}^{dq}$ will undergo a transient process with a stepwise drop due to measurement errors introduced by T/4 delay behaviors. The amplitude of $\dot{U}_{PCC,m(1)}^{dq}$ decreases from 1.0 pu to 0.8 pu and then drops from 0.8 pu to 0.6 pu. A 5 ms dynamic error between measured PCC voltage and the actual value after the fault will affect transient characteristics of the fault current of IIDGs, which is theoretically identical to the fault analysis.

4.1.2. Comparison with Traditional Method

The output current of IIDGs and the amplitude-frequency characteristics of an A-phase current are compared when a three-phase symmetrical fault occurs at f_1 and the system voltage drops to 20%. Comparison results between the method proposed in this paper and the traditional algorithm [39] are shown in Figure 9.

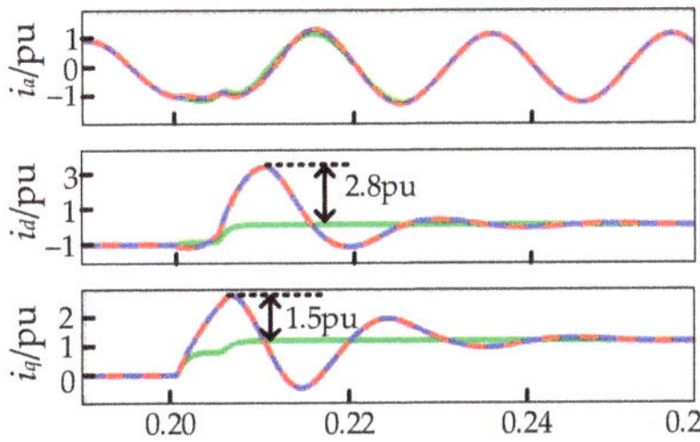

(a) IIDG output current

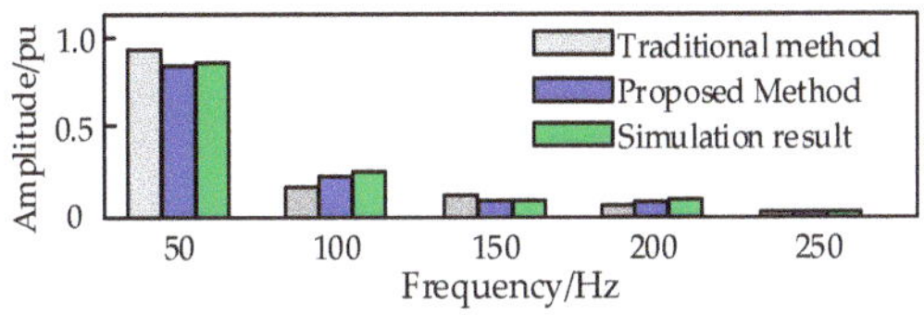

(b) Amplitude–frequency characteristics of A-phase current

Figure 9. Comparative results of fault currents.

In Figure 9a, the transient process of the IIDG fault current under the system voltage drop of 20% is divided into two stages. Due to the voltage feedforward error and LVRT control strategy, the d-axis and q-axis current increases within the first 5 ms of stage I. As for stage II, feedforward errors between measured voltages and actual voltages gradually become 0 and the reference current in the q-axis increases again. Compared with the traditional method, the q-axis current significantly overshoots with a maximum of 2.39 pu and then reaches the command value. From Figure 9b, it is found that all the low-order harmonic components of the IIDG fault current will increase. The second-order harmonic components of the transient current are the dominant components, which is consistent with the previous analysis.

From Figure 9a,b, it can be seen that the traditional equivalent model does not correctly reflect the transient characteristics of the IIDG fault currents within 0–20 ms after the fault, whereas the model used in this paper has a better match result. Furthermore, this section quantitatively compares the accuracy of the two models under different levels of voltage reductions, as shown in Tables 2 and 3. In the case of voltages which dropped by 20%, the calculation accuracy of the traditional methods is between 1% and 3% within 20 ms after a fault, while the accuracy of the model in this study is less than 0.01%. When the system voltage drops by 80%, the computational error of the traditional method exceeds 30% within 20 ms after a fault, which can reach 71.61% at maximum. The calculation accuracy of the proposed model is still below 0.1%, the computational accuracy is improved by more than 30%, and the narrower the time window, the more obviously the computational accuracy is improved.

Table 2. Comparison of the two models when the system voltage drops to 80%.

Method	Current	5 ms	10 ms	20 ms
Proposed Method	d-axis	<0.01%	<0.01%	<0.01%
	q-axis	<0.01%	<0.01%	<0.01%
Method in [39]	d-axis	1.74%	1.64%	1.36%
	q-axis	2.03%	1.93%	1.56%

Table 3. Comparison of the two models when the system voltage drops to 20%.

Method	Current	5 ms	10 ms	20 ms
Proposed Method	d-axis	0.02%	0.03%	0.05%
	q-axis	0.03%	0.06%	0.05%
Method in [39]	d-axis	71.61%	49.38%	33.22%
	q-axis	3.99%	15.31%	7.86%

Therefore, the proposed method in this study has high accuracy in fault calculations which is identical to the theoretical analyses and demonstrates the accuracy of the fault transient analysis of IIDG considering T/4 delay behaviors, as summarized in this paper.

4.2. Different Locations and Fault Resistances

To verify the performance of the proposed protection scheme under various locations and fault resistances, 27 groups of two-phase (A-phase and B-phase) grounding faults are simulated in the DN model shown in Figure 7, where K is set to 0.2 and I_{set} is set to 0.1 kA. The fault points are f_1, f_2 and f_3, respectively. The fault resistances are set to 10 Ω, 100 Ω and 500 Ω, respectively. The operation results of relays are shown in Table 4. I_{act} is the two-terminal differential currents of faulty sections for the action criterion in (4), which is equal to $\left|\left|\dot{I}_m\right| - \left|\dot{I}_n\right|\right| - F_\eta K\left(\left|\dot{I}_m\right| + \left|\dot{I}_n\right|\right)$.

Table 4. Operation results of the relays for the proposed protection scheme under different fault locations and resistances.

Fault Point	Fault Location	Fault Resistance (Ω)	η_A	η_B	F_η	I_{act} (kA)	Relay AB Tripped or Not	Relay
f_1	10%	10	1	0.8457	0.8398	10.5620	YES	AB
		100	0.9990	0.8171	0.8180	10.6497	YES	AB
		500	0.9991	0.8185	0.8192	10.6496	YES	AB
	50%	10	0.9979	0.9081	0.9385	7.9419	YES	AB
		100	0.9981	0.9216	0.9233	7.9289	YES	AB
		500	0.9977	0.9316	0.9338	7.9563	YES	AB
	90%	10	0.9980	0.8625	0.8643	6.4316	YES	AB
		100	0.9979	0.8723	0.8741	6.1900	YES	AB
		500	0.9973	0.8758	0.8181	6.5142	YES	AB
f_2	10%	10	0.9977	0.9975	0.9999	7.5339	NO	BC
		100	0.9978	0.9972	0.9993	7.3486	NO	BC
		500	0.9966	0.9963	0.9999	7.5341	NO	BC
	50%	10	0.9863	0.9862	0.9999	5.9441	NO	BC
		100	0.9866	0.9863	0.9997	5.0407	NO	BC
		500	0.9863	0.9864	0.9999	5.9212	NO	BC
	90%	10	0.9869	0.9866	0.9999	4.7786	NO	BC
		100	0.9858	0.9857	0.9999	5.4212	NO	BC
		500	0.9870	0.9869	0.9999	5.2130	NO	BC
f_3	10%	10	0.8853	0.8867	0.9984	9.9337	NO	AD
		100	0.8879	0.8864	0.9983	10.2895	NO	AD
		500	0.8871	0.8857	0.9984	9.9200	NO	AD
	50%	10	0.8742	0.8753	0.9987	6.6909	NO	AD
		100	0.8065	0.8015	0.9938	6.7521	NO	AD
		500	0.7953	0.7989	0.9955	6.6799	NO	AD
	90%	10	0.9111	0.9139	0.9969	4.9530	NO	AD
		100	0.8848	0.8863	0.9984	4.9544	NO	AD
		500	0.8081	0.7962	0.9853	4.9296	NO	AD

In the results shown in Table 4, the proposed protection scheme can correctly identify faults at different locations and fault resistances. When a fault f_2 occurs downstream of the protected section, the short-circuit current flowing through relays on both sides of AB is supplied by the system. The calculated FCRs on both sides are similar, that is, F_η close to 1. In the case of the fault occurring at a point, f_3, that is located at the outlet of a neighboring feeder on the same bus, a backward power flow may occur. The short-circuit currents at the relays on both sides of AB are contributed by the IIDG. The FCRs on both sides are small but approximately equal, that is, F_η is still close to 1. In summary, the relay can reliably trip to eliminate faults with good sensitivity and generality.

Furthermore, to show the correctness of the reconstructed electrical quantity of relays, the performance of the Prony algorithm implemented in the proposed protection method is assessed in Table 4 (f_1, 50%, 10 Ω). A comparison between fitted curves based on the Prony algorithm and simulated curves from PSCAD is given in Figure 10, where $\dot{I}_m$ and $\dot{I}_n$ are terminal currents of the section AB of Figure 7. From Figure 10, it is evident that the terminal currents $\dot{I}_m$ and $\dot{I}_n$ have been successfully fitted by the Prony algorithm, with a high fitting accuracy. The above retrieved results are correctly applied to the protection criterion, which can indirectly validate the effectiveness of the proposed methodology.

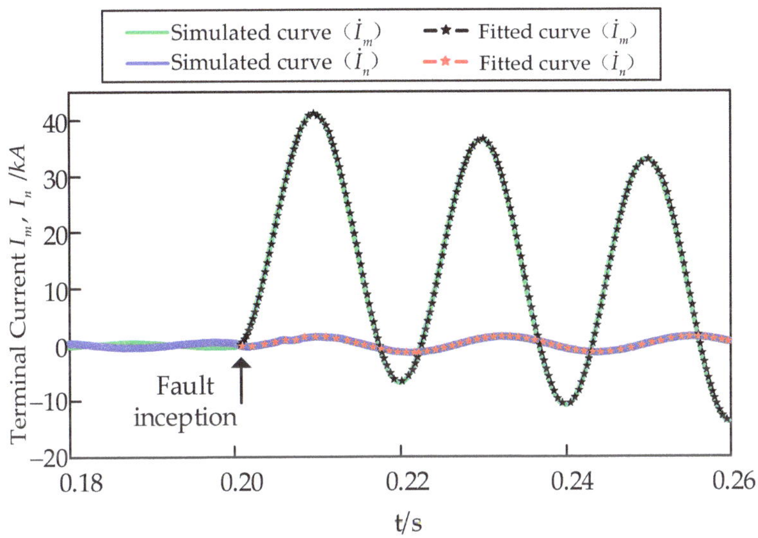

Figure 10. Comparison between fitted curves based on the Prony algorithm and simulated curves from PSCAD.

4.3. Different Fault Types

To verify the correctness of the proposed protection scheme in case of different fault types, two-phase short-circuit faults (A-phase and B-phase), two-phase to ground faults (A-phase and B-phase) and three-phase short-circuit faults are simulated at the midpoint of the AB section shown in Figure 7, respectively. The fault resistances are set to 1 Ω, 5 Ω, 50 Ω and 100 Ω, respectively. The simulation results are shown in Table 5. It can be seen from Table 5 that the proposed protection scheme can identify different fault types and each relay trips correctly. In the event of a serious fault, IIDG may provide a large number of harmonic currents. η calculated by relays on the IIDG side and F_η are close to 0, which can maximize the protection sensitivity.

Table 5. Operation results of the relays for the proposed protection scheme under different fault types.

Fault Type	Fault Resistance (Ω)	η_A	η_B	F_η	I_{act} (kA)	Trip or Not
three-phase short-circuit fault	—	0.9961	0.0001	0.0001	8.6650	YES
two-phase short-circuit fault	—	0.9984	0.9647	0.9662	8.3815	YES
two-phase to ground fault	1	0.9978	0.9131	0.9151	7.8778	YES
	5	0.9981	0.8846	0.8863	8.0605	YES
	50	0.9979	0.8814	0.8832	8.0672	YES
	100	0.9979	0.9102	0.9121	8.0038	YES

4.4. IIDG Output Power and Location

As the IIDG output power increases, the voltage support capabilities provided by the IIDG to the grid will become stronger after a fault, according to the LVRT control strategy. It ensures that the DN can continue to operate stably and weakens the fault current characteristics of the grid, which may result in traditional fixed restraint coefficients refusing to trip. To assess the impact of IIDGs on the proposed method, different IIDG output powers and locations are tested based on the model shown in Figure 7. The operation results of relays are obtained, as shown in Table 6, where the AB two-phase ground fault occurs at the fault point f_1, IIDG1 output powers are set to 1 MW, 2 MW, 3 MW and 4.5 MW, respectively, and the locations are 0.5 km, 1.5 km and 2.5 km from bus A, respectively; the fault resistance is set to 10 Ω. From the results shown in Table 6, it can be found that the proposed protection schemes can be operated accurately after a fault under different IIDG output power and locations. The proposed CDP scheme is immune to the output powers and locations of IIDGs.

Table 6. Operation results of the relays for the proposed protection scheme under different IIDG output powers and locations.

IIDG Output Power (MVA)	IIDG Location (km)	η_A	η_B	F_η	I_{act} (kA)	Trip or Not
1	0.5	0.9977	0.9431	0.9453	8.3592	YES
	1.5	0.9976	0.9491	0.9515	8.3423	YES
	2.5	0.9976	0.9378	0.9401	8.3371	YES
2	0.5	0.9976	0.9301	0.9323	8.3432	YES
	1.5	0.9977	0.9377	0.9399	8.3379	YES
	2.5	0.9977	0.9269	0.9290	8.4387	YES
3	0.5	0.9975	0.9208	0.92311	8.2625	YES
	1.5	0.9979	0.9141	0.9160	8.3865	YES
	2.5	0.9980	0.9163	0.9181	8.3367	YES
4.5	0.5	0.9976	0.9026	0.9048	8.2436	YES
	1.5	0.9974	0.9038	0.9061	8.2242	YES
	2.5	0.9979	0.9011	0.9030	8.3819	YES

4.5. DNs without IIDGs

To evaluate the performance of the proposed scheme on the traditional DN without IIDGs, IIDG is not considered in the model shown in Figure 7. The fault locations, fault resistances and other fault conditions are the same as those in Section 4.2. Table 7 shows the operation results for the relays in the proposed protection scheme.

Table 7. Operation results for the relays in the proposed protection scheme for the traditional DN without IIDGs.

Fault Point	Fault Location	Fault Resistance (Ω)	η_A	η_B	F_η	I_{act} (kA)	Relay AB Tripped or Not	Relay
f_1	10%	10	0.9893	0.9998	0.9895	10.3600	YES	AB
		100	0.9998	0.9999	0.9999	10.7699	YES	AB
		500	0.9899	0.9971	0.9928	10.3574	YES	AB
	50%	10	1.0000	1.0000	1.0000	8.1853	YES	AB
		100	0.9993	0.9990	0.9997	8.4347	YES	AB
		500	0.9995	0.9996	0.9999	8.2247	YES	AB
	90%	10	0.9996	0.9975	0.9979	6.4943	YES	AB
		100	0.9991	0.9994	0.9997	6.4898	YES	AB
		500	0.9986	0.9982	0.9996	6.5035	YES	AB
f_2	10%	10	1.0000	1.0000	1.0000	6.8398	NO	BC
		100	0.9978	0.9978	1.0000	6.8314	NO	BC
		500	0.9963	0.9962	0.9999	6.8742	NO	BC
	50%	10	1.0000	1.0000	1.0000	4.6295	NO	BC
		100	0.9888	0.9901	0.9987	4.6328	NO	BC
		500	0.9984	0.9985	0.9999	5.4846	NO	BC
	90%	10	0.9952	0.9989	0.9964	4.8398	NO	BC
		100	0.9968	0.9986	0.9982	4.6977	NO	BC
		500	0.9954	0.9958	0.9996	5.5084	NO	BC
f_3	10%	10	1.0000	1.0000	1.0000	10.1800	NO	AD
		100	0.9998	0.9998	1.0000	10.1804	NO	AD
		500	0.9997	0.9998	0.9999	10.1761	NO	AD
	50%	10	1.0000	1.0000	1.0000	6.8431	NO	AD
		100	0.9996	0.9992	0.9996	6.8430	NO	AD
		500	1.0000	0.9998	0.9998	6.8427	NO	AD
	90%	10	1.0000	1.0000	1.0000	5.0166	NO	AD
		100	0.9982	0.9987	0.9995	5.0142	NO	AD
		500	0.9999	0.9999	1.0000	5.0158	NO	AD

As can be seen in Table 7, the proposed protection schemes can accurately judge the fault and disconnect the faulty line for the traditional DN without IIDGs. Furthermore, a comparison of Tables 6 and 7 also indicates that IIDGs will contribute low-order harmonic components in the post-fault transient process for DNs, which confirms the validity of the previous conclusion. The method is efficient for a DN without IIDGs or with IIDGs, with good adaptability.

5. Discussion

Steady-state-based fault analyses of IIDGs have been studied by different scholars in the past [4,27,38–40]. To analyze and calculate fault currents of IIDGs more accurately, delay behaviors of sequence-component extractors have to be considered, which is an important prerequisite for the study of fault features and relay protection. The results in Tables 2 and 3 show that the maximum error of the proposed fault analysis method is below 0.06%, with higher accuracy in fault calculations than the above methods.

The protection system in [11] is capable of clearing ground faults only, while the performance under different fault types needs a deeper examination. Tables 4 and 5 demonstrate the good performance of the proposed method during different fault types. Referring to the existing literature [14,38], the adaptive scheme needs to obtain the IIDG information; however, electrical quantities or parameters of IIDGs are not readily available through relays and instrument transformers in DNs due to construction costs [23]. The active methodology [42] that relies on injecting the frequency signal may be costly and has a potential risk to the utility grid, which limits the application in real-life engineering. Fortunately, there is no need to obtain or exchange electrical information of each IIDG, so the proposed protection scheme is an economically feasible solution for DNs incorporating multiple IIDGs, without the requirements for the supplementary signal equipment and physical injections.

In addition, the flexible access of IIDGs, including the T-connection and bus-connection in the protection zone, will decrease the reliability and sensitivity of the conventional CDP [27] and could trigger process failure, while the proposed method can adapt to the flexible access of IIDGs with different output powers and locations, as shown in Tables 6 and 7, with good practicality and performance. In particular, the technique applied in this paper has low arithmetic complexity compared to those heuristic algorithms [22,26]. The noteworthy transient feature shown in Figures 8 and 9 is used to identify faults in DNs with IIDGs. Considering today's relay protection configurations, it is not difficult to achieve the requirements of the extraction of transient features and may be successfully implemented in the actual DNs.

The variety of DGs will increase with more diverse power electronics interfaces and control strategies in future DNs. Although the protection scheme here is described for a DN with PQ-IIDGs, the fault analysis method and protection philosophy in this paper can provide a theoretical basis and guidance for further research. Moreover, the proposed principle can be carried out on the existing communication-based protection system, which has a certain requirement for the communication of DNs. If the communication links between the relays at both terminals fail to function, it may lead to failure identification for faults in DNs due to missing or time-delayed information. Such a communication failure or latency is beyond the scope of this paper but will be addressed in future work.

6. Conclusions

A novel CDP framework using delay features that are contributed by the sequence component extractor has been proposed in this paper for protecting IIDG-dominated DNs. The T/4 delay behaviors associated with the grid-connected PQ-IIDGs are added to the fault analysis of DNs and the required mathematical model in the frequency domain is presented. It is found that, the two-stage delay behavior in the fault transient process for IIDGs will generate low-order harmonic components where the second-order harmonic components increase significantly. For the current differential relays, FCRs have been obtained using the Prony technique to construct the protection criteria.

The study shows that the proposed fault analysis method has high accuracy in fault calculations for transient electrical quantities for IIDGs. The feasibility of the FCRs-based current differential principle is substantiated by the simulation results under various fault types, fault locations and fault resistances, which is well adaptable to different IIDG locations and capacities. It is clear that the proposed protection does not require additional equipment to obtain electrical information from the PCCs of IIDGs, with favorable economic benefits. Consequently, this method is simple and could be a good solution for DNs with or without IIDGs.

Future work that involves various power electronics interfaces and control strategies of DGs connected to DNs, as well as communication failures, should be performed to show the adaptability and validity of the proposed fault analysis method and CDP philosophy.

Author Contributions: Conceptualization, G.W. and M.H.; methodology, J.L. and M.H.; software, C.L.; validation, M.H. and J.L.; formal analysis, G.W.; investigation, H.B.; resources, H.B. and R.Y.; data curation, H.W.; writing—original draft preparation, G.W. and M.H.; writing—review and editing, J.L.; supervision, R.Y.; funding acquisition, J.L. and H.B. All authors have read and agreed to the published version of the manuscript.

Funding: This research was funded by the science and technology project of China's southern power grid, grant number GZHKJXM20200081, and the science and technology program of Guangzhou, China, grant number 2023A04J1589.

Data Availability Statement: Data are contained within the article.

Conflicts of Interest: The authors declare no conflict of interest.

References

1. Medhat, W.; Azzouz, M.A. Triple Current Control of Four-Wire Inverter-Interfaced DGs for Correct Fault Type Identification. *IEEE Trans. Smart Grid* **2022**, *13*, 3607–3618. [CrossRef]
2. Gunawardane, K.; Bandara, N.; Subasinghage, K.; Kularatna, N. Extending the Input Voltage Range of Solar PV Inverters with Supercapacitor Energy Circulation. *Electronics* **2021**, *10*, 88. [CrossRef]
3. Alatrash, H.; Mensah, A.; Mark, E.; Haddad, G.; Enslin, J. Generator Emulation Controls for Photovoltaic Inverters. *IEEE Trans. Smart Grid* **2012**, *3*, 996–1011. [CrossRef]
4. Wang, Q.; Zhou, N.; Ye, L. Fault Analysis for Distribution Networks with Current-Controlled Three-Phase Inverter-Interfaced Distributed Generators. *IEEE Trans. Power Deliv.* **2015**, *30*, 1532–1542. [CrossRef]
5. Wang, C.; Pang, K.; Shahidehpour, M.; Wen, F. MILP-Based Fault Diagnosis Model in Active Power Distribution Networks. *IEEE Trans. Smart Grid* **2021**, *12*, 3847–3857. [CrossRef]
6. Mahamedi, B.; Eskandari, M.; Fletcher, J.E.; Zhu, J. Sequence-Based Control Strategy with Current Limiting for the Fault Ride-Through of Inverter-Interfaced Distributed Generators. *IEEE Trans. Sustain. Energy* **2020**, *11*, 165–174. [CrossRef]
7. Haj-ahmed, M.A.; Illindala, M.S. The Influence of Inverter-Based DGs and Their Controllers on Distribution Network Protection. *IEEE Trans. Ind. Appl.* **2014**, *50*, 2928–2937. [CrossRef]
8. Chen, L.-H. Overcurrent Protection for Distribution Feeders with Renewable Generation. *Int. J. Electr. Power Energy Syst.* **2017**, *84*, 202–213. [CrossRef]
9. Andruszkiewicz, J.; Lorenc, J.; Staszak, B.; Weychan, A.; Zięba, B. Overcurrent Protection against Multi-Phase Faults in MV Networks Based on Negative and Zero Sequence Criteria. *Int. J. Electr. Power Energy Syst.* **2022**, *134*, 107449. [CrossRef]
10. Rahmati, A.; Adhami, R. An Overcurrent Protection Relay Based on Local Measurements. In Proceedings of the 2013 IEEE Industry Applications Society Annual Meeting, Lake Buena Vista, FL, USA, 6–11 October 2013.
11. Li, J.; Feng, J.; Bai, H.; Wang, H.; Li, W.; Huang, M.; Wang, G. An Adaptive Protection Scheme for Multiple Single-Phase Grounding Faults in Radial Distribution Networks with Inverter-Interfaced Distributed Generators. *Int. J. Electr. Power Energy Syst.* **2023**, *152*, 109221. [CrossRef]
12. Qiang, J. Study on Adaptive Current Instantaneous Trip Protection Scheme for Distribution Network with Inverter Interfaced DG. *Autom. Electr. Power Syst.* **2009**, *31*, 71–76.
13. Yousaf, M.; Muttaqi, K.M.; Sutanto, D. A Control Strategy to Mitigate the Sensitivity Deterioration of Overcurrent Protection in Distribution Networks with the Higher Concentration of the Synchronous and Inverter-Based DG Units. *IEEE Trans. Ind. Applicat.* **2021**, *57*, 2298–2306. [CrossRef]
14. Zeng, D.; Wang, G.; Guo, J.; Sun, X. Adaptive Current Protection Scheme for Distribution Network with Inverter-Interfaced Distributed Generators. *Autom. Electr. Power Syst.* **2017**, *41*, 86–92.
15. Salem, M.M.; Elkalashy, N.I.; Atia, Y.; Kawady, T.A. Modified Inverter Control of Distributed Generation for Enhanced Relaying Coordination in Distribution Networks. *IEEE Trans. Power Deliv.* **2017**, *32*, 78–87. [CrossRef]
16. Zeineldin, H.H.; Sharaf, H.M.; Ibrahim, D.K.; El-Zahab, E.E.-D.A. Optimal Protection Coordination for Meshed Distribution Systems with DG Using Dual Setting Directional Over-Current Relays. *IEEE Trans. Smart Grid* **2015**, *6*, 115–123. [CrossRef]
17. Saldarriaga-Zuluaga, S.D.; López-Lezama, J.M.; Muñoz-Galeano, N. An Approach for Optimal Coordination of Over-Current Relays in Microgrids with Distributed Generation. *Electronics* **2020**, *9*, 1740. [CrossRef]
18. Ukil, A.; Deck, B.; Shah, V.H. Current-Only Directional Overcurrent Protection for Distribution Automation: Challenges and Solutions. *IEEE Trans. Smart Grid* **2012**, *3*, 1687–1694. [CrossRef]
19. Jia, K.; Yang, Z.; Fang, Y.; Bi, T.; Sumner, M. Influence of Inverter-Interfaced Renewable Energy Generators on Directional Relay and an Improved Scheme. *IEEE Trans. Power Electron.* **2019**, *34*, 11843–11855. [CrossRef]
20. Zidan, A.; El-Saadany, E.F. A Cooperative Multiagent Framework for Self-Healing Mechanisms in Distribution Systems. *IEEE Trans. Smart Grid* **2012**, *3*, 1525–1539. [CrossRef]
21. Yinger, R.J. Self-Healing Circuits at Southern California Edison. In *PES T&D 2012*; IEEE: New York, NY, USA, 2012; pp. 1–3.
22. Casagrande, E.; Woon, W.L.; Zeineldin, H.H.; Svetinovic, D. A Differential Sequence Component Protection Scheme for Microgrids with Inverter-Based Distributed Generators. *IEEE Trans. Smart Grid* **2014**, *5*, 29–37. [CrossRef]
23. Gao, H.; Li, J.; Xu, B. Principle and Implementation of Current Differential Protection in Distribution Networks with High Penetration of DGs. *IEEE Trans. Power Deliv.* **2017**, *32*, 565–574. [CrossRef]
24. Nikolaidis, V.C.; Michaloudis, G.; Tsimtsios, A.M.; Tzelepis, D.; Booth, C.D. A Coordinated Multi-Element Current Differential Protection Scheme for Active Distribution Systems. *IEEE Trans. Power Deliv.* **2022**, *37*, 4261–4271. [CrossRef]
25. Zhou, C.; Zou, G.; Zang, L.; Du, X. Current Differential Protection for Active Distribution Networks Based on Improved Fault Data Self-Synchronization Method. *IEEE Trans. Smart Grid* **2022**, *13*, 166–178. [CrossRef]
26. Zhou, C.; Zou, G.; Zhang, S.; Zheng, M.; Tian, J.; Du, T. Mathematical Morphology-Based Fault Data Self-Synchronization Method for Differential Protection in Distribution Networks. *IEEE Trans. Smart Grid* **2023**, *14*, 2607–2620. [CrossRef]
27. Han, B.; Li, H.; Wang, G.; Zeng, D.; Liang, Y. A Virtual Multi-Terminal Current Differential Protection Scheme for Distribution Networks with Inverter-Interfaced Distributed Generators. *IEEE Trans. Smart Grid* **2018**, *9*, 5418–5431. [CrossRef]
28. Chen, G.; Liu, Y.; Yang, Q. Impedance Differential Protection for Active Distribution Network. *IEEE Trans. Power Deliv.* **2020**, *35*, 25–36. [CrossRef]

29. Hossain, M.; Leevongwat, I.; Rastgoufard, P. Revisions on Alpha Plane for Enhanced Sensitivity of Line Differential Protection. *IEEE Trans. Power Deliv.* **2018**, *33*, 3260–3262. [CrossRef]
30. Nsengiyaremye, J.; Pal, B.C.; Begovic, M.M. Low-Cost Communication-Assisted Line Protection for Multi-Inverter Based Microgrids. *IEEE Trans. Power Deliv.* **2021**, *36*, 3371–3382. [CrossRef]
31. Nsengiyaremye, J.; Pal, B.C.; Begovic, M.M. Microgrid Protection Using Low-Cost Communication Systems. *IEEE Trans. Power Deliv.* **2020**, *35*, 2011–2020. [CrossRef]
32. Li, X.; Lu, Y. Improved Amplitude Differential Protection Scheme Based on the Frequency Spectrum Index for Distribution Networks with DFIG-Based Wind DGs. *IEEE Access* **2020**, *8*, 64225–64237. [CrossRef]
33. Lin, X.; Ma, X.; Wang, Z.; Sui, Q.; Li, Z.; Ye, Y.; Wu, Y.; Cao, S.; Wang, G. A Novel Current Amplitude Differential Protection for Active Distribution Network Considering the Source-Effect of IM-Type Unmeasurable Load Branches. *Int. J. Electr. Power Energy Syst.* **2021**, *129*, 106780. [CrossRef]
34. Chowdhury, A.; Paladhi, S.; Pradhan, A.K. Adaptive Unit Protection for Lines Connecting Large Solar Plants Using Incremental Current Ratio. *IEEE Syst. J.* **2022**, *16*, 3272–3283. [CrossRef]
35. Zheng, X.; Chao, C.; Weng, Y.; Ye, H.; Liu, Z.; Gao, P.; Tai, N. High-Frequency Fault Analysis-Based Pilot Protection Scheme for a Distribution Network with High Photovoltaic Penetration. *IEEE Trans. Smart Grid* **2023**, *14*, 302–314. [CrossRef]
36. Pathirana, A.; Rajapakse, A.; Perera, N. Development of a Hybrid Protection Scheme for Active Distribution Systems Using Polarities of Current Transients. *Electr. Power Syst. Res.* **2017**, *152*, 377–389. [CrossRef]
37. Dong, J.; Pan, J.; Mao, X. Low Voltage Ride-through Strategy of VSG Based on Adaptive Positive and Negative Sequence Composite Control. *Power Syst. Technol.* **2023**, *47*, 815–822.
38. Li, H.; Deng, C.; Zhang, Z.; Liang, Y.; Wang, G. An Adaptive Fault-Component-Based Current Differential Protection Scheme for Distribution Networks with Inverter-Based Distributed Generators. *Int. J. Electr. Power Energy Syst.* **2021**, *128*, 106719. [CrossRef]
39. Li, J.; Bi, H.; Li, C.; Li, J.; Zeng, D.; Wang, G. Analysis and Calculation Method for Multiple Faults in Low-Resistance Grounded Systems with Inverter-Interfaced Distributed Generators Based on a PQ Control Strategy. *Int. J. Electr. Power Energy Syst.* **2022**, *138*, 107980. [CrossRef]
40. Pan, G.; Zeng, D.; Wang, G.; Zhu, G.; Li, H. Fault Analysis on Distribution Network with Inverter Interfaced Distributed Generations Based on PQ Control Strategy. *Proc. CSEE* **2014**, *34*, 555–561.
41. Yao, R.; Sun, K.; Qiu, F. Vectorized Efficient Computation of Padé Approximation for Semi-Analytical Simulation of Large-Scale Power Systems. *IEEE Trans. Power Syst.* **2019**, *34*, 3957–3959. [CrossRef]
42. Soleimanisardoo, A.; Kazemi Karegar, H.; Zeineldin, H.H. Differential Frequency Protection Scheme Based on Off-Nominal Frequency Injections for Inverter-Based Islanded Microgrids. *IEEE Trans. Smart Grid* **2019**, *10*, 2107–2114. [CrossRef]

electronics

MDPI

Article

Optimization Decomposition of Monthly Contracts for Integrated Energy Service Provider Considering Spot Market Bidding Equilibria

Chen Wu [1,*], Zhinong Wei [1,*], Xiangchen Jiang [1], Yizhen Huang [1] and Donglou Fan [2]

1 School of Electrical and Power Engineering, Hohai University, Nanjing 211100, China; 231606030018@hhu.edu.cn (X.J.); 231306080020@hhu.edu.cn (Y.H.)

2 Wiscom System Co., Ltd., Nanjing 211100, China; dlfan@wiscom.com.cn

* Correspondence: 170206030002@hhu.edu.cn (C.W.); 19870038@hhu.edu.cn (Z.W.)

Abstract: Under the current power trading model, especially in the context of the large-scale penetration of renewable energy and the rapid integration of renewable energy into the power system, reasonable medium- and long-term decomposition can reduce the fluctuation in the energy price when the integrated energy service provider (IESP) participates in the spot market. It helps to avoid the price risk of the spot market. Additionally, it promotes the optimization of the operation of the regional energy day-ahead scheduling. At the present stage, most of the medium- and long-term contract decomposition methods focus on the decomposition of a single power and take less consideration of the bidding space in the spot market. This limitation makes it challenging to achieve efficient interaction and interconnection among multi-energy resources and smooth integration between the medium- and long-term market and the spot market. To address these issues, this paper proposes an optimal monthly contract decomposition method for IESPs that takes into account the equilibrium of spot bidding. First, the linking process and rolling framework of multi-energy transactions between the medium- and long-term market and the spot market are designed. Second, an optimal decomposition model for monthly contracts is constructed, and a daily decomposition method for monthly medium- and long-term contracts that accounts for the spot bidding equilibrium is proposed. Then, the daily preliminary decomposition result of medium- and long-term multi-energy contracts is used as the boundary condition of the day-ahead scheduling model, and the coupling characteristics of the multi-energy networks of electricity, gas, and heat are taken into account, as well as the operational characteristics. Then, considering the coupling characteristics and operating characteristics of electricity, gas, and heat networks, the optimal scheduling model of a multi-energy network is constructed to minimize the sum of cumulative daily operating costs, and the monthly final contract decomposition value and daily spot bidding space are derived. Finally, examples are calculated to verify the validity of the decomposition model, and the examples show that the proposed method can reduce the variance in spot energy purchase by about 4.64%, and, at the same time, reduce the cost of contract decomposition by about USD 0.33 million.

Keywords: spot bidding equilibrium; monthly homogeneous decomposition; medium- and long-term markets

1. Introduction

In recent years, the global energy sector has been undergoing profound changes with the large-scale penetration of renewable energy sources and the growing emphasis on the operation of a low-carbon economy. The rapid development of renewable energies has given new impetus to the power system; however, the challenges that come with it are becoming increasingly apparent. Power grids are faced with higher demand for renewable energy integration, increased uncertainty in energy supply, and the complexity of energy trading markets.

In this context, effective trading strategies and optimization methods are essential to achieving a balance between efficient operation, low carbon emissions, and economic viability in the new power system. It is necessary to steadily increase the scale of medium- and long-term power trading, solidly promote the trial operation of spot pilot settlements, actively and steadily push forward the construction of the power spot market, and strengthen the organic connection between the medium- and long-term, spot, and auxiliary service markets for power [1–4]. However, the complexity and variety of energy types in the integrated energy system (IES) and its participation in the spot market transactions are susceptible to the impact of the energy decomposition results of the medium- and long-term market; therefore, it is necessary to take into full account the demand for spot market bidding and the coupling and complementary study of multi-energy resources compatible with the medium- and long-term market and the contract decomposition of the spot market to alleviate the fluctuations in the spot market volume and price and to ensure that the market has a stable and economic operation [5–7].

Compared with the spot market, most of the power transactions conducted in the medium- and long-term markets are longer time scales such as annual or monthly markets [8,9]. To realize the effective connection between the medium- and long-term market and the spot market, its minimum time scale should be consistent with that of the spot market; therefore, the medium- and long-term contract decomposition methods at the present stage are all power decomposition with curves [10,11]. The current research on the decomposition of medium- and long-term contracts mainly focuses on the research of medium- and long-term curve decomposition methods and the research of the method of connecting the medium- and long-term market with the spot market.

In terms of medium- and long-term market curve decomposition methods, current studies mainly focus on the decomposition of power [12–14]. The decomposition of power refers to a method used in the context of energy markets, specifically for breaking down medium- and long-term power contracts into smaller, more manageable timeframes, typically for operational or trading purposes. This method is primarily concerned with decomposing the energy from a single type of power source, such as electricity, without considering the interplay or conversion between different types of energy resources. Literature [15] delves into the medium- and long-term trading mechanism within the electricity spot market. Market players customize bilateral and listed transactions to shape trading curves. Centralized competitive transactions are decomposed based on typical curves like the average curve and the peak curve. Literature [16] designs the bilateral spot market mechanism and proposes that the medium- and long-term contracted electricity on both the generation side and the user side should be curve-decomposed to form medium- and long-term transaction settlement curves. Literature [17], for the new type of power market and scheduling characteristics, constructs a two-stage wind power consumption model considering the connection of medium- and long-term transactions and short-term scheduling to realize the effective connection between the market and schedule. Literature [18] constructs a decomposition model for thermal electricity in the context of the Turkish electricity market by integrating the derivatives market, the bilateral contract market, and the spot market under the curve decomposition for single electricity. In literature [19], the Nordic region performs a curve decomposition of the contracted power to continue to organize the coupled clearing of the Nordic market, which results in the clearing of the coupled market as a basis for the settlement and execution of the coupled market. The above study proposes the decomposition method of medium- and long-term contracts considering both supply and demand and short-term scheduling, which is more flexible and diversified, but the above decomposition method is power decomposition, and it fails to consider the conversion of energy between electricity, gas, heat, and other energy sources and is unable to realize the synergistic supply of multiple energies in the day-to-day situation.

In terms of the convergence of the medium- and long-term market and spot market transactions, the traditional decomposition method is mostly for the average decomposition of the medium- and long-term contract volume, and the decomposition result of the medium- and long-term market is used as the clearing boundary of the spot market to obtain the final actual operating curve [20–22]. Literature [23] proposes a quadratic planning method for the decomposition of the medium- and long-term contracted power volume and takes the decomposed daily power volume as the constraint to ensure the fairness of the bidding unit scheduling. Literature [24] addresses the problem of the autonomous decomposition of medium- and long-term contracted power for power producers and investigates the medium- and long-term contracted power decomposition method based on distributional robust optimization for power producers in the environment of serious uncertainty in spot market price. Literature [25] adopts the method of monthly bidding space equalization to decompose the annual winning contract into months, and the unfinished monthly decomposition of the total annual winning power in actual operation is corrected to the subsequent part on a rolling basis while maintaining the monthly bidding space equalization in the remaining months. Literature [26], concerning the Nordic region, performs a curve decomposition of the contracted power to continue to organize the coupled clearing of the Nordic market, which results in the clearing of the coupled market as a basis for the settlement and execution of the coupled market. In literature [27], the marginal node tariff mechanism is commonly used in the US electricity market to differentially reflect the value of electrical energy in time and space by selecting a series of nodes that are less affected by blockages and awarding them with a fixed weight, and then calculating a representative weighted average price to realize the transaction convergence. The above studies are mostly on the monthly or weekly decomposition of the medium- and long-term contracted electricity, without considering the bidding space and bidding fluctuations in the spot market, which makes it difficult to realize the smooth connection from the medium- and long-term market to the spot market.

Aiming to address the aforementioned issues, this paper proposes an optimal decomposition method for monthly contracts of IESPs. This method takes into account the equilibrium of spot bidding and considers the coupling and complementarity of multiple types of energy sources within the IES, as well as the effective connection between the medium- and long-term market and the spot market. First, a daily decomposition model of monthly medium- and long-term contracts is constructed, with weekly equilibrium decomposition of the centralized bidding volume and the remaining multi-energy demand supplemented by the contracted volume, and the objective of minimizing the variance in the residual load after the decomposition of the medium- and long-term contracts and minimizing the cost of decomposition is to perform the daily decomposition of the medium- and long-term contracted volume. Secondly, based on the results of the daily decomposition of the medium- and long-term contracts, the day-ahead scheduling model of the IES is constructed by considering the coupling and complementarity of the multi-energy streams to minimize the sum of the cumulative daily operating costs of the network. A reasonable medium- and long-term energy decomposition method can reduce the volatility and uncertainty of the IESP's participation in the spot market transactions and ensure the effective execution of the medium- and long-term contracted energy at the day-ahead level.

To distinguish this paper from other studies, its main contributions can be summarized in the following three areas:

(1) Given the difficulties in the connection between the medium- and long-term market and the spot market and the volatility and uncertainty of the price and quantity of the spot market transactions, this paper designs a medium- and long-term monthly contract optimization decomposition strategy that can stabilize the bidding space in the spot market and reduce the fluctuation in the demand of spot bidding by considering the equilibrium of the bidding space in the spot market to ensure the smooth connection between the medium- and long-term market and the spot market;

(2) To ensure the effective execution of the mid-and long-term contract decomposition results daily and reduce the contract decomposition deviation, this paper constructs a day-ahead scheduling model of the IES that accounts for the coupling of multiple energy flows, takes into account the substitution and cross-complementary character-

istics of electricity, gas, and heat of multi-energy resources, and carries out the optimal scheduling of various types of resources based on the daily decomposition results of the mid-and long-term contracts;

(3) To effectively alleviate the pressure of energy supply at the energy supply point and realize the efficient interconnection and interaction of electricity, gas, and heat networks in the IES, this paper considers the dynamic pipe storage characteristics of the gas network and realizes the low storage and high generation of the gas network. The results show that the method can effectively reduce the dispatch operation cost.

2. The Connection Process of Multi-Energy Transactions between the Medium- and Long-Term Market and the Spot Market

Given the high flexibility of the spot market, the IESP, acting as a third-party entity, engages in supplemental multi-energy trading in the spot market after procuring the bulk of energy in the medium- and long-term market. A rational approach to decomposing medium- and long-term energy can mitigate the volatility and unpredictability of the IESP's involvement in spot market transactions, thereby fostering a seamless transition between the medium- and long-term market and the spot market. The challenge lies in effectively optimizing the daily decomposition of IESP's medium- and long-term traded energy while ensuring the stability of the bidding space in the spot market. On one hand, it is crucial to ensure that the decomposed medium- and long-term energy can be effectively implemented on the spot time scale. On the other hand, it is equally important to minimize the fluctuation in spot bidding demand to ensure a smooth bidding space for IESPs in the spot market. Moreover, the IESP must also thoroughly consider fairness among units when decomposing the capacity of units and gas source points. The interface between the medium- and long-term and the spot market is illustrated in Figure 1.

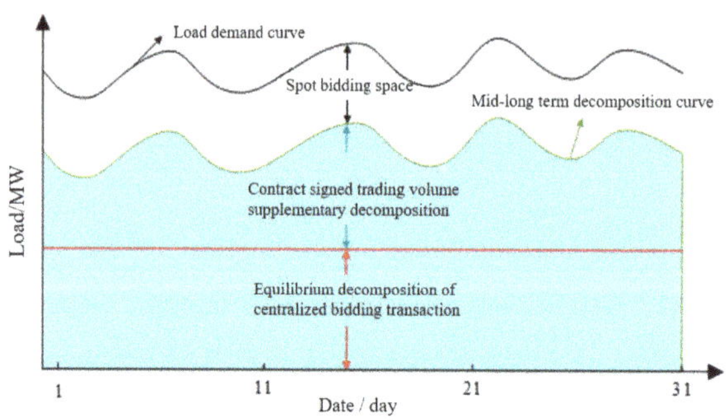

Figure 1. The decomposition and connection relationship between long-term contracts and spot markets of IESPs.

As illustrated in Figure 1, the IESP's decomposition of medium- and long-term contract volumes encompasses the equilibrium decomposition of centralized bidding transactions and supplementary decomposition of the contracted volume. The centralized bidding transactions are primarily decomposed through weekly equilibrium decomposition, ensuring a stable level of daily transaction decomposition. The contracted volume, on the other hand, serves as a supplementary decomposition to meet the remaining multi-energy demand, thereby mitigating the fluctuating amplitude of the medium- and long-term bidding transactions' tariffs. When conducting the monthly and weekly decomposition of multi-energy trading volumes, the decomposition schematic is demonstrated in Figure 2.

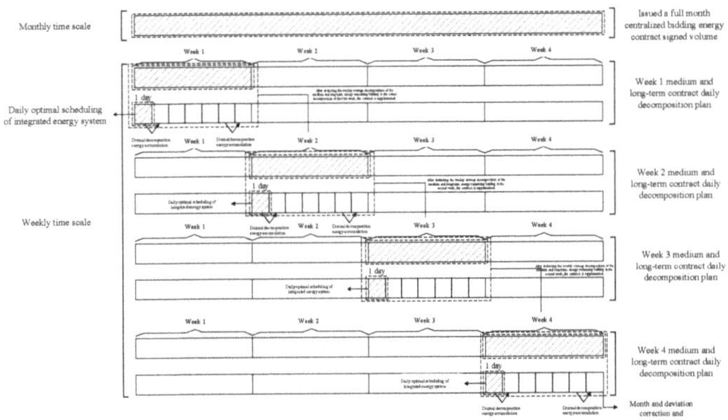

Figure 2. Optimal decomposition of long-term time scale contract considering spot bidding equilibrium of IESP.

When the IESP executes the optimized decomposition of the medium- and long-term contract equilibrium, it initiates a weekly decomposition of the total monthly contracted energy sourced from the transactions. Throughout the decomposition process, the centralized bidding trading volume is consistently maintained at a weekly equilibrium. The contracted volume serves as a supplement to the remaining medium- and long-term demand portion, and the weekly balance energy is computed into the contracted energy for the ensuing week, continuing in this manner until the final week of the month. Specifically, for the daily contract volume decomposition within a week, the IESP takes into account the daily IES dispatch optimization. The daily contract decomposition volume is accumulated to the medium- and long-term contracted trading volume allocated in the corresponding week, and the balance volume accumulated by the end of the week is counted as the allocable energy for the remaining week.

3. Daily Decomposition of Monthly Medium- and Long-Term Contracts Considering Spot Bid Equalization

3.1. Monthly Contract Optimization Decomposition Model

3.1.1. Objective Function

Drawing from the daily optimal scheduling model, this paper introduces an optimal decomposition method for the energy of medium- and long-term contracts. This method uses a week as the time window, decomposes the medium- and long-term energy from month to day, conducts rolling decomposition from week to week, and performs deviation assessment every month. The method takes into account the deviation in contract execution, load demand, network security checks, and the balance of spot bidding space, thus ensuring that the decomposition of medium- and long-term contracts is as efficient and balanced as possible. The objective is to minimize the variance in the residual load following the decomposition of the medium- and long-term contract and to minimize the cost of decomposition, as represented in Equations (1) and (2):

$$minR = \frac{1}{M_{mth}}[\sum_{w=1}^{W}\sum_{d=1}^{D}(P_{buye,d,w} - \overline{P}_{buye,d,w})^2 + (f_{buyg,d,w} - \overline{f}_{buyg,d,w})^2] \tag{1}$$

$$\begin{aligned} minOF &= \sum_{d=1}^{D}[\sum_{i=1}^{N_{gene}}(E_{e,i,d}^{bid}q_{e,i,d}^{bid} + E_{e,i,d}^{cone}q_{e,i,d}^{cone} + E_{e,i,d}^{buye}q_{e,i,d}^{buye}) \\ &+ \sum_{m=1}^{N_{gas}}(E_{g,m,d}^{bid}q_{g,m,d}^{bid} + E_{g,m,d}^{cone}q_{g,m,d}^{cone} + E_{g,m,d}^{buye}q_{g,m,d}^{buye})] \end{aligned} \tag{2}$$

where R represents the sum of the variances of the monthly residual load; OF stands for the cost of contract decomposition; M_{mth} is the number of days in the month; $P_{buye,d,w}$ is the total spot electricity purchase volume on day d of week w; $\overline{P}_{buye,d,w}$ is the average monthly spot electricity purchase volume on day d of week w; $f_{buyg,d,w}$ is the total spot gas purchase volume on day d of week w; $\overline{f}_{buyg,d,w}$ is the average monthly spot gas purchase volume on day d of week w.

3.1.2. Restrictive Condition

This section discusses the constraints of the monthly contract optimisation decomposition model, which are essential to ensure that the model is practical and meets the actual operational requirements. The specific constraints are listed below:

$$P_{buye,d,w} = \sum_{t=1}^{T}[\sum_{n=1}^{N_{EL}}(P_{e,n,t} - P_{WT,n,t} - P_{EH,n,t})] - \sum_{i=1}^{N_{gene}} E_{e,i,d,w} \tag{3}$$

$$P_{re,d,w} = \sum_{t=1}^{T}\sum_{n=1}^{N_{EL}}(P_{e,n,t} - P_{WT,n,t} - P_{EH,n,t}) \tag{4}$$

$$\overline{P}_{buye,d,w} = \frac{1}{M_{mth}}\sum_{w=1}^{W}\sum_{d=1}^{D} P_{buye,d,w} \tag{5}$$

$$E_{e,i,d,min} \leq E_{e,i,d,w} \leq E_{e,i,d,max}, \forall i \in N_{gene} \tag{6}$$

$$f_{buyg,d,w} = \sum_{t=1}^{T}[\sum_{k=1}^{N_{GL}}(f_{g,k,t} + f_{EH,k,t})] - \sum_{m=1}^{N_{gas}} E_{g,m,d,w} \tag{7}$$

$$f_{rg,d,w} = \sum_{t=1}^{T}\sum_{k=1}^{N_{GL}}(f_{g,k,t} + f_{EH,k,t}) \tag{8}$$

$$\overline{f}_{buyg,d,w} = \frac{1}{M_{mth}}\sum_{w=1}^{W}\sum_{d=1}^{D} f_{buyg,d,w} \tag{9}$$

$$E_{g,m,d,min} \leq E_{g,m,d,w} \leq E_{g,m,d,max}, \forall m \in N_{gas} \tag{10}$$

where $E^{bid}_{e,i,w}$ is the upper and lower bounds of the monthly contract decomposition to the daily contract energy; $P_{re,d,w}, f_{rg,d,w}$ is the simulated electricity/gas load demand on day d of week w.

Building upon the decomposition of the monthly contract, it is also essential for the IESP to consider the fulfilment of said contract. This paper contemplates the weekly decomposition of the monthly contract to ensure its effective decomposition and execution. Notably, the weekly decomposition energy comprises both the decomposition of centralized bidding energy and contract signing energy.

$$\sum_{d=1}^{D} E_{e,i,d,w} = E^{bid}_{e,i,w} + E^{con}_{e,i,w}, \forall i \in N_{gene} \tag{11}$$

$$\sum_{d=1}^{D} E_{g,m,d,w} = E^{bid}_{g,m,w} + E^{con}_{g,m,w}, \forall m \in N_{gas} \tag{12}$$

where $E^{bid}_{e,i,w}$ represents the centralized bidding electricity volume of generator set I in week w; $E^{con}_{e,i,w}$ is the monthly contract signing electricity volume of generator set I in week w; $E^{bid}_{g,j,w}$ is the centralized bidding gas volume of gas source point j in week w; $E^{con}_{g,m,w}$ is the contract signing gas volume of gas source point m in week w.

Based on the literature [25,28], in this paper, in the weekly optimal decomposition of the monthly energy, the bidding energy balanced decomposition is used to decompose the centralized bidding energy in a balanced way, and the remaining part is supplemented by using the contracted volume. The calculations are as follows:

$$E_{e,i,w}^{bid} = \frac{1}{W} E_{me,i}^{bid}, \forall i \in N_{gene} \tag{13}$$

$$E_{g,m,w}^{bid} = \frac{1}{W} E_{mg,m}^{bid}, \forall m \in N_{gas} \tag{14}$$

where $E_{g,m,w}^{bid}$ is the monthly bidding power of generating unit I; $E_{g,m,w}^{bid}$ is the monthly bidding gas volume of gas source point m.

For the remaining weeks, the centralized bidding energy decomposition and the contract signing volume decomposition are as follows:

$$E_{e,i,w}^{bid} = \frac{1}{4-j}[\sum_{w=j+1}^{W}\sum_{d=1}^{D} P_{re,d,w} - (E_{me,i}^{con} - \sum_{w=1}^{j} E_{e,i,w}^{con,real})], \forall i \in N_{gene} \tag{15}$$

$$E_{e,i,w}^{con} = \sum_{d=1}^{D} P_{re,d,w} - [\frac{1}{4-j}(E_{me,i}^{bid} - \sum_{w=1}^{j} E_{e,i,w}^{bid,real})], \forall i \in N_{gene} \tag{16}$$

$$E_{g,m,w}^{bid} = \frac{1}{4-j}[\sum_{w=j+1}^{W}\sum_{d=1}^{D} f_{rg,d,w} - (E_{mg,m}^{con} - \sum_{w=1}^{j} E_{g,m,w}^{con,real})], \forall m \in N_{gas} \tag{17}$$

$$E_{g,m,w}^{con} = \sum_{d=1}^{D} f_{rg,d,w} - [\frac{1}{4-j}(E_{mg,m}^{bid} - \sum_{w=1}^{j} E_{g,m,w}^{bid,real})], \forall m \in N_{gas} \tag{18}$$

where j is the week number;$E_{e,i,w}^{con,real}$ is the actual disaggregated bid and contracted power obtained in week w of generating unit I; $E_{g,m,w}^{con,real}$ is the actual decomposition of bidding gas and contracted gas obtained in week w at the gas source point m.

In addition to this, to ensure the stable operation of the system, the unit's mandatory energy constraints, network security constraints, and end-of-month deviation energy assessment constraints need to be taken into account, as shown in the following equation:

$$\begin{cases} 0 \le |\sum_{w=1}^{W} (E_{e,i,w}^{bid} - E_{e,i,w}^{bid,real})| \le 0.3 \times \sum_{w=1}^{W} E_{e,i,w}^{bid} \\ 0 \le |\sum_{w=1}^{W} (E_{e,i,w}^{con} - E_{e,i,w}^{con,real})| \le 0.3 \times \sum_{w=1}^{W} E_{e,i,w}^{con} \end{cases} \tag{19}$$

$$\begin{cases} 0 \le \sum_{w=1}^{W} (E_{g,i,w}^{bid} - E_{g,i,w}^{bid,real}) \le 0.1 \times \sum_{w=1}^{W} E_{g,i,w}^{bid} \\ 0 \le \sum_{w=1}^{W} (E_{g,i,w}^{con} - E_{g,i,w}^{con,real}) \le 0.1 \times \sum_{w=1}^{W} E_{g,i,w}^{con} \end{cases} \tag{20}$$

4. A Day-Ahead Scheduling Model for IESs Accounting for Multiple Energy Flow Coupling

4.1. Objective Function

After completing the medium- and long-term contract transactions, the IESP needs to perform a daily decomposition of the traded monthly electricity to facilitate the optimal operation of the electrical and thermal multi-energy network under its jurisdiction as well as to improve the interface relationship between the spot market and the medium- and long-term market. The optimal dispatch model of the multi-energy network is based on the objective of minimizing the sum of cumulative daily operating costs of the network, as shown in the following equation:

$$minC = \sum_{w=1}^{W}\sum_{d=1}^{D}(C_{gen,d,w} + C_{\text{grid},d,w} + C_{\text{buy},d,w}) \tag{21}$$

$$C_{gen,d,w} = \sum_{t=1}^{T}\sum_{i=1}^{N_{gene}} c_{start,i}P_{gene,i,t}\mu_{gene,i,t} \tag{22}$$

$$C_{\text{grid},d,w} = \sum_{t=1}^{T}[\sum_{i=1}^{N_{gene}}(a_{e,i}P_{gene,i,t}^2 + b_{e,i}P_{gene,i,t} + c_{e,i}) + \sum_{i=1}^{N_{EH}} c_{EH,i}P_{EH,i,t} + \sum_{i=1}^{N_{gas}} c_{gas,i}f_{gas,i,t}] \tag{23}$$

$$C_{\text{buy},d,w} = \sum_{t=1}^{T}(P_{buye,t}\rho_{spot,e,t} + f_{buyg,t}\rho_{spot,g,t}) \tag{24}$$

where C is the total monthly cost of the IESP's operating region; W, D, and T are the number of weeks, days, and moment numbers; $C_{gen,d,w}$ is the unit startup cost on day d of week w; $C_{grid,d,w}$ is the network operating cost on day d of week w; $C_{buy,d,w}$ is the cost of purchasing power in the spot market on day d of week w; $C_{dr,d,w}$ is the cost of the demand response to load on day d of week w; N_{gene}, N_{EH}, N_{gas}, N_{EL}, N_{GL}, N_{HL} are the number of generating units, the number of energy hubs, the number of gas points, the number of electrical load nodes, the number of gas load nodes, and the number of heat load nodes, respectively; $c_{start,I}$ is the unit startup cost of the ith unit; $\mu_{gene,i,t}$ is a 0–1 variable indicating the unit's operating status; $P_{gene,i,t}$ is the output of the ith unit in time period t; $a_{e,I}$,$b_{e,I}$,$c_{e,i}$ are the output cost coefficients for generating unit i, respectively; $c_{EH,i}$ is the unit cost of output for the energy hub; $P_{EH,i,t}$ is the injected power for the energy hub; $c_{gas,i}$ is the unit cost of gas purchased at the point of origin; $f_{i,t}$ is the output flow rate at time period t at the gas source point; $P_{buye,t}$, $\rho_{buye,t}$ are the spot market purchased power and purchased price in time period t; $f_{buyg,t}$, $\rho_{buyg,t}$ are the volume and price of gas purchased in the spot market at time t.

4.2. Restrictive Condition

In the context of an integrated energy network, gas pipelines exhibit certain storage characteristics during day-ahead optimized scheduling due to the relatively slow gas flow [29,30]. Such characteristics will have a large impact on the daily energy decomposition of the IESP in the region. This impact is likely to cause uneven energy decomposition and may lead to the failure of some energy decomposition plans. Therefore, it is necessary to consider the daily network-related constraints, including grid operation constraints, gas network operation constraints, heat network operation constraints, energy hub constraints, etc., in the energy decomposition.

4.2.1. Grid Operation Constraints

Based on the references [30], the current constraints of the power network are as follows:

$$P_{mn,t} = \frac{\theta_{m,t} - \theta_{n,t}}{X_{mn}} \tag{25}$$

$$P_{e,n,t} + \sum_{n\in m} P_{mn,t} = P_{EH,n,t} + \sum_{i=1}^{N_{gene}}(P_{buye,i,t} + P_{gene,i,n,t}) + P_{WT,n,t} \tag{26}$$

$$\mu_{gene,i,t}P_{gene,i,min} \le P_{gene,i,t} \le P_{gene,i,max}\mu_{gene,i,t} \tag{27}$$

$$RD_{gene,i} \le P_{gene,i,t} - P_{gene,i,t-1} \le RU_{gene,i} \tag{28}$$

$$\sum_{t=1}^{T} P_{gene,i,t} \le E_{gene,i,d}^{real} \tag{29}$$

$$(\mu_{gene,i,t-1} - \mu_{gene,i,t})(T_{\text{on},i,t-1} - T_{\text{on},i,min}) \geq 0 \quad (30)$$

$$(\mu_{gene,i,t} - \mu_{gene,i,t-1})(T_{\text{off},i,t-1} - T_{\text{off},i,min}) \geq 0 \quad (31)$$

where $P_{mn,t}$ is the line transmission power of the branch mn; $\theta_{mn,t}$ is the phase angle difference between node m and node n at time t; X_{mn} is the reactance of the branch mn; r_{mn} is the resistance of the branch mn; $P_{e,n,t}$ is the electrical load power of node n in time period t; $P_{gene,i,n,t}$ is the sum of the outputs of the generating units decomposed to node n in time period t; $P_{PV,n,t}$ is the outgoing power of the PV at node n for time period t; $P_{WT,n,t}$ is the output of wind power at node n in time period t; $P_{gene,i,min}$, $P_{gene,i,max}$ are the upper and lower output limits of generator set i, respectively; $RU_{gene,i}$, $RD_{gene,i}$ are the upward and downward climbing power of genset i; $E^{real}_{gene,i,d}$ is the actual total daily electricity for the medium- and long-term power breakdown to day d; $T_{on,i,t}$,t, $T_{off,i,t}$ are the number of hours the generating unit i has been out of service and the number of hours it has been shut down, respectively; $T_{on,i,min}$, $T_{off,i,max}$ are the minimum output duration and minimum downtime, respectively.

4.2.2. Gas Network Operational Constraints

Based on the reference [31], the dynamic properties of the gas network model are mainly reflected in the pipeline inventory. Therefore, the tidal flow constraints of the gas network are as follows:

$$f_{kl,t} = \tau_{kl,t} \times 5.72 \times 10^{-4} \sqrt{\frac{(p_{k,t}^2 - p_{l,t}^2)D^5}{FSL_{kl}}} \quad (32)$$

$$F = 0.0044(1 + \frac{12}{0.276D}) \quad (33)$$

$$\sum_{l \in k} f_{kl,t} = \sum_{k \in l} f_{kl,t} f_{lk,t} + f_{gas,k,t} + f_{buyg,k,t} - f_{EH,k,t} - (f_{g,k,t} - f_{drg,k,t}) \quad (34)$$

$$f_{kl,min} \leq f_{kl,t} \leq f_{kl,max} \quad (35)$$

$$-f_{gas,k,max} \leq f_{gas,k,t} \leq f_{gas,k,max} \quad (36)$$

$$M_{kl,t} = M_{kl,t-1} + f_{kl,t} + f_{lk,t} \quad (37)$$

$$\frac{C_{kl,con}}{L_{kl}} M_{kl,t} = p_{kl,t} \quad (38)$$

$$p_{k,min} \leq p_{k,t} \leq p_{k,max} \quad (39)$$

$$M_{kl,0} = M_{kl,T} \quad (40)$$

$$\sum_{t=1}^{T} f_{gas,k,t} \leq E^{real}_{gas,i,d} \quad (41)$$

where $f_{kl,t\,t}$ is the gas flow rate through pipe kl at time t; $\tau_{kl,t}$ is the flow of natural gas in pipeline kl from k to l. Then, flowing from l to k, $p_{k,t}$, $p_{l,t}$ are the pressures at nodes k and l; D is the pipe diameter; F is the pipe friction coefficient; Sis is the specific gravity of the gas; L_{kl} is the length of pipe kl; $f_{gas,k,t}$ is the output gas flow rate of the gas source at node k in period t; $f_{EH,k,t}$ is the gas flow rate output/consumed by the energy hub at node k in period t; $f_{g,k,t}$ is the gas load flow at node k in period t; $f_{kl,\,max}$, $f_{kl,\,min}$ are the upper and lower limits of the pipe flow, respectively; $f_{gas,k,\,max}$, $f_{gas,k,\,min}$ are the upper and lower limits of the output gas flow at the gas source point, respectively; $M_{kl,t}$ is the pipe inventory of pipe kl at period t; $C_{kl,\,the\,con}$ is a constant related to the length, diameter, etc., of the pipe; $p_{k,\,max}$, $p_{k,\,min}$ are the upper and lower limits of the air pressure at the pipe node; $E^{real}_{gas,i,d}$ is the actual gas production broken down to day d of the medium- and long-term contract signed by the IESP.

4.2.3. Thermal Network Operational Constraints

The thermal network model is divided into a hydraulic model and a thermal model, and the hydraulic model is as follows:

$$Am = m_q \tag{42}$$

$$Bh_f = 0 \tag{43}$$

$$h_f = Km|m| \tag{44}$$

where A is the network correlation matrix; m is the flow rate of water flow in the heat network pipe; m_q is the flow rate of water flow into the load node or out of the heat source node; B is the loop correlation matrix; h_f is the pressure drop in the pipe caused by friction loss; and K is the pipe resistance coefficient.

The thermal modeling is as follows:

$$H - H_{drh} = C_p m_q (T_s - T_o) \tag{45}$$

$$T_{out,ij,t} = (T_{s,t} - T_{a,t}) e^{-\lambda L_{ij}/C_p m_{ij,t}} + T_{a,t} \tag{46}$$

$$\left(\sum m_{out} T_{out}\right) = \sum (m_{in} T_{in}) \tag{47}$$

$$T_{s,min} \leq T_s \leq T_{s,max} \tag{48}$$

$$T_{o,min} \leq T_o \leq T_{o,max} \tag{49}$$

where H is the heat load matrix; H_{drh} is the load power matrix consisting of the tangential heat load; T_s is the nodal water supply temperature; $T_{s,min}$, $T_{s,max}$ are the upper and lower bounds of the nodal water supply temperature; T_o is the nodal return temperature; $T_{o,min}$, $T_{o,max}$ are the upper and lower bounds of the nodal return temperature; $T_{out,ij,t}$ is the exit temperature of the pipeline end; λ is the heat transfer coefficient; C_p is the specific heat capacity of water; m_{in} is the flow rate of the pipeline into the node; m_{out} is the flow rate of the pipeline out of the node; T_{in} is the temperature of the input pipeline end; T_{out} is the nodal mixing temperature.

4.2.4. Source Hub Constraint

The energy hub, as a key support for connecting the multi-energy network, internally involves cogeneration equipment to realize the conversion of natural gas to electricity and heat, and the external input–output model is shown below:

$$P_{EH,i,t} = f_{EH,i,t} H_{vg} \lambda_e \tag{50}$$

$$h_{EH,i,t} = f_{EH,i,t} H_{vg} \lambda_h \tag{51}$$

where H_{vg} is the calorific value of natural gas, taken as 7.8 kW·h/m^3; λ_e, λ_h are the natural gas to electricity and heat transfer efficiencies, respectively.

4.3. Model Solving Steps

The solution flow of the IESP monthly contract daily optimization decomposition method, constructed in this paper to account for the dynamic characteristics of the network, is illustrated in Figure 3. First, input the monthly total bidding volume and total contract signing volume, make $w = 1$, calculate the total bidding volume of the first week as the boundary of the contract supply energy within the week, make $d = 1$, input the daily load forecast value, spot market clearing price, and integrated energy network parameters, use the CPLEX solver to optimize and solve the IES scheduling model, and derive the daily output of each unit and gas source point. Make $d = d + 1$, solve the optimization model for D times of daily scheduling in a loop, superimpose the power of units and gas source points derived from D times, and determine whether the variance in the spot bidding volume reaches the minimum; if it meets the convergence condition, then make $w = w + 1$. Determine whether it meets the convergence condition corresponding to whether the monthly integrated energy optimization scheduling result reaches the minimum and whether the variance in the spot trading volume reaches the minimum of the weekly; if it meets the condition, output the day-to-day energy decomposition value of the medium- and long-term contract and the corresponding daily spot bidding space.

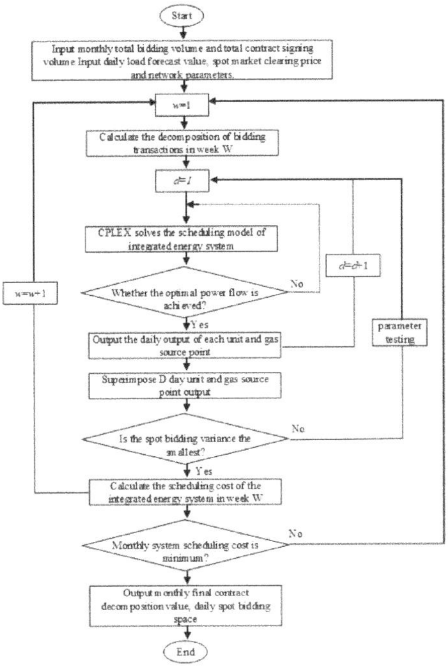

Figure 3. Solving process.

5. Calculus Analysis

5.1. Basic Data

To validate the effectiveness of the contract optimization decomposition method proposed in this chapter, the model is solved using the EMP + MINLP algorithm package in GAMS 43 with the YALMIP + CPLEX solver in MATLAB 2020a. In this chapter, an IES consisting of an IEEE 39-node electric system, a 20-node natural gas system, and a 6-node thermal system is used, as shown in Figure 4, with the specific parameters described in reference [31]. The system is connected with two EHs and the grid's 36 nodes are connected to the gas grid's 20 nodes with gas turbines. The power system consists of seven generating units, the gas system consists of four gas source points, and the heat system is supplied by the EHs. The EHs adopt the heat-determined power method, in which the electric efficiency of the EHs is 0.375 and the thermal efficiency is 0.5. The monthly contracted price is set at 13.3 USD/MWh, the centralized bidding price is set as a step offer, the power interval is [20,000, 60,000], the interval band is [20,000, 600], the initial band offer is set at 12.5 USD/MWh, the gas interval is set at [15, 500], and the gas interval is set at [15, 500]. Based on the data of a typical month, the initial electricity/gas/heat load curve is shown in Figure 5. The monthly spot electricity and gas prices and the 24 h electricity and gas prices for a particular day of the month are shown in Figure 6.

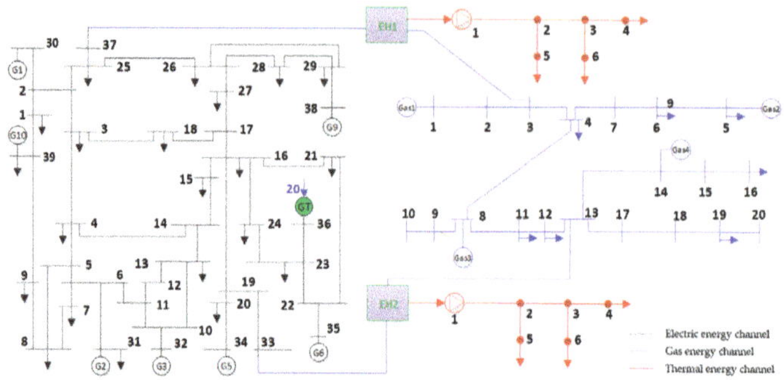

Figure 4. Integrated energy example system diagram.

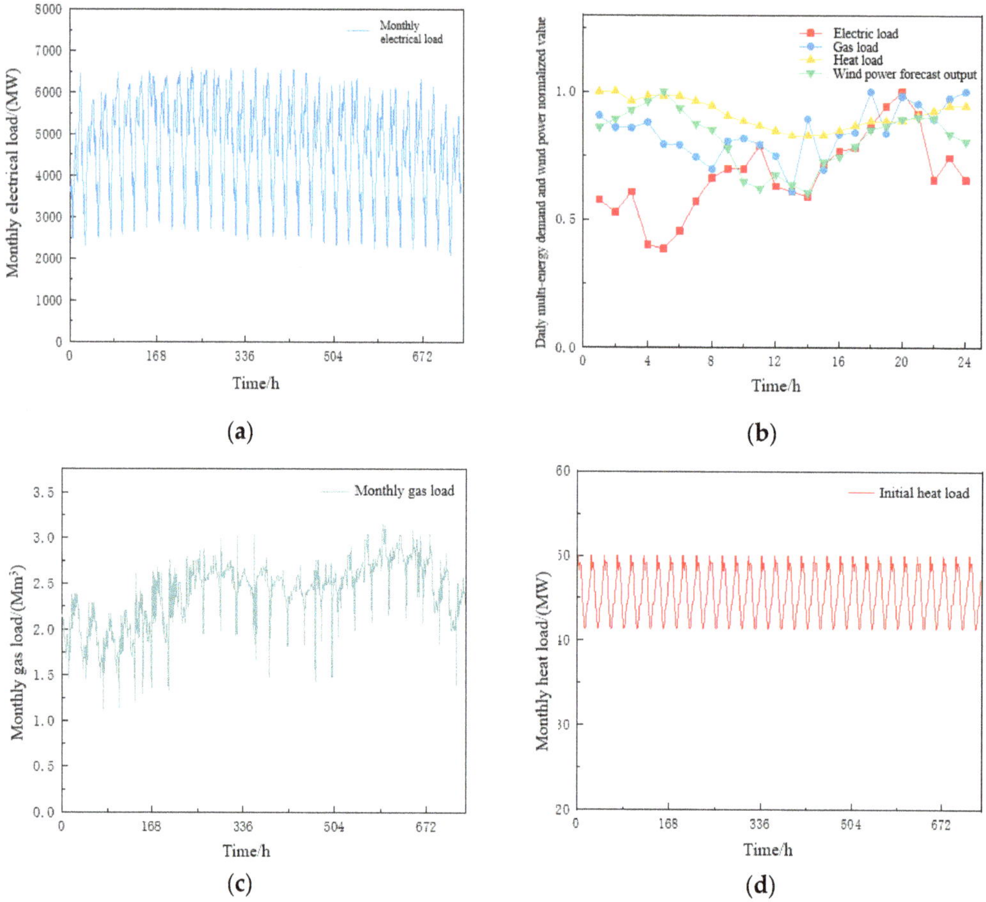

Figure 5. Initial load curve (**a**); (**b**) daily wind power and demand; (**c**) monthly gas demand; (**d**) monthly calorie requirements.

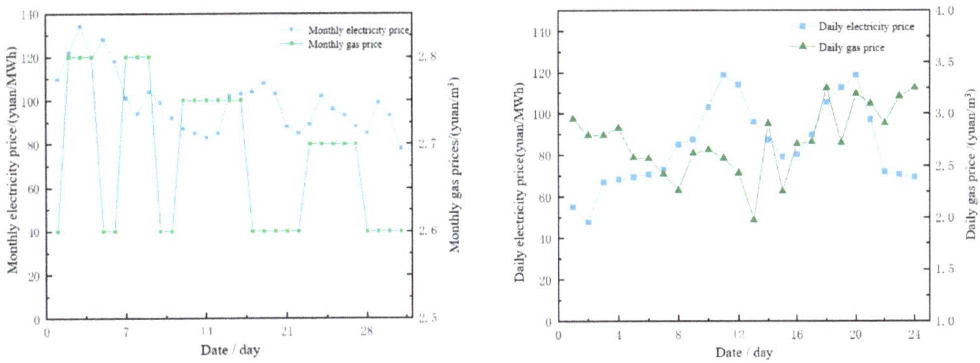

(**a**) (**b**)

Figure 6. Electricity price and gas price diagram. (**a**) Monthly spot energy price curve; (**b**) daily spot energy price curve.

5.2. Analysis of the Effectiveness of the Monthly Contract Equalization Decomposition Method

To verify the rationality and effectiveness of the spot bidding equilibrium decomposition method proposed in this paper, three scenarios are set up for comparative analysis, respectively:

S1: Contract decomposition using the traditional decomposition method, i.e., average contract volume decomposition;

S2: A contractual decomposition using the proposed centralized bidding volume equilibrium decomposition, but not considering the spot bidding equilibrium;

S3: A contractual decomposition using the proposed centralized bidding volume equalization decomposition and taking into account the spot bidding equalization.

Considering that the heat load demand of the IESP is all converted to energy via an EH and decomposed to electricity demand and gas demand, the heat load demand is converted to electricity/gas demand every month to arrive at the electricity demand and gas demand and the initial decomposition is carried out by using the methodology proposed in this chapter, as shown in Figure 7.

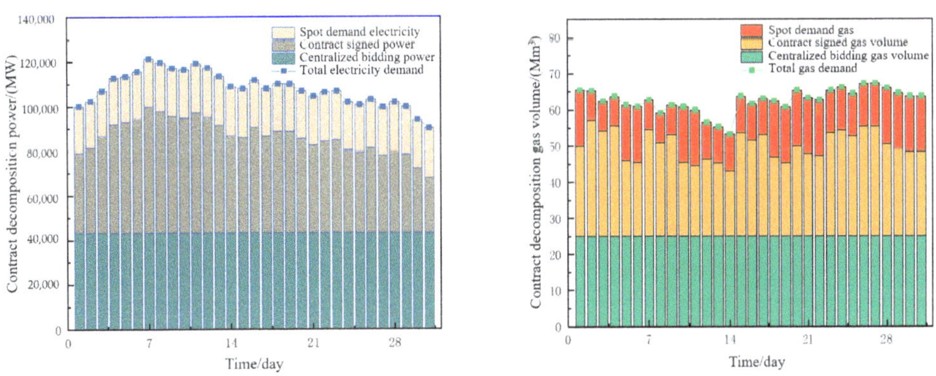

(**a**) (**b**)

Figure 7. Contract decomposition results in S3 scenario. (**a**) Preliminary breakdown of monthly electricity contracts; (**b**) preliminary breakdown of monthly gas contracts.

As can be seen from Figure 7, the trend of contracted electricity and gas volume is consistent with the monthly electricity/gas demand while the centralized bidding electricity and gas volume is in a constant state, and the daily decomposition of the spot purchasing energy obtained is basically in an equilibrium state due to the consideration of the equilibrium of the space of the spot bidding, but it also shows a certain tendency to increase or decrease with the fluctuation in the price.

The contract disaggregation costs for the three scenarios are shown in Table 1, and the fluctuations in spot purchase energy for each scenario are shown in Figure 8.

Table 1. Contract decomposition cost comparison under different scenarios.

Scene	Contract Decomposition Cost/(USD)	Spot Purchase Energy Variance
S1	45,871,627	24.5681
S2	44,302,449	87,193.68
S3	45,534,180	23.4273

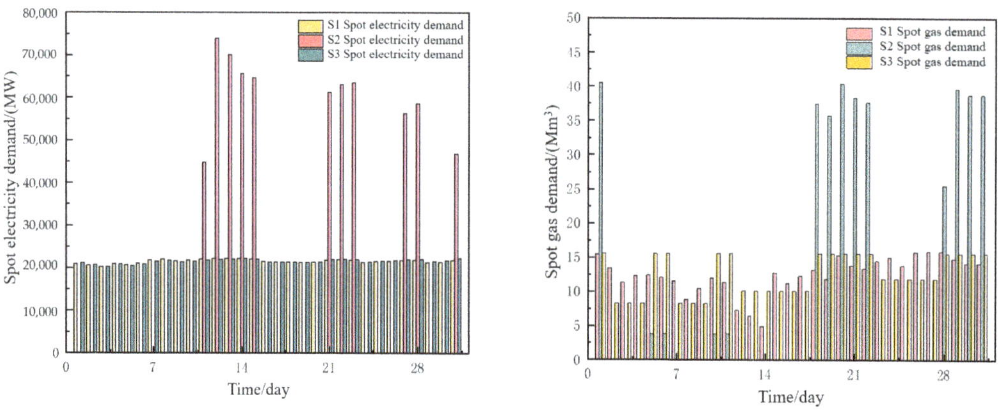

(**a**) (**b**)

Figure 8. Spot purchase energy in different scenarios. (**a**) Spot power contract breakdown results; (**b**) spot gas contract breakdown results.

As can be seen from Table 1, after S3 takes into account the bidding equilibrium decomposition, the cost of contract decomposition is reduced by about USD 0.33 million compared with S1, and the variance in spot energy purchase also decreases by about 4.64%; this leads to the conclusion that the equilibrium decomposition is less costly compared with the average decomposition method of the monthly contract, and the smaller variance in the spot energy purchase indicates the smaller fluctuation in the daily spot price, which is conducive to the stabilization of the intra-month spot market trading. As shown in Figure 8, S2 does not take into account the equilibrium of the spot space and is combined with the monthly energy price curve in Figure 6a. The IESP prefers to participate in the spot market to purchase energy at the time when the price is cheaper, and thus the contract decomposition cost of S3 is improved by about 2.7% relative to that of S2. Yet, the impact of S2 on the variance in spot energy purchased surpasses that of S3. This is evident in Figure 8, illustrating the daily differences in spot energy purchases. Consequently, daily spot bidding transactions throughout the month experience significant price–volume fluctuations. This instability hampers the smooth functioning of the spot market and adversely affects the profitability of participating market players' bids.

5.3. Analysis of the Actual Daily Performance of Monthly Contracts

After arriving at the preliminary decomposition results of the monthly contract, considering the execution effect of the actual decomposed quantity daily, the execution of the decomposed quantity is verified by utilizing the day-ahead contract decomposition model proposed in this paper, which results in the output of the generating units, the gas source points, and the EHs in the IES under the management of the IESP with the actual execution of the contract as shown in Figures 9 and 10.

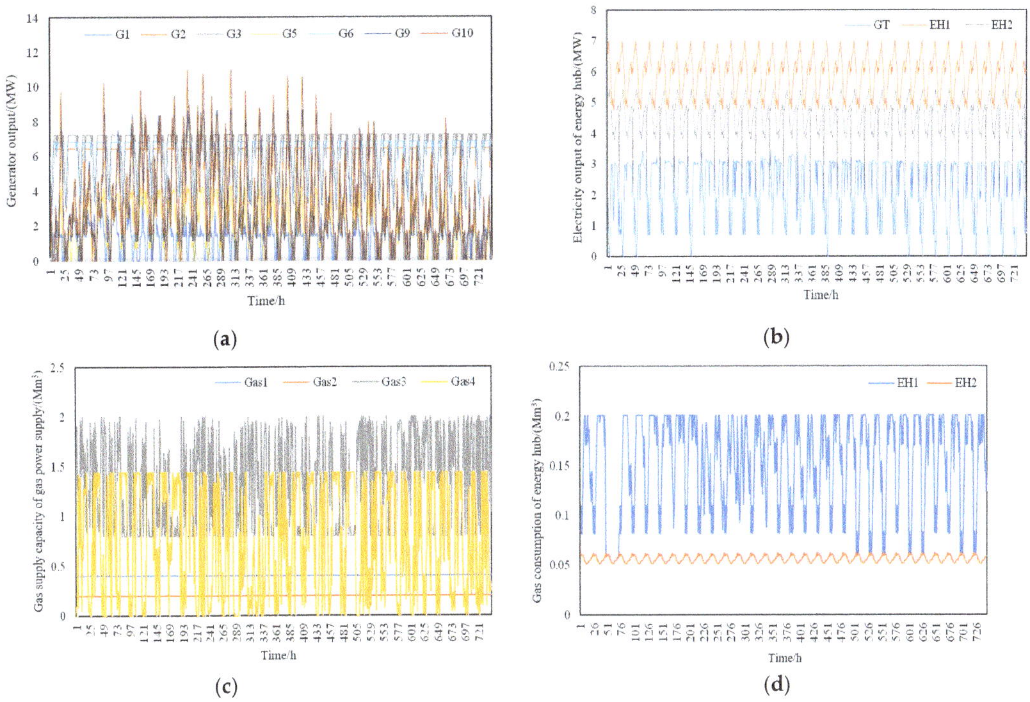

Figure 9. Daily actual power/gas output of IES in typical months. (**a**) Actual power output of generating units; (**b**) actual power output of energy conversion hub; (**c**) actual gas supply of the gas source point; (**d**) actual gas consumption of the energy conversion hub.

As can be seen from Figure 9a,b, the daily trend of the generating unit and gas point output is more consistent, which is mainly due to the overall load trend of each node of the system being similar, and the difference is mainly reflected in the difference in the amount of load, so the output curve of each unit is in a roughly consistent state. At the same time, due to the large difference in the operating cost of each unit, such as generators G1 and G5 with higher operating costs, the output of G1 and G5 is in a lower state in most cases while the remaining part of the unit's output cost is relatively low, such as G3 and G6, and therefore it has been in a full output state after startup. When the daily load is in high demand, the original unit output cannot meet the load demand, and then G9 and G10 units enhance their output to help the system reach the trend balance. From Figure 9c,d, it can be seen that the gas supply at the gas source point is basically in the full state, and the gas supply of Gas3 and Gas4 is in the complementary state. This is mainly because, on the one hand, the capacity of these two gas source points is larger, and the cost of the gas supply of Gas3 and Gas4 is comparatively smaller and its regulation ability is more flexible; on the other hand, due to the existence of the EH, the heat load demand of the two heat networks need to be supplied indirectly by the gas network, which makes the gas supply at the gas source point larger than that at the gas supply point. On the other hand, due to the existence of the EH, the heat load demand of both heat networks has to be indirectly supplied by the gas network, which makes the gas supply capacity of the gas source points larger, and, combined with the pipe storage characteristics of the gas network, the load demand of the heat network and the power grid can be effectively satisfied.

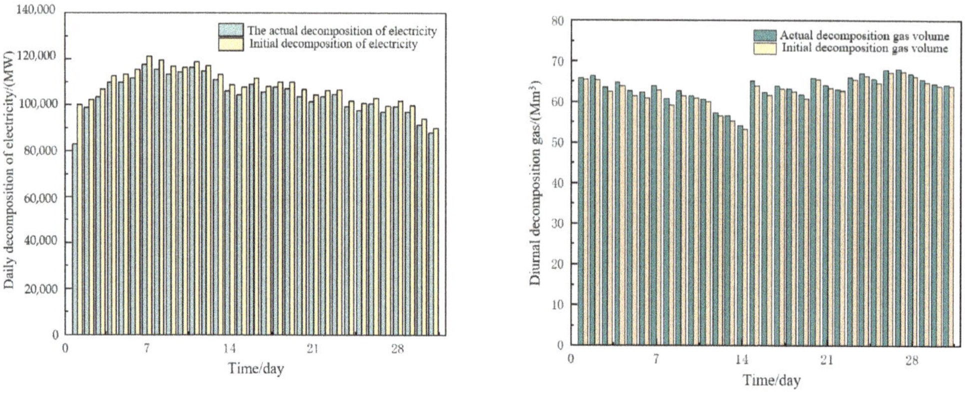

(a) (b)

Figure 10. Contract decomposition daily total execution. (**a**) Actual execution and preliminary decomposition of electric quantity; (**b**) actual execution and preliminary decomposition of gas quantity.

From Figure 10, it can be found that, after adopting the decomposition method proposed in this chapter, the gap between the actual daily execution and the initial contractual decomposition amount is small, and the actual electric/gas output execution deviations are all within 3% by constraints (49) and (50). And, comparing Figure 10a,b, it can be found that the actual gas supply exceeds the projected gas supply from the gas source point while the actual electricity output is smaller than the projected electricity output, which is mainly due to the heat-to-electricity method in the EH and the presence of a gas turbine in the grid. In this setup, the gas grid needs to provide a part of the additional gas to meet the natural gas demand of the GT and the EH, and the cost of the energy conversion equipment's output is smaller than the generating unit's operating costs, hence the positive and negative deviations described above.

The day of the month was selected for the attribution determination of the specific breakout contract volumes, and the breakout contract volumes attributed to each generating unit, gas source point, and EH are shown in Figure 11.

As depicted in Figure 11, the contract decomposition attribution for a day of the month reveals a distinct distribution trend. In Figure 11a, the three units on the left tend to derive revenue from centralized bidding contracts, driven by higher marginal costs. Consequently, there is an increase in the volume of centralized bidding in decomposition to enhance revenue from centralized bidding offers. Concurrently, in the period of higher spot power prices as illustrated in Figure 6b spot power sales contracts are pursued for profit maximization, aiming to capitalize on revenue opportunities. According to Figures 3–11a, since the marginal cost of G9, G10, and other units is small, they can choose the stable medium- and long-term physical contract accordingly and tend to sell electricity in the period of higher spot prices to improve their income from electricity sales and further reduce their operating costs. For the decomposition of the power output of the energy hub in Figure 11c, as the energy hub adopts the heat-determined power mode, its power output trend is approximately the same as the trend of the heat network load, and the trend is opposite to the price of electricity, which makes the decomposition of the power output of the energy hub not able to participate in the spot transaction effectively and obtain large profits from it. Therefore, in the relatively high price period, such as 4:00~8:00, the spot contract will be chosen, while, for the rest of the period, the centralized bidding and physical contract will be chosen to ensure their interests.

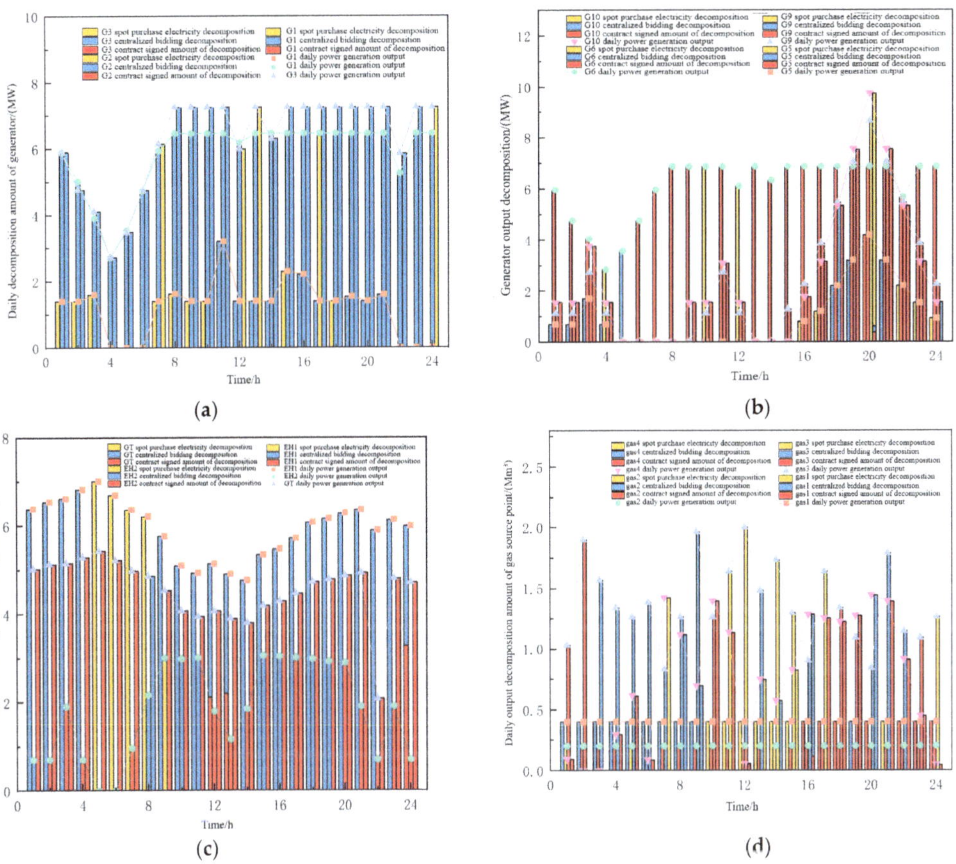

Figure 11. Daily output decomposition of generator set and gas source point. (**a**) Daily output decomposition of generator set; (**b**) daily output decomposition of generator set; (**c**) daily power decomposition of energy conversion hub; (d) daily output decomposition amount of gas source point.

Figure 11d shows the decomposition of gas supply at the gas source points, combining the high and low cost of gas supply at each gas source point: Gas1 and Gas2 are higher, Gas3 and Gas4 are lower, and, from the figure, it can be seen that Gas1 and Gas2 are more inclined to select spot bidding contracts to obtain more revenue from the sale of gas during the time of high gas prices, whereas Gas3 and Gas4 are more inclined to select centralized bidding contracts versus physical contracts. Gas3 and Gas4 prefer centralized bidding contracts and physical contract contracts, and spot contracts are only identified during a few peak gas price periods. The attribution of generating units and gas source points can effectively improve their operating income through the decomposition of their output and ensure the effective execution of medium- and long-term trading contracts and balanced trading in the spot market.

The dynamic pipe storage characteristics of the gas network are considered in the IES constructed in this paper, and, to verify the influence of the pipe storage characteristics on the optimal operation of the system, the changes in the gas supply volume at the gas source point and the changes in the network operation cost are analyzed with and without considering the pipe storage. The IES incurs an operating cost of USD 0.54 million without factoring in the pipe storage characteristic. With this, the gas network's operating cost amounts to USD 123.14 million. Conversely, considering that the pipe storage characteristic lowers the IES's operating cost to USD 0.525 million, with the gas network's operating cost reduced to USD 115.14 million, the network's operating cost decreases by approximately 2.01% when incorporating the pipe storage characteristic, while the gas network's cost drops by about 6.4%. Therefore, the operating cost of the system can be effectively reduced after considering the pipe storage characteristics of the gas network to promote the economic and stable operation of the IES. Specifically, the changes in the gas supply volume of each gas source point in the gas network are shown in Figure 12, and the changes in the gas supply volume of Gas3 and Gas4 are given in Figure. Since the maximum capacity of Gas1 and Gas2 in the gas network is small and is at full output in both scenarios, it is not discussed here.

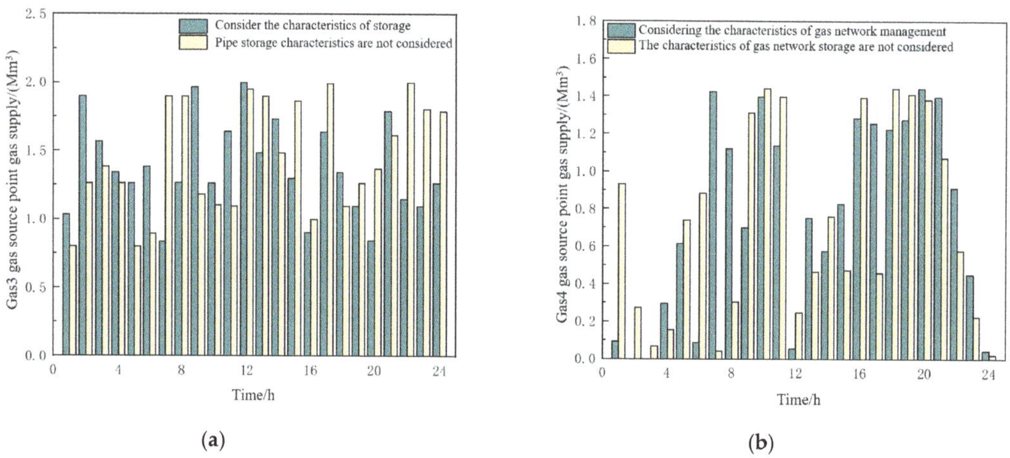

Figure 12. Change trend of gas supply at gas source point considering pipe storage characteristics. (**a**) Gas3 gas supply change; (**b**) Gas4 gas supply change.

As can be seen in Figure 12, after considering the dynamic pipe storage characteristics of the gas network, the gas supply at Gas3 is significantly lower during peak demand hours, such as 8:00~12:00 and 20:00~24:00, compared with the gas supply at Gas3 without considering the pipe storage, and the gas output at Gas3 is relatively higher during the trough hours of the gas load, when the pipeline gas storage is used to release gas in the peak hours to alleviate the pressure of gas supply at Gas3. The gas supply at Gas3 and Gas4 is the same during 0:00~4:00 and 8:00~12:00, and Gas3 and Gas4 are also the same during 0:00~4:00 and 8:00~12:00. The pressure of gas supply at the gas source point is also the same for gas supply at Gas4 during the hours of 0:00 to 4:00 and 8:00 to 12:00, and Gas3 and Gas4, due to the small difference in cost, will have a relatively lower output of the other gas source point when the output of one of the gas source points is higher on the basis of satisfying the balance of the trend of the gas network. Specifically, the total gas network inventory changes are shown in Figure 13.

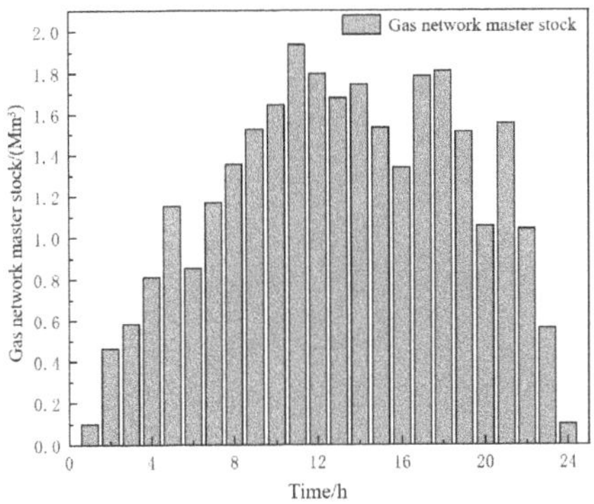

Figure 13. Change in the trend of gas network master stock.

As can be seen from Figure 13, natural gas pipelines store gas during the gas load trough time and supply gas during the gas load demand peak time, forming a complementary trend with the output of the gas source point, effectively alleviating the gas supply pressure at the gas source point and stabilizing the pipeline air pressure of the gas network to a certain extent, preventing the risk of overrun of the pipeline air pressure. In summary, the pipe storage characteristics of the gas network can help the IES to alleviate the energy supply pressure at the energy supply point.

6. Conclusions

Aiming at the problem that there are fewer medium- and long-term decomposition methods for electricity/gas multi-energy contracts under multi-energy coupling and that the connection between integrated energy medium- and long-term and spot markets is unclear, this chapter proposes an optimal decomposition method for IESPs' monthly contracts that takes into account the equilibrium of spot bidding, and the specific conclusions are as follows:

(1) The proposed method in this chapter considers the equilibrium decomposition of the centralized bidding volume as well as the equalization of the spot bidding space, which reduces the contract decomposition cost by about USD 0.33 million compared to the contract decomposition cost without considering the equilibrium decomposition, and the spot energy purchase variance also decreases by about 4.64%; at the same time, the consideration of the spot bidding variance can also effectively alleviate the fluctuation in the spot market and achieve smooth convergence of the spot market with the medium- and long-term market;

(2) In this chapter, the daily operation plan of the IES is formulated based on the preliminary decomposition of the contract, which ensures that the deviation in daily energy supply and output does not exceed 3% in response to the assessment index at the end of the month, and then the optimized decomposition of the attributed amount of the medium- and long-term contract and the spot purchase plan is carried out so that the daily decomposition of the contract can maximize the operating income of each supplier;

(3) Considering that the dynamic pipe storage characteristics of the gas network trim the overall daily operating cost of IES by 2.01% compared to the scenario without this consideration, the noteworthy cost reduction is predominantly in the gas network, lowered by approximately 6.4%. Simultaneously, the gas source point's output exhibits a smoother trend. This substantiates the effectiveness of gas network pipe storage characteristics in easing energy supply pressure at the energy supply point and achieving efficient interconnection of the electricity, gas, and heat networks in the IES.

Author Contributions: Conceptualization, C.W.; methodology, C.W.; data curation, Z.W.; writing—original draft, X.J., Y.H. and D.F.; writing—review and editing, X.J.; visualization, D.F.; supervision, C.W. and D.F.; project administration, C.W., Z.W. and X.J.; funding acquisition, C.W. and Z.W. All authors have read and agreed to the published version of the manuscript.

Funding: This work was supported by the Postdoctoral Research Funding Program of Jiangsu Province Under Grant No. 2021K622C.

Data Availability Statement: Data are contained within the article.

Conflicts of Interest: Author Donglou Fan was employed by the company Wiscom System Co., Ltd. The remaining authors declare that the research was conducted in the absence of any commercial or financial relationships that could be construed as a potential conflict of interest.

References

1. Haaskjold, K.; Pedrero, R.A. Long-term optimization of the Norwegian energy system under the influence of the European power market. In Proceedings of the 2023 19th International Conference on the European Energy Market (EEM), Lappeenranta, Finland, 6–8 June 2023; pp. 1–6. [CrossRef]
2. Gomez, T.; Herrero, I.; Rodilla, P.; Escobar, R.; Lanza, S.; de la Fuente, I.; Llorens, M.L.; Junco, P. European Union Electricity Markets: Current Practice and Future View. *IEEE Power Energy Mag.* **2019**, *17*, 20–31. [CrossRef]
3. Lu, T.; Zhang, W.; Wang, Y.; Xie, H.; Ding, X. Medium and Long-term Trading Strategies for Large Electricity Retailers in China's Electricity Market. *Energies* **2022**, *15*, 3342. [CrossRef]
4. Wang, K.; Wang, X.; Jia, R.; Dang, J.; Liang, Y.; Du, H. Research on Coupled Cooperative Operation of Medium and Long-term and Spot Electricity Transaction for Multi-Energy System: A Case Study in China. *Sustainability* **2022**, *14*, 10473. [CrossRef]
5. Manjunatha, H.M.; Purushothama, G.K.; Nanjappa, Y.; Deshpande, R. Auction-Based Single-Sided Bidding Electricity Market: An Alternative to the Bilateral Contractual Energy Trading Model in a Grid-Tied Microgrid. *IEEE Access* **2024**, *12*, 48975–48986. [CrossRef]
6. Kohansal, M.; Samani, E.; Mohsenian-Rad, H. Understanding the Structural Characteristics of Convergence Bidding in Nodal Electricity Markets. *IEEE Trans. Ind. Inform.* **2021**, *17*, 124–134. [CrossRef]
7. Cui, M.Y.; Xuan, M.Y.; Lu, Z.G.; He, L.C. Operation optimization strategy of multi-integrated energy service companies based on cooperative game theory. *Proc. CSEE* **2022**, *42*, 3548–3564. [CrossRef]
8. Li, T.; Gao, C.; Chen, T.; Jiang, Y.; Feng, Y. Medium and long-term electricity market trading strategy considering renewable portfolio standards in the transitional period of electricity market reform in Jiangsu, China. *Energy Econ.* **2022**, *107*, 105860. [CrossRef]
9. Huang, Y.; Qu, K.; Li, C.; Si, G. Research on Modeling Method of Medium-and long-term Wind Power Time Series Based on K-means MCMC Algorithm. *Power Syst. Technol.* **2019**, *43*, 2469–2476.
10. Li, C.; Hu, Y.; Zhao, H.; Li, X. General Model and Algorithm for Contract Energy Decomposition. *Power Syst. Autom.* **2007**, *31*, 26–30.
11. Wang, Y.; Zhang, P.; Shu, J.; Yang, Y. The economic distribution of contract trading volume and bidding trading volume in the daily plan. *Power Syst. Autom.* **2002**, *26*, 45–48.
12. Huang, Y.; Wang, X.; Zhang, W.; Cao, C. Energy decomposition of long and middle term contract in multi-energy system. In Proceedings of the 2019 IEEE Innovative Smart Grid Technologies-Asia (ISGT Asia), Chengdu, China, 21–24 May 2019; pp. 3735–3740.
13. Guntermann, C.; Tebbenjohanns, P.; Moser, A. Price convergence in an integrated simulation of the electricity and the gas market. In Proceedings of the 2023 19th International Conference on the European Energy Market (EEM), Lappeenranta, Finland, 6–8 June 2023; pp. 1–5. [CrossRef]
14. Sotiropoulos, E.; He, Y.; Hildmann, M.; Andersson, G. Modeling of electricity load for forward contract pricing. In Proceedings of the 2013 IEEE Power & Energy Society General Meeting, Vancouver, BC, Canada, 21–25 July 2013; pp. 1–5. [CrossRef]
15. Li, L.; Wang, N.; Chen, Y. Research on the Trial Operation of Mid-term and Long-term Transactionsthe Southern Electricity Spot Market Environment. *Price Theory* **2020**, *2*, 47–50. [CrossRef]
16. Chen, Z.; Yang, C.; Zhang, B.; Han, J.; Yang, Y.; Zhang, T.; Chen, Y. Design of Bilateral Trading Mechanism for Gansu Electricity Spot Market. *Glob. Energy Interconnect.* **2020**, *3*, 441–450.
17. Cao, H.; Qiu, Z.; Xiang, J.; Hao, Q.; Gui, W. Wind Power Accommodation Model Considering the Link of Medium and Long-term Transactions with Short-term Dispatch. *Power Grid Technol.* **2020**, *44*, 4200–4210. [CrossRef]
18. Yucekaya, A. Electricity trading for coal-fired power plants in Turkish power market considering uncertainty in spot, derivatives and bilateral contract market. *Renew. Sustain. Energy Rev.* **2022**, *159*, 112189. [CrossRef]
19. EPEX SPOT. PCR Project Main Features. [EB/OL]. [2020-4-1]. Available online: https://www.EPEXSPOT.com (accessed on 3 January 2024).

20. Xu, C.; Pang, K.; Wen, F.; Palu, I.; Gong, J.; Xie, Y.; Chen, C. Decomposition model of contract for difference considering market power mitigation. In Proceedings of the2020 IEEE International Conference on Environment and Electrical Engineering and 2020 IEEE Industrial and Commercial Power Systems Europe(EEEIC/I& CPS Europe), Madrid, Spain, 9–12 June 2020; pp. 1–5.
21. Nie, Z.; Gao, F.; Wu, J.; Guan, X.; Liu, K. Contract for difference energy decomposition model for maximizing social benefit in the electricity market. In Proceedings of the World Congress on Intelligent Control and Automation, Guilin, China, 12–15 June 2016; pp. 2449–2454.
22. Samani, E.; Mohsenian-Rad, H. A Data-Driven Study to Discover, Characterize, and Classify Convergence Bidding Strategies in California ISO Energy Market. In Proceedings of the 2021 IEEE Power & Energy Society Innovative Smart Grid Technologies Conference (ISGT), Washington, DC, USA, 16–18 February 2021; pp. 1–5. [CrossRef]
23. Miao, S.; Luo, B.; Shen, J.; Cheng, C.; Li, G.; Sun, Y. Short-Term Multi-Objective Hydro-Thermal Generation Dispatch Considering Electricity Market Transition and Mid-and Long-Term Contract Decomposition. *Power Grid Technol.* **2018**, *42*, 2221–2231. [CrossRef]
24. Zhang, S.; Liu, S.; Wang, Y.; Wu, B. A distributionally robust optimization model for power generators' medium and long-term contracted energy decomposition. *Power Syst. Prot. Control.* **2023**, *51*, 71–80. [CrossRef]
25. Wang, Y.; Yu, J.; Liu, Z. Decomposition of Yearly Bided Volume Based on Rol-uniformization of Monthly Competitive Bidding Spaces. *Power Syst. Autom.* **2006**, *30*, 24–2764.
26. Chen, Q.; Zhang, W.; Teng, F.; Guo, H.; Liu, X.; Jiang, N. Transmission Mechanisms and Coupling Approaches in European Transnational Electricity Markets. *Glob. Energy Interconnect.* **2020**, *3*, 423–429. [CrossRef]
27. Borisovsky, P.A.; Eremeev, A.V.; Grinkevich, E.B.; Klokov, S.A.; Vinnikov, C.V. Trading hubs construction for electricity markets. In *Optimization in the Energy Industry*; Springer: Berlin/Heidelberg, Germany, 2010.
28. Xu, S.; Liu, J. Decomposition of Yearly Bided Volume Based on Roll-uniformization of Monthly Competitive Bidding Spaces. *Modern Electric Power.* **2023**, *40*, 505–513. [CrossRef]
29. Mengelkamp, E.; Staudt, P.; Garttner, J.; Weinhardt, C. Trading on local energy markets: A comparison of market designs and bidding strategies. In Proceedings of the 2017 14th International Conference on the European Energy Market (EEM), Dresden, Germany, 6–9 June 2017; pp. 1–6. [CrossRef]
30. Xu, J.; Hu, J.; Liao, S.; Ke, D.P.; Wang, Y.L.; Sun, R.F. Collaborative optimization of integrated energy system considering network dynamic characteristics and integrated demand response. *Power Syst. Autom.* **2021**, *45*, 40–48.
31. Wu, G.; Liu, J.; Xiang, Y.; Shen, X.D.; Ma, Y.H. Day-ahead optimal scheduling of integrated electricity and natural gas system with medium and long-term e electricity contract decomposition and wind power uncertainties. *Power Syst. Autom. Equip.* **2019**, *39*, 246–253.

electronics

Article

Optimal Capacity Configuration of Energy Storage in PV Plants Considering Multi-Stakeholders

Fan Sun *, Weiqing Wang and Dongliang Nan

School of Electrical Engineering, Xinjiang University, Urumqi 830047, China; wwq59@xju.edu.cn (W.W.); nandongliang@dky.xj.sgcc.com.cn (D.N.)
* Correspondence: sunfan@stu.xju.edu.cn; Tel.: +86-152-7656-8077

Abstract: With the integration of large-scale renewable energy generation, some new problems and challenges are brought for the operation and planning of power systems with the aim of mitigating the adverse effects of integrating photovoltaic plants into the grid and safeguarding the interests of diverse stakeholders. In this paper, a methodology for allotting capacity is introduced, which takes into account the active involvement of multiple stakeholders in the energy storage system. The objective model for maximizing the financial proceeds of the PV plant, the system for the storage of energy, and a power grid company is studied. Then, in order to maximize the benefit of three stakeholders, a modified particle swarm optimization algorithm is devised, employing the prevailing typical allocation strategy. Finally, a case study is provided based on the modified IEEE 14-bus and the actual power grid from South Xinjiang, China. The simulation results and findings of the case study conclusively illustrate that the proposed methodology adeptly ensures the maximization of interests for the triad of stakeholders.

Keywords: multi-stakeholders; photovoltaic (PV) plant; optimal capacity of energy storage; modified particle swarm optimization

1. Introduction

To address global electricity demand in an environmentally sustainable manner, one pivotal approach involves the establishment of an innovative power system predominantly fueled by the robust advancement of renewable energy sources, encompassing hydropower, solar power, and wind power [1,2]. However, the grid characterized by a substantial share of photovoltaic (PV) faces great risks as PV has strong randomness and volatility owing to the intrinsic characteristic associated with PV generation [3,4]. The integration of energy storage has the capability to mitigate the variability in renewable energy output without necessitating alterations to the current grid infrastructure. Also, it can alleviate the strain caused by the peaking of general units while harnessing the potential of sustainable energy [5–7]. The collaborative undertaking of PV and stored energy can enhance the flexibility and ensure the stability of the sophisticated optical data archival system, which can bring maximum benefits to the grid. Hence, investigating the storage capability of the energy reservoir is crucial given the substantial investment costs associated with energy storage.

Over the past few years, an abundance of research has focused on the configuration to optimize the energy storage capacity of PV plants. Bullichthe-Massagué et al. (2020) and Zhang et al. (2021) summarized and analyzed different characteristics of energy storage. The conclusion was the establishment of an optimal energy storage solution for providing diverse services in large PV plants [8,9]. Shi et al. (2020) introduced an innovative hierarchical optimization approach to enhance the efficiency of both sustainable energy and energy reservoir [10]. Li et al. (2021) and Liu et al. (2021) both proposed optimization models for energy storage systems (ESS) aimed at enhancing PV consumption efficiency in the distribution network [11,12]. Zhang et al. (2020) proposed an advanced heuristic algorithmic approach of determining capacity for regulatory control in an integrated incremental hydropower PV-pumped storage system [13]. Zhu et al. (2023) developed a profitable energy storage capacity optimization model [14]. Zhang et al. (2019) and Chaima et al. (2021) proposed fast configuration methods for energy storage derived from the forecasting of PV and an energy reservoir topologyed hydro storage–PV plant system [15,16]. Deng et al. (2021) and An et al. (2019) devised an established optimization scheduling model, which ensured the steadiness and dependable functioning of an expansive multi-energy coupling system [17,18]. The above-mentioned results mainly focus on the siting and capacity-setting methods for ESS to promote PV consumption. However, few studies discuss the optimization capacity allocation in terms of the benefits.

Energy reservoirs can also be applied as a flexible adjustable resource to gain some benefits from participation in grid operations. Contreras-Ocana et al. (2017) and Wang et al. (2018) explored mechanisms to enhance the role of energy storage in grid operations [19,20]. Contreras-Ocana et al. developed an optimization model for aggregators to participate in grid peaking and dispatching through competitive bidding, while Wang et al. introduced a capacity contract mechanism for power sellers, facilitating optimal trading strategies. Also, it was argued that power sellers can achieve real-time regulation of distributed energy storage resources through a certain trading mechanism. Mortaz et al. (2019) and Croce et al. (2020) proposed advanced strategies to optimize energy distribution and investment [21,22]. Mortaz et al. utilized a tri-level stochastic optimization program to assess the impact of zonal diversification on investment portfolios. Croce et al. introduced a machine learning-enhanced multi-criteria optimization methodology to optimize energy zone flexibility, aiming to reduce peak power consumption and prevent reverse power flow. By taking into account the cost of the whole life cycle and the benefit of the ESS, Li et al. (2020) established a bilayer optimization model to solve the challenge of voltage fluctuations of distributed power sources in microgrids [23]. Wu et al. (2019) and Mediwaththe et al. (2020) both focused on optimizing smart grid operations [24,25]. Wu et al. developed a distributed hierarchical equivalence algorithm to solve economic challenges in smart grids, while Mediwaththe et al. explored integrating energy storage with PV for demand management within acceptable voltage levels, employing a Stackelberg game based on a linearized model to enhance energy trading benefits for storage providers. To solve the impossibility of quantifying the indirect benefits of energy storage, Wang et al. (2020) established an economic benefit and evaluation approach taking into account a reduction in unitary deficits and postponed returns from investment [26]. Zhang et al. (2022) proposed a capacity allocation method with maximum energy storage benefit with the life cycle as the optimization objective [27]. The energy management strategy considering peak and valley regulation was proposed for the mixed energy storage system of the electrified railroad. Cui et al. (2017) established a two-stage energy storage arbitrage model to address the chronic energy storage arbitrage problem of maximizing annual returns in a grid characterized by a substantial share of wind energy [28]. Zhao et al. (2021) developed a model to assess the benefits of energy storage system scheduling strategies based on time-of-use tariffs, focusing on maximizing economic gains [29]. Ying et al. (2023) targeted the uncertainties of wind power with a capacity configuration for pneumatic ESS, aiming to enhance economic outcomes [30]. Meanwhile, Sun et al. (2020) proposed an energy flow-based optimization strategy for the integrated energy system, adding a different dimension to optimizing energy resources [31]. All three studies aim at optimizing energy storage for economic benefits. Yang et al. (2021) proposed a framework of analyzing battery operation in an Australian National Electricity Market electricity spot; an operating strategy was proposed based on battery degradation costs [32]. However, most of the aforementioned results mainly focus on analyzing the benefits of energy storage without taking into account other stakeholders. Therefore, an important topic is the optimization of the capacity configuration of ESS from the perspective of multiple interests, which is worth studying.

Building upon the preceding discussions, in this paper, a capacity optimization allocation infrastructure is proposed based on maximizing the net benefit by considering multiple interests (such as PV plants, energy storage systems, and grid companies). The auxiliary service benefits of energy storage are fully exploited by an improved particle swarm optimization (PSO). Finally, simulation and analysis are carried out in the adapted IEEE 14-bus network and a real-world electricity grid in a region of South Xinjiang, China, which are expected to demonstrate the efficacy and utility of the proposed approach.

2. Photovoltaic-Storage Architecture

The structure of a typical multi-source interaction system is shown in Figure 1. The integration of large-scale PV into the electricity network, with the inherent randomness of PV generation, will significantly affect the stability of energy supply and power quality in the grid. The energy storage system has the characteristics of rapid bidirectional power regulation. It can significantly enhance the negative impact of PV volatility on the grid. The configuration of the combined ESS for PV is depicted in Figure 1. The energy reservoir device can be selectively connected to either the DC or AC side of the PV inverter as needed. It is convenient to retrofit the existing PV generator set when connected from the AC side. The investment can be reduced when connected from the DC side.

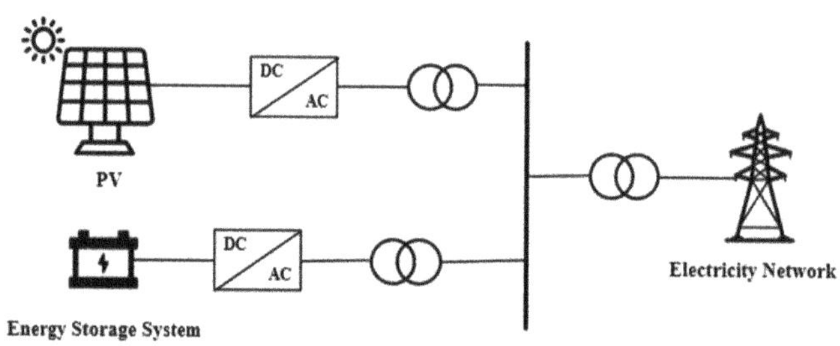

Figure 1. A typical multi-source interaction system.

3. Multi-Stakeholder Model

The reservoir of energy equipment is integrated into the pre-existing power grid to control the storage system to store and release energy according to the change in load and PV plant output. In this paper, three parties' revenue (PV plant revenue, energy storage system revenue, and grid company revenue) is taken as the optimization object and an intricate model for optimizing multiple objectives is developed.

3.1. *Revenue Model of PV Plant*

Revenues from PV plants can be expressed as

$$F_{1,\text{income}} = W_{1.PSI} + W_{1.NES} + W_{1.RBI}\frac{\tau}{1-(1+\tau)^{-T_1}} \tag{1}$$

$$W_{1.RBI} = P_1 D_1 \tag{2}$$

where τ represents the prevailing discount rate, $\tau/[1-(1+\tau)^{-T_1}]$ is the equivalent annual coefficient of the device, P_1 is the total installed capacity of the PV plant, D_1 is the scrap price of PV cells per unit of power, $W_{1.PSI}$ is the annual electricity sales revenue, $W_{1.NES}$ is the renewable energy subsidies of government, and $W_{1.RBI}$ is the PV battery scrap income.

$$F_{1,\text{cost}} = C_{1.ICC}\frac{\tau}{1-(1+\tau)^{-T_1}} + C_{1.OMC} \tag{3}$$

$$C_{1.ICC} = P_1 U_1 \tag{4}$$

$$C_{1.OMC} = P_1 M_1 \tag{5}$$

where U_1 is the cost of PV cells per unit of power and M_1 is the annual maintenance cost of the PV plant per unit of power. T_1 is the lifespan of the photovoltaic power plant.

Therefore, the net income model of the PV plant can be expressed as

$$F_1 = F_{1,\text{income}} - F_{1,\text{cost}} \tag{6}$$

3.2. Revenue Model of ESS

The advantages of ESS mainly include peak saving and backup, which can be represented as

$$F_{2.\text{income}} = \sum_{t=1}^{T} c_{PE} E_{2.PE}^{t} + \sum_{t=1}^{T} c_{RE} E_{2.RE}^{t} \tag{7}$$

where c_{PE} represents the unit peak saving income, c_{RE} is the unit standby income, and $E_{2,PE}^{t}$ and $E_{2,RE}^{t}$ are the peak saving power and spare capacity delivered by ESS during the specified time period t, respectively.

The expend of storage system can be represented as

$$F_{2,\text{cost}} = C_{2.PUR} + C_{2.ICC}\frac{\tau}{1-(1+\tau)^{-T_2}} + C_{2.OMC} \tag{8}$$

$$C_{2.PUR} = \sum_{t=1}^{T} E_{2.PUR}^{t} \lambda_2 \tag{9}$$

where $E_{2.PUR}^{t}$ is the energy storage purchased from the grid power in period t, λ_2 is the price of unit power, $C_{2.PUR}$ is the power purchase cost, $C_{2.ICC}$ is the primal capital expenditure, and $C_{2.OMC}$ is the daily expense of operation and maintenance. T_2 is the lifespan of ESS.

Therefore, the net profit model of storage system can be represented as

$$F_2 = F_{2,\text{income}} - F_{2,\text{cost}} \tag{10}$$

3.3. Grid Company Revenue Model

The income of grid company can be expressed as

$$F_{3.\text{income}} = W_{3.SEL} + W_{3.TGC} \tag{11}$$

$$W_{3.TGC} = \lambda_{TGC}(M-N) \tag{12}$$

where λ_{TGC} is the green certificate unit electricity trading price, M is the actual holding of green certificates, N is the renewable energy quota of the grid company, $W_{3.SEL}$ is the electricity sales income, and $W_{3.TGC}$ is the green certificate transaction income.

The cost of the grid company can be expressed as

$$F_{3.\text{cost}} = C_{3.PUR} + C_{3.GTR}\frac{\tau}{1-(1+\tau)^{-T_3}} \tag{13}$$

where $C_{3.PUR}$ is the purchase cost and $C_{3.GTR}$ is the cost of network expansion. T_3 is the lifespan of the grid structure.

The equation for computing the augmentation expense of the grid is as follows:

$$C_{3,GTR} = \sum_{k=1}^{N_L} C_n L_n \tag{14}$$

where N_L represents the total number of candidate amplification lines, C_n represents average expense of line n, and L_n represents the length of alternative amplification line.

Therefore, the net income model of grid company can be expressed as

$$F_3 = F_{3,\text{income}} - F_{3,\text{cost}} \tag{15}$$

3.4. Constraint Conditions

3.4.1. Power Balance Constraint

The balance of power can be expressed as

$$P_{gen}^{t} + P_{PV}^{t} + P_{h}^{t} = P_{load}^{t} + P_{bess}^{t} + P_{loss}^{t} \tag{16}$$

where P_{gen}^{t} represents the power of the output for unit i, P_{PV}^{t} represents the power of output for the PV field in time interval t, P_{h}^{t} represents the power of output for the hydroelectric facility during the specified time span t, P_{load}^{t} represents the load in time interval t, P_{bess}^{t} represents the power of output for the storage system during the specified time span t, and P_{loss}^{t} is the active power of network loss.

3.4.2. Constraints on Output of Conventional Unit within the Upper and Lower Bounds

The output of conventional unit can be expressed as

$$P_{gi}^{\min} \leq P_{gi,t} \leq P_{gi}^{\max} \tag{17}$$

where $P_{gi}^{\max}$ and $P_{gi}^{\min}$ represent the upper and lower bounds of the output, respectively.

3.4.3. Conventional Unit Ramp Constraint

The ramp of conventional unit can be expressed as

$$-\Delta P_{i}^{down} \leq P_{gi,t} - P_{gi,t-1} \leq \Delta P_{i}^{up} \tag{18}$$

where ΔP_{i}^{down} is the downhill climbing rate limit of conventional unit i and ΔP_{i}^{up} is the limit of rising ramp rate for conventional unit i.

3.4.4. Reserve Capacity Constraint

To mitigate the unpredictability in grids, a certain amount of reserve capacity should be reserved in the system. However, there will be decreased economy of grid operation due to excessive spare capacity.

$$\begin{cases} \sum\limits_{i=1}^{N_g} \mu_{i,t}\left(P_{gen,i}^{\max} - P_{gen,i}^{t}\right) \geq \lambda_1 P_{h}^{t} + \lambda_2 P_{PV}^{t} + \lambda_3 P_{i}^{t} \\ \sum\limits_{i=1}^{N_g} \mu_{i,t}\left(P_{gen,i}^{t}(t) - P_{gen,i}^{\min}\right) \geq \lambda_1 P_{h}^{t} + \lambda_2 P_{PV}^{t}(t) + \lambda_3 P_{i}^{t} \end{cases} \tag{19}$$

where N_g represents the number of power plants, $\lambda_1 = \lambda_2 = 10\%$, λ_3 = 2~5%, $\mu_{i,t}$ represent the operational cycling status of a conventional power facility i, $\mu_{i,t} = 1$ represents power-on, $\mu_{i,t} = 0$ indicates downtime, and $P_{gen,i}^{\max}$ and $P_{gen,i}^{\min}$ represent the upper and lower boundaries of the conventional unit's output, respectively.

3.4.5. Voltage Constraint

The safety constraint of voltage can be expressed as

$$\begin{cases} V_{i}^{\min} \leq V_i \leq V_{i}^{\max} \\ \theta_{i}^{\min} \leq \theta_i \leq \theta_{i}^{\max} \end{cases} \tag{20}$$

3.4.6. Line Flow Constraint

The safety constraint of line flow can be expressed as

$$P_{t}^{\text{flow}} \leq P_{t}^{\text{flow,max}}, (t = 1, \ldots, T) \tag{21}$$

where P_t^{flow} and $P_t^{\text{flow,max}}$ represent the actual power and maximum power on the line, respectively.

3.4.7. Energy Storage State of Charge Constraint

The constraint of energy storage state can be expressed as

$$S_t^{\text{soc}} = \begin{cases} S_{t-1}^{\text{soc}} + \frac{P_t^{\text{bess}} \times \eta_{bc} \times \Delta t}{E_{bess}(t)}, P_t^{\text{bess}} > 0 \\ S_{t-1}^{\text{soc}} + \frac{P_t^{\text{bess}} \times \eta_{bd} \times \Delta t}{E_t^{\text{bess}}}, P_t^{\text{bess}} < 0 \end{cases} \tag{22}$$

where P_t^{bess} represents the grid charge and discharge power of ESS during the specified time span t, $P_t^{\text{bess}} > 0$ represents the charging state of the ESS, $P_t^{\text{bess}} < 0$ represents the discharge state of ESS, $E_{bess}(t)$ represents the remaining capacity of the ESS at the previous moment, S_t^{soc} represents the state of charge of ESS, Δt represents the duration of ESS charging, and η_{bc} and η_{bd} represent the efficiency during charging and discharging in ESS, respectively.

4. Model Solving

4.1. Improved Particle Swarm Optimization

The conventional PSO showcases the benefits of a broad optimization range and straightforward coding. It is often used to obtain the optimal solution, which is given by [33].

$$\begin{cases} V_{i+1}^d = \omega V_i^d + c_1 r_1 \cdot (P_i^d - X_i^d) + c_2 r_2 \cdot (P_i^d - X_i^d) \\ X_{i+1}^d = X_i^d + V_i^d \end{cases} \tag{23}$$

where V_i^d and X_i^d are the speed and position before the update, V_{i+1}^d and X_{i+1}^d are the updated speed and location, d represents the iteration of population i in X_i^d, ω is weight, and c_1 and c_2 represent learning parameters of population, which are the random values within the range of (0, 1).

However, the standard PSO is susceptible to local optima and premature. In this paper, a modified PSO from two aspects is established.

(1) Improvement in Inertia Weight

In the standard PSO, the weight ω affects the optimization accuracy of the particles. The slower particle flight speed at the smaller weight is beneficial to the local optimum in the later stage. The faster particle flight speed at the larger weight is conducive to a global optimum. Therefore, compared to the weights of traditional PSO, the improved PSO in this paper adopts dynamic weight ω. The improved update formula is as follows:

$$D(x) = \frac{n_{\text{var}}}{x_{\max} - x_{\min}} \sum_{n=1}^{n_{\text{var}}} (x_{gest,n} - x_n) \tag{24}$$

$$\omega = \omega_{\min} + (\omega_{\max} - \omega_{\min}) \times (D(x) - 1) \tag{25}$$

where n_{var} is the number of decision variables, $x_{\min}$ is the particle position lower limit, and $x_{\max}$ is the particle position and decision variable upper limit.

(2) Improvement in the Learning Factor

PSO has better global optimum ability in the early stage when c_1 is large and c_2 is small. Also, PSO has better local optimum ability in the later stage when c_2 is large and c_1 is small. Therefore, by adopting a non-linear change method, the improvement in the learning factor can be formulated as

$$c_1 = 2 - \frac{r}{r_{\max}} \tag{26}$$

$$c_2 = 1 + \frac{r}{r_{\max}} \tag{27}$$

where $r \in (0,1)$ is the random number.

4.2. Solution Process

The solution process of storage system capacity optimization configuration based on modified PSO is depicted in Figure 2. It is important to note that the variable to be optimized in the multi-objective optimal configuration model is the installed capacity of ESS. The spatial positions of the particles are determined through the modified PSO.

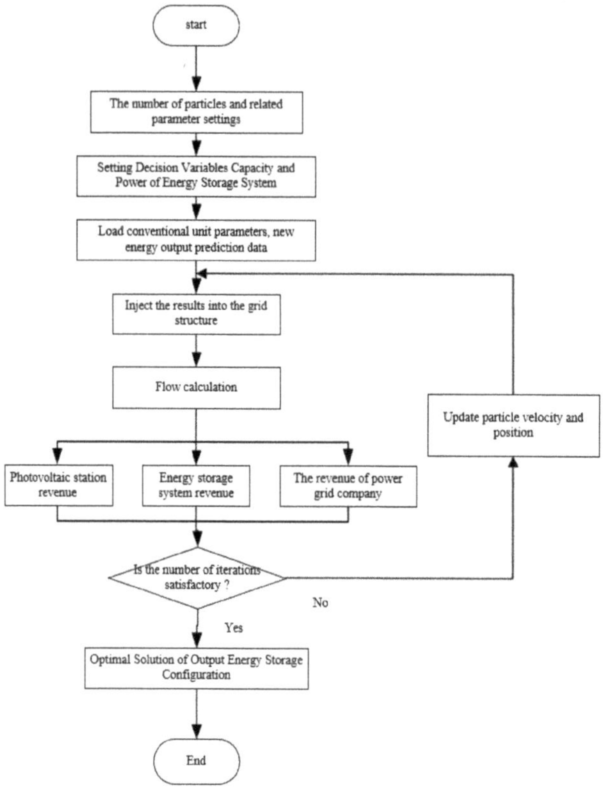

Figure 2. Flowchart of storage capacity optimization configuration.

The depicted flowchart outlines an optimization protocol for an energy system incorporating photovoltaic stations and storage units. Initially, the algorithm's parameters are established, including the number of solutions modeled as particles. Decision variables for the storage system's capacity and power are set, followed by the introduction of parameters for conventional energy units and predictions for renewable outputs. These inputs inform the energy distribution to the grid. Subsequently, energy flow is calculated, and an iterative optimization loop begins, updating the solution candidates' attributes to maximize the system's revenue components. The loop continues until a satisfactory solution is reached based on convergence criteria or the iteration count, culminating in the identification of an optimal configuration for the energy storage system output, thereby concluding the optimization process.

5. Case Analysis

5.1. Parameter Setting

The adapted IEEE 14-bus network and the actual power grid in a certain area of Southern Xinjiang are taken as the case models for simulation verification. The load information and PV output curve in the case are shown in Figure 3. The improved network topologies are depicted in Figures 4 and 5, respectively. The operating parameters of thermal units, ESS and hydropower units, the electricity price information, and equipment cost information are shown in Tables 1–4, respectively.

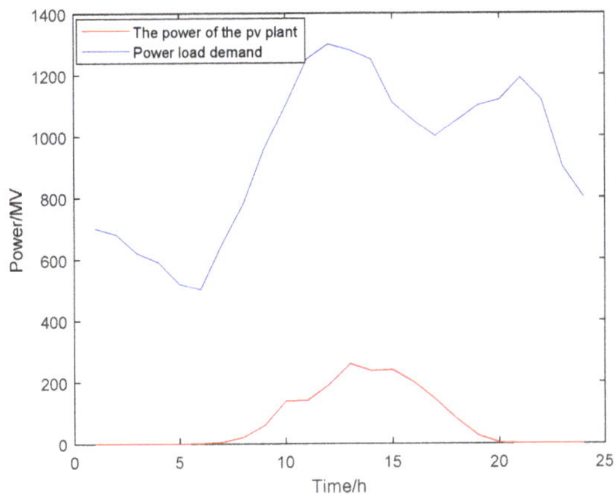

Figure 3. Typical daily operation curve.

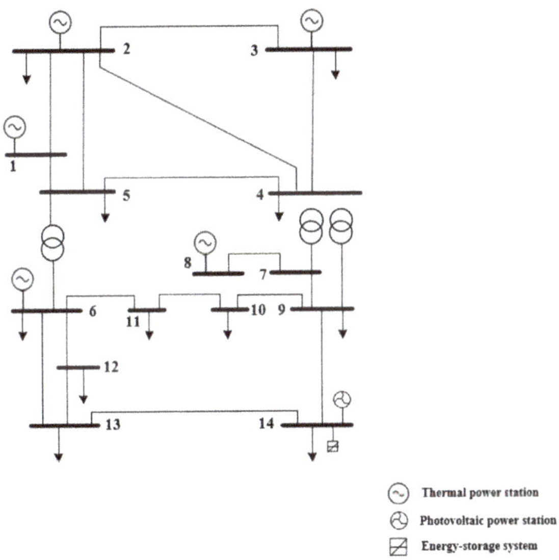

Figure 4. The adapted IEEE 14-bus network.

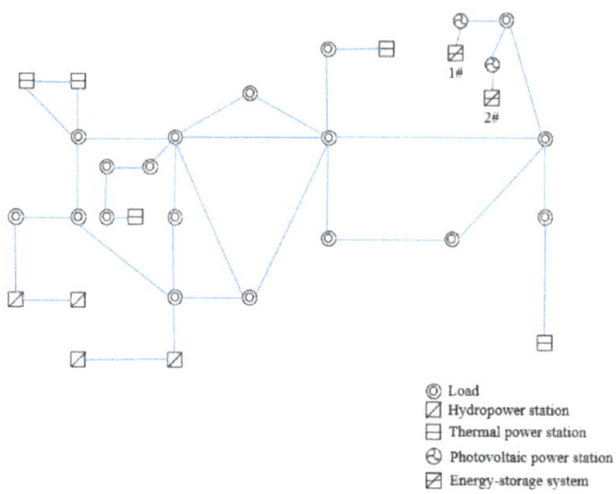

Figure 5. Actual power grid topology of a region in Southern Xinjiang.

Table 1. Parameter settings for the case system.

Parameters	Data	Parameters	Data
PV plant construction cost	11,340 yuan/kW	Number of ESS	2
PV plant maintenance cost	1120 yuan/kW per year	SOC upper limit of energy storage battery	0.9
PV panel scrap income	120 yuan/kW per year	SOC lower limit of energy storage battery	0.1
PV panel life	20 years	Battery life	15 years
Energy storage system construction cost	3800 yuan/kW	Grid expansion cost	1 million yuan/KM
Maintenance cost of energy storage system	300 yuan/kW per year	Government's expected photovoltaic subsidies	0.3 yuan/kWh
Energy storage unit peaking income	30 yuan/MWh	Photovoltaic grid electricity price	0.7 yuan/kWh
Energy storage unit standby revenue	0.55 yuan/kWh	Power grid selling price	0.5 yuan/kWh

Table 2. Relevant parameters of the thermal power unit.

Unit Number	1	2	3	4	5
Min (MW)	120	120	20	20	20
Max (MW)	455	455	130	130	162
Coal consumption coefficient a	0.00048	0.00031	0.000212	0.00211	0.00398
Coal consumption coefficient b	16.19	17.26	16.61	16.5	19.72
Coal consumption coefficient c	1000	970	700	680	450

Table 3. Relevant parameters of the hydropower unit.

Unit Number	1	2	3	4
V_{min} (10,000 m^3)	80	60	100	70
V_{max} (10,000 m^3)	150	120	240	160
Q_{min} (10,000 m^3/s)	5	6	10	6
Q_{max} (10,000 m^3/s)	15	15	30	20
PH_{min} (MW)	10	15	35	20
PH_{max} (MW)	500	350	600	450

Table 4. Parameter settings for the algorithm.

Parameters	Data	Parameters	Data
Iterations	100	Learning factor c_1	0.1
Number of populations	50	Learning factor c_2	0.2
Initial value of inertia weight	0.2	Concluding inertia weight parameter	0.8

The simulation time span is 1 day, the annualized equivalent factor of equipment is 0.08, the upper limit of capacity of PV plant and storage system is set as the maximum load of the system, the trading price of green certificate is set as 0.2 yuan/KWh, and the renewable energy quota obligation of power grid company is set as 15% of the annual load power. The scratch income from photovoltaic panels in Table 1 represents annual data. Therefore, it is necessary to prorate this to daily data when optimizing the established model.

5.2. Case Analysis

In order to validate the viability of the model presented in this paper, three scenarios are formulated for analysis. All scenarios operate under identical system parameters.

Scenario 1: Only the power grid company's revenue;
Scenario 2: Only the energy storage system benefits;
Scenario 3: The income of grid companies, ESS, and PV plants.

Figures 6 and 7 show the state of ESS operation throughout the day under three optimization scenarios, utilizing the adapted IEEE 14-bus network.

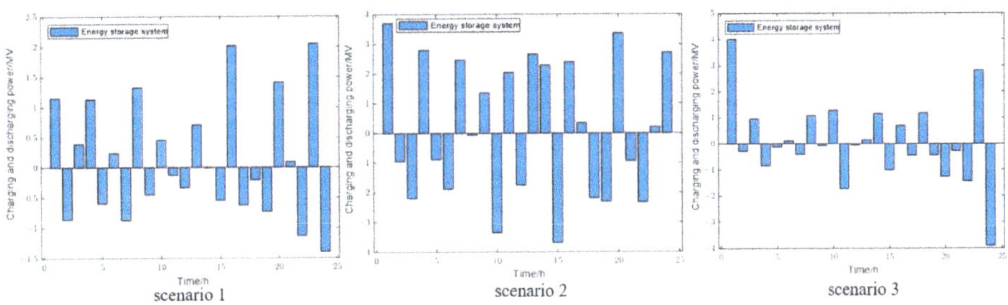

Figure 6. Power flow into and out of the ESS for scenario 1 to 3.

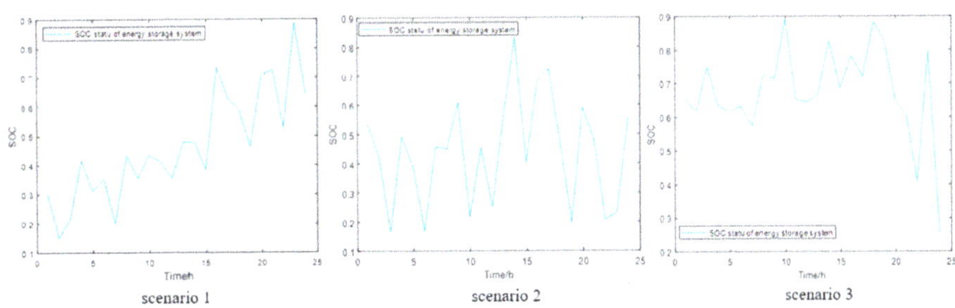

Figure 7. SOC status of ESS for scenarios 1 to 3.

Table 5 lists the peak saving capacity and reserve capacity that ESS can provide to the grid in a cycle of 24 h.

ESS is equivalent to the provision of energy to the grid when it is in the discharged state again. At the same time, it can also be charged on demand, which is equivalent to a load.

Table 5. Energy storage capacity configuration results.

Scene No.	Configure Capacity/MWh	Peaking Capacity/MWh	Standby Capacity/MWh
1	5.71	7.88	48.9
2	8.51	22.54	68.83
3	7.22	12.35	54.36

Scenario 3's energy management profile, as illustrated by Figure 6, suggests a highly strategic and dynamic system operation, with sharp peaks in charging and pronounced troughs in discharging, indicating aggressive energy turnover within the 24-h cycle. Upon comparative analysis, Scenario 3 emerges as superior for an energy storage system due to its dynamic operation. Unlike Scenario 1, which shows moderate charging and discharging, and Scenario 2, which indicates higher utilization, Scenario 3 demonstrates the most significant range of power fluctuations. With peaks nearly reaching 5 MW and troughs around −4 MW, Scenario 3 suggests an aggressive and responsive energy management strategy. It indicates that the system robustly adapts to varying energy supply and demand conditions, optimizing its storage capacity. The pronounced activity in Scenario 3 likely reflects strategic operational decisions to store energy during periods of low cost or excess supply and discharge during high demand, potentially leading to improved efficiency and cost-effectiveness. Thus, Scenario 3's utilization pattern may be indicative of a more economically savvy and operationally efficient system.

The variability in the SOC graph points to a system that responds actively to the changing energy landscape, possibly accommodating high levels of renewable energy generation and rapid shifts in electricity demand. Scenario 3 may offer advantages in scenarios where energy prices fluctuate widely, allowing operators to buy low during periods of surplus and sell high during shortages, effectively leveraging market dynamics. The pronounced dips in SoC also hint at the potential for deep discharge cycles. When juxtaposed with the charging and discharging profiles, they suggest that ESS is not just a backup but a pivotal component of the energy strategy, engaging in both load leveling and peak saving.

To ascertain the enhanced particle swarm algorithm's efficacy and utility in addressing the issue delineated herein, a comparative analysis of the algorithm's performance pre- and post-enhancement toward solving the designated problem was conducted, with the computational outcomes depicted in Figure 8. An examination of the convergence rates of both algorithms reveals that the post-improvement version exhibits accelerated convergence, thereby underscoring the selected algorithm's superiority.

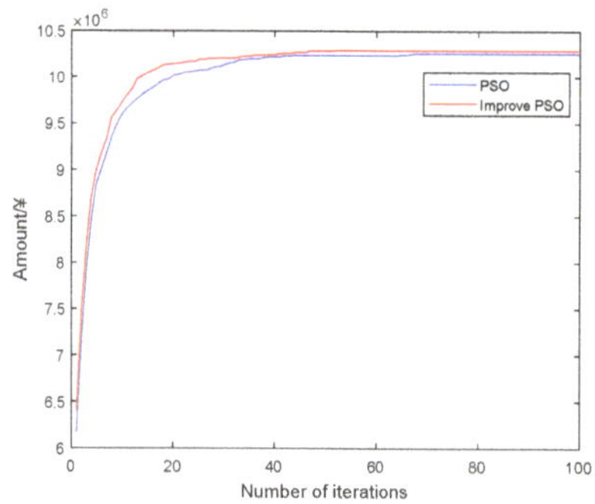

Figure 8. Comparison of optimization results between PSO and improved PSO algorithms.

During the execution of the actual Simulation, three scenarios were systematically configured. Figures 9 and 10 show the state of ESS operation throughout the day under three optimization scenarios, utilizing the actual power grid topology of a region in Southern Xinjiang.

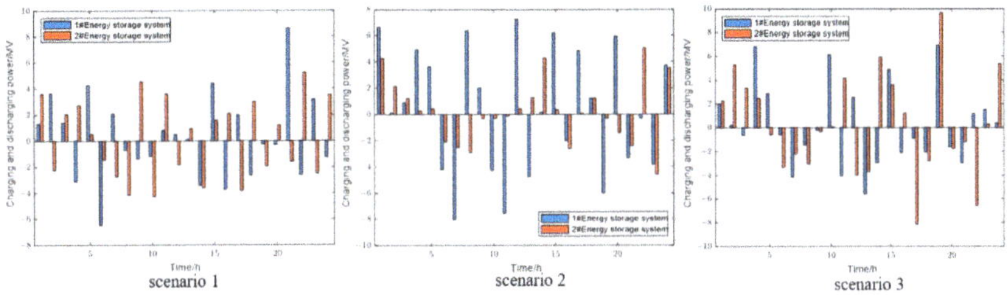

Figure 9. Power flow into and out of the ESS for scenario 1 to 3.

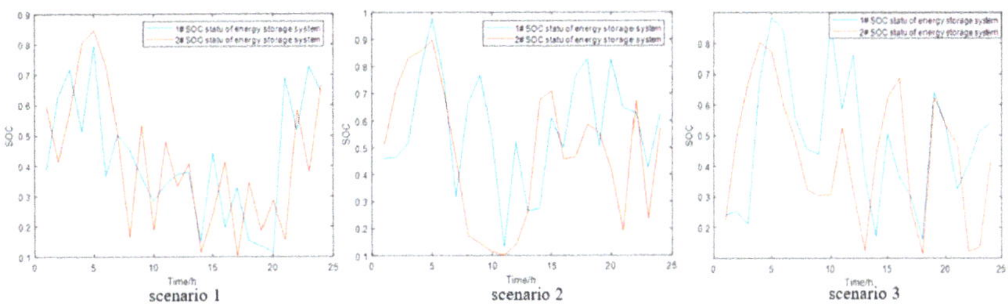

Figure 10. SOC status of ESS for scenarios 1 to 3.

This paper configures ESS in the place where the PV farms gather more densely according to the actual grid structure of South Xinjiang, China. The final determination of energy storage capacity allocation is 14.4 MWh and 19.1 MWh, respectively, taking into account the interests of the three parties.

The peaking capacity and standby capacity that the two energy storage systems can provide to the grid in one cycle of 24 h are presented in Table 6. In analyzing the ESS across three scenarios, it becomes evident that Scenario 3 offers a more balanced and efficient configuration. This scenario showcases System 1 and System 2 with configured capacities of 14.4 MWh and 191 MWh, peaking capacities of 29.1 MWh and 36.9 MWh, and substantial standby capacities of 138.7 MWh and 168.6 MWh, respectively. The distribution of capacities in Scenario 3 suggests an optimized balance between the immediate availability of energy for peak demand (peaking capacity) and the strategic reserve for unforeseen circumstances (standby capacity), without excessively prioritizing one over the other. This balanced approach not only ensures a reliable energy supply during peak times but also maintains a higher level of preparedness for emergency situations. Furthermore, the relatively high configured and standby capacities of both systems in Scenario 3 indicate a robust framework designed to accommodate fluctuations in demand and supply, enhancing the overall efficiency and stability of the energy grid. Consequently, Scenario 3 emerges as the superior configuration, demonstrating an effective integration of energy storage capabilities to meet diverse operational demands while ensuring grid reliability and efficiency. Meanwhile, the above results indicate that considering multiple interests simultaneously will result in a more balanced outcome.

It can be concluded that the total revenue of the established model is the highest in the PV plant energy storage system from Table 7. Scenario 2 has a higher charging and discharging capacity than Scenarios 1 and 3, as it considers the maximum benefit of ESS. However, scenario 3 has a relatively flat charging and discharging capacity because it integrates the revenue objectives of multiple parties. The revenues of the three parties of interest in the simulation case and actual case are enumerated in Table 7. The findings indicate that the revenue of PV plants mainly comes from their power sales revenue and government subsidy policies. Auxiliary services are the main source of revenue for ESS. Grid companies' revenues are closely related to the installed renewable energy capacity in the region. A higher price of green certificate trading can increase the incentive for energy storage and grid companies to invest. ESS can increase the utilization of sustainable energy resources.

Table 6. Energy storage capacity configuration results.

Scene No.	Energy Storage System No.	Configured Capacity/MWh	Peaking Capacity/MWh	Standby Capacity/MWh
1	1	15.1	27.1	128.7
	2	12.2	30.1	101.1
2	1	18.4	44.2	217.7
	2	10.3	19.52	102.5
3	1	14.4	29.1	138.7
	2	19.1	36.9	168.6

Table 7. Third party income of three scenarios.

Scene No.	Subject of Interest	Total Revenue in 14-Bus Network/¥	Total Revenue in Actual Power Grid/¥
1	PV plant	499,722	240,898
	ESS	70,878	16,628
	Grid company	9,701,638	1,333,863
2	PV plant	504,338	247,063
	ESS	113,620	17,869
	Grid company	9,602,817	1,314,823
3	PV plant	504,603	246,790
	ESS	101,236	17,392
	Grid company	9,683,937	1,328,108

6. Conclusions

The energy storage system can efficaciously mitigate a range of issues arising from large-scale PV into grids. By the combination of multi-stakeholders and an improved PSO, a multi-faceted optimization model was introduced for the PV of ESS. The simulation results demonstrated that the reasonable allocation of ESS in the grid could enhance the harnessing of alternative energy sources. The rational arrangement of the ESS can achieve the maximum benefit for multiple parties. Simultaneously, the adapted PSO devised in this study could efficiently compute multi-objective optimization models and achieve good performance.

Author Contributions: F.S. and W.W. did the methodology, simulation, and validation. F.S. did the analysis and wrote the paper. Conceptualization, F.S. and W.W.; Software, F.S. and D.N.; Investigation, D.N. and W.W.; Resources, F.S. and W.W.; Data Curation, F.S. and D.N.; Writing—Original Draft Preparation, F.S. and D.N.; Writing—Review and Editing, F.S., W.W. and D.N.; Visualization, F.S., W.W. and D.N.; Supervision, W.W.; Project Administration, W.W.; Funding Acquisition, D.N. and W.W. All authors have read and agreed to the published version of the manuscript.

Funding: This work was supported by the National Natural Science Foundation of China (No. 51667020).

Data Availability Statement: Data are contained within the article.

Conflicts of Interest: The authors declare no conflicts of interest.

References

1. Gielen, D.; Boshell, F.; Saygin, D.; Bazilian, M.D.; Wagner, N.; Gorini, R. The role of renewable energy in the global energy transformation. *Energy Strategy Rev.* **2019**, *24*, 38–50. [CrossRef]
2. Kim, Y.; Wilson, C. Analyzing future change in the EU's energy innovation system. *Energy Strategy Rev.* **2019**, *24*, 279–299. [CrossRef]
3. Qiu, Y.; Lin, J.; Liu, F.; Song, Y.; Chen, G.; Ding, L. Stochastic online generation control of cascaded run-of-the-river hydropower for mitigating solar power volatility. *IEEE Trans. Power Syst.* **2020**, *35*, 4709–4722. [CrossRef]
4. Ding, Y.; Xie, D.; Hui, H.; Xu, Y.; Siano, P. Game-theoretic demand side management of thermostatically controlled loads for smoothing tie-line power of microgrids. *IEEE Trans. Power Syst.* **2021**, *36*, 4089–4101. [CrossRef]
5. Li, X.; Wang, L.; Yan, N.; Ma, R. Cooperative dispatch of distributed energy storage in distribution network with PV generation systems. *IEEE Trans. Appl. Supercond.* **2021**, *31*, 1–4. [CrossRef]
6. Li, F.; Li, C.; Sun, K.; Zhang, J.; Li, H. Capacity configuration of hybrid CSP/PV plant for economical application of solar energy. *Chin. J. Electr. Eng.* **2020**, *6*, 19–29. [CrossRef]
7. Wang, Q.; Zhang, X.; Yi, C.; Li, Z.; Xu, D. A novel shared energy storage planning method considering the correlation of renewable uncertainties on the supply side. *IEEE Trans. Sustain. Energy* **2022**, *13*, 2051–2063. [CrossRef]
8. Bullich-Massagué, E.; Cifuentes-García, F.-J.; Glenny-Crende, I.; Cheah-Mañé, M.; Aragüés-Peñalba, M.; Díaz-González, F.; Gomis-Bellmunt, O. A review of energy storage technologies for large scale photovoltaic power plants. *Appl. Energy* **2020**, *274*, 115213. [CrossRef]
9. Zhang, D.; Mubaarak, S.; Jiang, S.; Wang, L.; Liu, J.; Chen, Y.; Li, M. Optimal allocation method of energy storage in PV station based on probabilistic power flow. *Energy Storage Sci. Technol.* **2021**, *10*, 2244–2251.
10. Shi, Z.; Wang, W.; Huang, Y.; Li, P.; Dong, L. Simultaneous optimization of renewable energy and energy storage capacity with hierarchical control. *CSEE J. Power Energy Syst.* **2020**, *8*, 95–104.
11. Yan, D.; Chen, Y. Distributed coordination of charging stations with shared energy storage in a distribution network. *IEEE Trans. Smart Grid.* **2023**, *14*, 4666–4682. [CrossRef]
12. Zhu, H.; Hou, R.; Jiang, T.; Lv, Q. Research on energy storage capacity configuration for PV power plants using uncertainty analysis and its applications. *Glob. Energy Interconnect.* **2021**, *4*, 608–618. [CrossRef]
13. Zhang, S.; Xiang, Y.; Liu, J.; Liu, J.; Yang, J.; Zhao, X.; Jawad, S.; Wang, J. A regulating capacity determination method for pumped storage hydropower to restrain PV generation fluctuation. *CSEE J. Power Energy Syst.* **2020**, *8*, 304–316.
14. Zhu, B.; Wang, Y.; Guo, H.; Yang, N.; Lu, L. Improved model of base station power system for the optimal capacity planning of photovoltaic and energy storage system. *Electronics* **2023**, *12*, 4826. [CrossRef]
15. Zhang, J.; Gao, H.; Wang, J.; Hong, J.H.; Liu, J.Y.; Zhuang, H.M. Economic allocation of energy storage in photovoltaic power station considering prediction accuracy and topological structure. *Electr. Power Autom. Equip.* **2019**, *39*, 115–121.
16. Chaima, E.; Lian, J.; Ma, C.; Zhang, Y.; Kavwenje, S. Complementary optimization of hydropower with pumped hydro storage-photovoltaic plant for all-day peak electricity demand in Malawi. *Energies* **2021**, *14*, 4948. [CrossRef]
17. Deng, X.; Huang, Y.; Chen, Y.; Chen, C.; Yang, L.; Gao, Q.; Chen, X.; Hou, W.; Lin, Z. Multi-scenario physical energy storage planning of integrated energy systems considering dynamic characteristics of heating network and demand response. *Front. Energy Res.* **2021**, *9*, 783588. [CrossRef]
18. An, Y.; Zhao, Z.; Wang, S.; Huang, Q.; Xie, X. Coordinative optimization of hydro-photovoltaic-wind-battery complementary power stations. *CSEE J. Power Energy Syst.* **2019**, *6*, 410–418.
19. Contreras-Ocana, J.; Ortega-Vazquez, M.; Zhang, B. Participation of an energy storage aggregator in electricity markets. *IEEE Trans. Smart Grid* **2017**, *10*, 1171–1183. [CrossRef]
20. Wang, K.; Chen, Q.; Guo, H.; Xia, Q. Mechanism design and trading strategy for capacity contract of energy storage towards transactive energy. *Autom. Electr. Power Syst.* **2018**, *42*, 54–60.
21. Mortaz, E. Portfolio diversification for an intermediary energy storage merchant. *IEEE Trans. Sustain. Energy* **2019**, *11*, 1539–1547. [CrossRef]
22. Croce, V.; Raveduto, G.; Verber, M.; Ziu, D. Combining machine learning analysis and incentive-based genetic algorithms to optimise energy district renewable self-consumption in demand-response programs. *Electronics* **2020**, *9*, 945. [CrossRef]
23. Li, T.; Xu, Y.; Chen, J.; Wang, K. Optimal configuration of energy storage for microgrid considering life cycle cost-benefit. *Proc. CSU-EPSA* **2020**, *32*, 46–51.
24. Wu, X.; Sun, Y.; Wei, Z.; Sun, G. Distributed hierarchical consensus algorithm for economic dispatch in smart grid. *IET Gener. Transm. Distrib.* **2019**, *13*, 5541–5549. [CrossRef]
25. Mediwaththe, C.P.; Blackhall, L. Network-aware demand-side management framework with a community energy storage system considering voltage constraints. *IEEE Trans. Power Syst.* **2020**, *36*, 1229–1238. [CrossRef]
26. Wang, J.; Liu, W.; Li, S.; Liu, Z.; Yang, M.; Guo, H.M. A method to evaluate economic benefits of power side battery energy storage frequency/peak regulation considering the benefits of reducing thermal power unit losses. *Power Syst. Technol.* **2020**, *44*, 4236–4245.
27. Wang, C.; Jiao, S.; Li, Y.; Zhang, Q. Capacity optimization of a hybrid energy storage system considering Wind-Solar reliability evaluation based on a novel Multi-strategy snake optimization algorithm. *Expert Syst. Appl.* **2023**, *231*, 120602. [CrossRef]

28. Cui, H.; Li, F.; Fang, X.; Chen, H.; Wang, H. Bilevel arbitrage potential evaluation for grid-scale energy storage considering wind power and LMP smoothing effect. *IEEE Trans. Sustain. Energy* **2017**, *9*, 707–718. [CrossRef]
29. Zhao, D.; Wang, H.; Huang, J.; Lin, X. Time-of-use pricing for energy storage investment. *IEEE Trans. Smart Grid* **2021**, *13*, 1165–1177. [CrossRef]
30. Ying, Y.; Tian, Z.; Wu, M.; Liu, Q.; Tricoli, P. Capacity configuration method of flexible smart traction power supply system based on double-layer optimization. *IEEE Trans. Transp. Electrif.* **2023**, *9*, 4571–4582. [CrossRef]
31. Sun, Y.; Zhang, B.; Ge, L.; Sidorov, D.; Wang, J.; Xu, Z. Day-ahead optimization schedule for gas-electric integrated energy system based on second-order cone programming. *CSEE J. Power Energy Syst.* **2020**, *6*, 142–151.
32. Yang, J.; Wang, W.; Ma, K.; Yang, B.; Dou, C.-X. Optimization and self-adaptive dispatching strategy for multiple shared battery stations of electric vehicles. *IEEE Trans. Ind. Inform.* **2021**, *17*, 1363–1374. [CrossRef]
33. Clerc, M.; Kennedy, J. The particle swarm—Explosion, stability, and convergence in a multidimensional complex space. *IEEE Trans. Evol. Comput.* **2002**, *6*, 58–73. [CrossRef]

electronics

MDPI

Article

Bad Data Repair for New Energy Stations in Power System Based on Multi-Model Parallel Integration Approach

Chenghao Li [1], Mingyang Liu [1], Ze Gao [1], Yi Wang [2,*] and Chunsun Tian [1]

1 State Grid Henan Electric Power Research Institute, Zhengzhou 450052, China; lichenghao@sgcc.com.cn (C.L.); liumingyang3@ha.sgcc.com.cn (M.L.); gaoze@ha.sgcc.com.cn (Z.G.); tianchunsun@ha.sgcc.com.cn (C.T.)
2 School of Electrical and Information Engineering, Zhengzhou University, Zhengzhou 450001, China
* Correspondence: yiwang@zzu.edu.cn

Abstract: The accurate and reliable acquisition of measurement information is very important for the stable operation of power systems, especially the operation status information of new energy stations. With the increasing proportion of new energy stations in power systems, the quality issues of data from these stations, caused by communication congestion, interference, and network attacks, become more pronounced. In this paper, to deal with the issue of low accuracy and poor performance of bad data restoration in new energy stations, a novel deep learning approach by combining the modified long short-term memory (LSTM) neural network and Wasserstein generative adversarial network with gradient penalty (WGAN-GP) is proposed. The proposed method can be implemented in a parallel ensemble way. First, the normal data set acquired from multiple sections of new energy stations is utilized to train the modified LSTM and WGAN-GP model. Secondly, according to the data characteristics and rules captured by each model, the two models are systematically integrated and the bad data repair model pool is constructed. Subsequently, the results of model repair are screened and merged twice by the parallel integration framework to obtain the final repair result. Finally, the extensive experiments are carried out to verify the proposed method. The simulative results of energy stations in a real provincial power grid demonstrate that the proposed method can effectively repair bad data, thereby enhancing the data quality of new energy stations.

Keywords: renewable energy station; long short-term memory neural network; Adam optimization algorithm; Wasserstein generative adversarial network; parallel ensemble learning; data repair

1. Introduction

In recent years, with the ongoing development of new energy systems, the integration of numerous new energy stations has led to a surge in the amount of data processed by power systems, making the data structures more complex [1]. In modern power systems, the tight integration of the physical and information layers has deepened the reliance of control centers on real-time measurement data. However, in addition to inevitable data noise, these real-time measurements frequently contain bad data. Such data not only distorts the system's true state and adversely affects the accuracy and convergence of state estimation, but also poses significant challenges to the performance of subsequent advanced applications [2,3]. For instance, voltage amplitude errors in line parameter identification can be amplified a thousandfold in resistance identification [4]. In addition, network and time synchronization attacks on synchronous phasor data pose a threat to the safe and stable operation of power systems, sometimes even triggering cascading failures [5,6]. To deal with issues, repairing the bad measurement data of the new energy stations in real time is not only crucial for improving data quality and reliability, but also enhances the situational awareness capabilities of power systems, supporting their safe and stable operation.

Generally speaking, the methods for repairing bad data at new energy stations fall into two main categories: model-dependent repair methods and data-driven repair methods.

Model-dependent repair methods require knowledge of the system's topology and detailed parameters, which repairs the bad data through state estimation models. Reference [7] proposed a synchronous phasor data conditioning algorithm in which Kalman filtering and smoothing techniques for synchronous phasor data conditioning preprocessed the data before they reached the linear estimator to ensure the quality of the data obtained from phasor measurement units (PMUs), achieving the accurate restoration of PMU data. Reference [8] proposed a state estimator (PSE) method based on phasor measurement, which used an augmented state vector approach to identify and correct angle deviations and current scaling errors in phasor data, thereby improving data consistency and facilitating the repair of PMU data.

Although the aforementioned model-based methods for repairing bad data can effectively address anomalies, it is important to note their high dependency on models, requiring detailed system topology and parameter information. However, acquiring the accurate topology and detailed parameters of actual power systems is often challenging. In contrast, data-driven strategies for bad data repair can establish mappings between normal data sets and target data by directly learning intrinsic data relationships using data mining and machine learning methods, thereby accomplishing data repair. Consequently, data-driven repair methods have gained widespread attention in recent years [9–12]. Reference [13] proposed a data repair model, which is constructed with a generative adversarial network (GAN), which combines a temporal convolutional network (TCN) and a bi-directional long short-term memory network (BiLSTM) to repair the AIS data. Reference [14] proposed a bidirectional recurrent imputation time series (BRITS) method, leveraging bi-directional recurrent neural networks to capture the dynamic properties of time series data from both directions and provide accurate predictions for missing values. Reference [15] proposed a method using artificial neural networks to estimate missing synchrophasor data, predicting missing data values from existing complete data sets. Reference [16] proposed a method using an improved generative adversarial network, which can learn the distribution of measurement data in the power system and realize the reconstruction of missing measurement data with higher accuracy.

Data-driven repair methods, not reliant on detailed system models, offer significant flexibility. For the above-mentioned data-driven methods, they mostly rely on a single model. However, this dependence on a single training method significantly reduces the model's generalization ability. Recently, ensemble learning has become a popular machine learning paradigm. Its core idea is to introduce diversity into models and appropriate combination strategies to overcome the limitations of single models, thereby enhancing overall repair accuracy [17]. Reference [18] proposed a multi-scale ensemble neural network method that utilizes long short-term memory (LSTM), gated recurrent units (GRU), and temporal convolutional networks (TCN) as the basic models. These networks are assembled on both single-model and multi-model scales to improve prediction accuracy. In dealing with large, variable, and nonlinear data sets, the heterogeneous base learning model can comprehensively capture the potential properties and associations of data with different distribution characteristics, and improve the overall performance of the model.

Therefore, addressing the issues of poor generalization and overfitting in single models, and considering the nonlinear time-series characteristics of new energy station data, in order to realize the reliable repair of abnormal data of new energy stations, a deep learning method combining modified LSTM with WGAN-GP is proposed in this paper. This method employs a parallel ensemble learning framework that integrates both modified LSTM and WGAN-GP models. Initially, the modified LSTM model is used to capture the temporal correlation features in the data of new energy stations, ensuring accurate fitting and prediction of time trends and patterns in the data. Simultaneously, the WGAN-GP model learns the distribution characteristics of the new energy station data, generating data similar to the actual data distribution, thereby enhancing robustness in real-world applications. Afterwards, by parallelly integrating the modified LSTM and WGAN-GP models, the strengths of both models are deeply fused to enhance the model's generalization capabilities. Finally, simulation tests on the actual operational dataset of a new provincial energy station demonstrate that, compared to single data models, the proposed multi-model parallel ensemble method exhibits stronger generalization abilities. It can achieve the high-precision repair of bad data under various conditions, thus improving the data quality level of new energy stations.

The rest of the paper is organized as follows. Section 2 introduces the two selected models, the modified LSTM and the WGAN-GP. Section 3 is dedicated to presenting the multi-model parallel ensemble model. Section 4 provides a detailed discussion of the simulation results. Finally, Section 5 concludes by summarizing the article and discussing the future scope for the proposed method.

2. Multi-Model Parallel Integrated Data Repair Method

In this section, a novel approach by combining the modified long short-term memory (LSTM) neural network and Wasserstein generative adversarial network with gradient penalty (WGAN-GP) is developed and will be introduced in detail.

2.1. Modified LSTM Algorithm

2.1.1. Traditional LSTM Algorithm

Long short-term memory is a special type of recurrent neural network (RNN) first proposed in [19], primarily designed to deal with the issues of gradient disappearance and gradient explosion during long sequence training. Different from a traditional RNN, LSTM introduces three gate mechanisms: the forget gate, the input gate, and the output gate, along with a cell state. These gate mechanisms, implemented through neural networks, enable effective memory retention and selective forgetting of information within measurement data, thereby capturing temporal dependencies among the measured data.

In general, LSTM can be represented in the following form:

$$\begin{cases} f_t = \sigma(w_f[x_t, h_{t-1}] + b_f) \\ i_t = \sigma(w_i[x_t, h_{t-1}] + b_i) \\ o_t = \sigma(w_o[x_t, h_{t-1}] + b_o) \\ c_t = (f_t \otimes c_{t-1}) \oplus (i_t \otimes \tanh(w_c[x_t, h_{t-1}] + b_c)) \\ h_t = o_t \otimes \tanh(c_t) \end{cases} \tag{1}$$

where f_t is the forgetting door; i_t indicates an input gate; o_t denotes an output gate; c_t represents a cell state; h_t is hidden state; x_t is the input vector at time t; t is a time step; b_f, b_i, b_o, b_c are the deviation of the corresponding gate control unit; w_f, w_i, w_o, w_c are the weight vector of the forget gate f_t, input gate i_t, output gate o_t, and the cell state c_t, respectively; $\sigma(\cdot)$ and $\tanh(\cdot)$ are the Sigmoid activation function and hyperbolic tangent function, respectively; $\otimes$ is the matrix multiplication operation; $\oplus$ is the matrix addition operation.

As shown in Figure 1, LSTM consists of input gates, control gates, forget gates, and output gates. Among these, the forget gate is responsible for filtering and discarding non-critical information from the cell state, allowing LSTM to disregard irrelevant information and retain valuable information for subsequent computations. The input gate regulates the input of new data information, enabling LSTM to update its internal cell state when processing new data. The output gate is in charge of transmitting information from the cell state to the next layer or the next time step, enabling LSTM to selectively output information that is useful for the current task. The cell state is the core of LSTM, spanning the entire time sequence and carrying essential information. The cell state can be updated through a combination of operations involving the forget gate and the input gate, ensuring the transmission of information in long sequences [19,20].

Despite its widespread application and robustness in handling sequential data, the traditional long short-term memory (LSTM) model is not without its limitations. One of the primary challenges associated with LSTM is its susceptibility to overfitting, especially in scenarios involving complex data sets with numerous parameters. Additionally, LSTM often faces difficulties in efficiently processing longer sequences, leading to a degradation in performance due to the vanishing gradient problem. In response to these shortcomings, our research focuses on further enhancements to the LSTM architecture. These improvements aim to address these specific limitations by optimizing the model's architecture and training process. The following sections detail the methodologies employed to refine the LSTM model, encompassing optimization algorithms and regularization techniques.

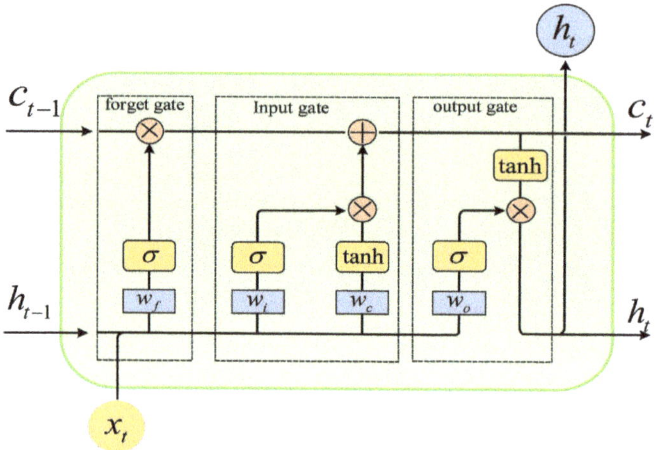

Figure 1. Schematic diagram of LSTM.

2.1.2. Adam Optimization Algorithm

Adam optimization algorithm is a learning rate adaptive optimization algorithm [21]. Adam algorithm can be understood as a learning rate adaptive optimizer with momentum method. It is an extension of stochastic gradient descent and can replace the classical stochastic gradient descent method to update network weights more effectively. It estimates the gradient first and second moments of each parameter according to the objective function, calculated using the exponential moving average. In order to solve the problem of high noise and gradient dilution during parameter space iteration, the feature scaling of the gradient of each parameter is kept constant. The formula is derived as follows:

$$\begin{cases} \theta_i^{(k+1)} = \theta_i^{(k)} - g_i^{(k)} \\ g_i^{(k)} = \frac{\eta \hat{v}_i^{(k)}}{\sqrt{\hat{s}_i^{(k)}} + \varepsilon} \end{cases} \quad (2)$$

$$\begin{cases} \hat{v}_i^{(k)} = \frac{v_i^{(k)}}{1-\beta_1^k} \\ \hat{s}_i^{(k)} = \frac{s_i^{(k)}}{1-\beta_2^k} \end{cases} \quad (3)$$

where k is the number of iterations; $\theta_i^{(k)}$ represents the ith characteristic parameter in the iterative process; $g_i^{(k)}$ denotes the descending distance value along the gradient direction; $s_i^{(k)}$ indicates the exponential decay average of the historical gradient; $v_i^{(k)}$ is the exponential decay average of the quadratic historical gradient; $\hat{s}_i^{(k)}$ and $\hat{v}_i^{(k)}$ are the deviation correction values of $s_i^{(k)}$ and $v_i^{(k)}$, respectively. Hyperparameter η = 0.001; ε is manually entered parameters; Hyperparameter β_1 = 0.9; β_2 = 0.999.

Adam optimization algorithm in LSTM has significant advantages over the traditional stochastic gradient descent (SGD). With Adam, the learning rate can be adjusted more easily, helping to overcome the problem of disappearing or exploding gradients. The adaptive performance of Adam makes it more suitable to deal with different gradient characteristics of different parameters, thus, improving the training efficiency and convergence speed.

2.1.3. Dropout Regularization

Dropout regularization is a technique commonly used in neural network training to prevent models from overfitting [22]. The core idea of dropout is to randomly turn off a subset of neurons in the network during each training iteration, which can reduce the complex co-adaptive relationships between neurons. Dropout regularization is introduced into the structure of the long short-term memory network (LSTM) [23] to enhance the generalization ability of the model. While LSTMs are more effective than traditional RNN at capturing long-term dependencies, they still run the risk of overfitting. By adding dropout regularization to the LSTM structure, a subset of connections or neurons in the network can be randomly dropped during training. This method can reduce the complex co-adaptive relationship between LSTM units and reduce the overfitting of the training data. Slightly different from standard dropout, when dropout is applied in LSTM, it is usually applied with different dropout proportions in different parts of the LSTM unit (such as input gate, forget gate, output gate) or between different layers to avoid excessive impact on the time dependence of LSTM. The modified LSTM model with dropout is shown in Figure 2.

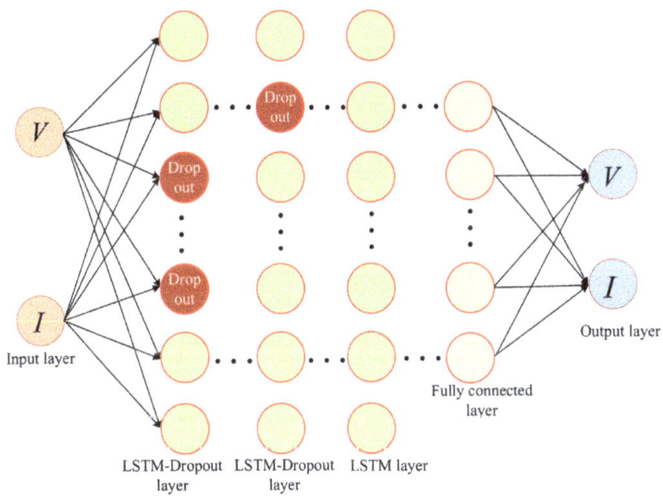

Figure 2. LSTM with dropout.

2.2. *WGAN-GP Algorithm*

Generative adversarial network (GAN) is a type of deep learning model [24]. This model is capable of generating data with similar characteristics to the training data. GAN consists of two opposing networks: a generator and a discriminator. The generator's task is to produce data samples resembling the distribution of real data, while the discriminator's task is to distinguish between generated data and real data. These two networks learn from each other in an adversarial environment. The generator continuously improves its ability to generate realistic data, while the discriminator aims to enhance its capability to distinguish between real and generated data. As the training progresses, the generator eventually becomes proficient at creating samples that exhibit characteristics similar to the real data.

The objective function of the original generated adversarial network is as follows:

$$L(G,D) = \min_G \max_D \{E_{x\sim p_r(x)}[\lg(D(x))] + E_{x\sim p_{g(z)}}[\lg(1-D(G(z)))]\} \quad (4)$$

where $L(\cdot)$ is the objective function of generator and discriminator; $E(\cdot)$ denotes the expectation function; $G(\cdot)$ represents the generator functions; $D(\cdot)$ is the discriminator function; $P_r(\cdot)$ indicates the distribution of the objective function x; $P_g(\cdot)$ is the noise data distribution; z is the input noise data vector.

The traditional GAN model often exhibits a certain instability during training, which can lead to the problem of mode collapse. This issue refers to the generator's tendency to produce highly similar samples, resulting in a lack of diversity in the generated samples [25]. In such cases, the generator struggles to capture the diversity present in the training data, thus failing to effectively generate new and diverse data samples. To address this problem, one approach is to introduce the Wasserstein distance to measure the discrepancy between generated data and real data. The definition of the Wasserstein distance is as follows:

$$W(P_r,P_g) = \frac{1}{K}\sup_{\|f\|\leq K}\left(\underset{x\sim P_r(x)}{E}[f(x)] - \underset{\tilde{x}\sim P_g(\tilde{x})}{E}[f(\tilde{x})]\right) \quad (5)$$

where K represents the Lipschitz constant; $\sup(\cdot)$ is the upper bound; x is the raw data; $\tilde{x}$ is the generated data.

During the actual operation of WGAN, to ensure training stability, it is necessary to perform weight clipping on the discriminator. However, this weight clipping approach might lead to issues such as exploding gradients and non-convergence. To address this problem, based on the WGAN, a gradient penalty is introduced to regularize the gradients of the discriminator. The gradient penalty term helps ensure smoothness during the training process and encourages the generator to produce more diverse and higher-quality samples. The loss function of the WGAN-GP model can be represented as follows:

$$L(G,D) = \min_G \max_D \left\{ \underset{x\sim P_r(x)}{E}[D(x)] - \underset{z\sim P_g(z)}{E}[D(G(z))] + \lambda \underset{\hat{x}\sim P_{\hat{x}}(\hat{x})}{E}[\|\nabla\hat{x}D(\hat{x})\|_p - 1]^2 \right. \quad (6)$$

where λ represents the gradient penalty coefficient; $\hat{x} = \alpha x + (1-\alpha)G(z)$ indicates a random interpolation between the real data and the generated data; α is the number randomly sampled from the uniform distribution [0, 1]; $P_{\hat{x}}(\cdot)$ is an interpolation distribution formed between the real data distribution and the generated data distribution. $\|\cdot\|_p$ is p norm; ∇ is the gradient operator.

By enhancing the discriminator, WGAN-GP not only overcomes the issue of mode collapse but also improves the stability of training in GANs. This enhancement enables the generator to produce more diverse and higher-quality data samples. Figure 3 depicts the architecture of the WGAN-GP network.

2.3. WGAN-GP Algorithm

Ensemble learning is a machine learning method that leverages the combination of multiple simple models to obtain a composite model with improved performance, thereby enhancing the accuracy and generalization capabilities of the model [26]. The primary strategies of ensemble learning involve generating diversity, model training, and model combination. Common ensemble learning techniques include bagging, boosting, and stacking. Ensemble learning not only enhances model performance but also allows researchers to design combination approaches tailored to specific machine learning problems to obtain more robust solutions.

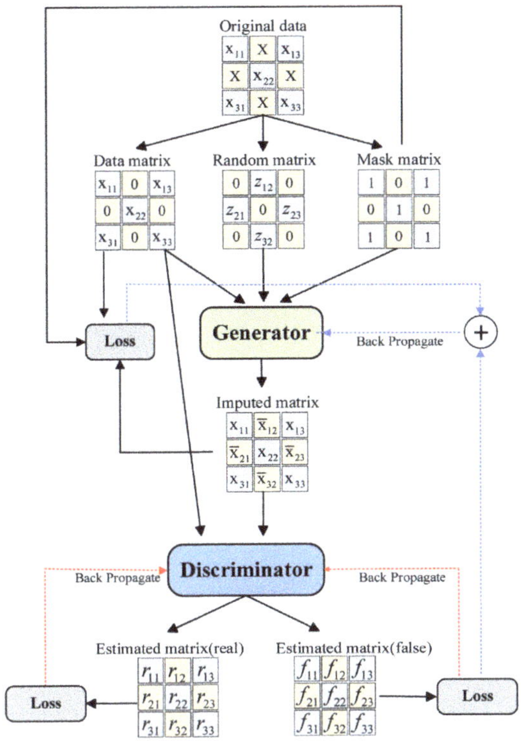

Figure 3. The architecture of WGAN-GP.

As an extension of ensemble learning, parallel ensemble learning can enhance the efficiency of ensemble learning through parallel computing [27]. In the conventional ensemble learning, the training and prediction of each base model are executed sequentially. However, in the parallel ensemble learning, multiple base models can be trained and make predictions simultaneously on different computing resources. Parallel ensemble learning can significantly reduce the time required for model training and prediction, thereby enhancing the model's performance. This is particularly significant for handling large-scale data sets and achieving real-time predictions.

This paper adopts the bootstrap aggregating (Bagging) algorithm from parallel ensemble learning to integrate LSTM and WGAN-GP. Bagging is a commonly used basic strategy in ensemble learning [28], and its fundamental principle is depicted in Figure 4. The core idea of the Bagging algorithm involves multiple rounds of resampling the original dataset to generate several diverse training subsets. These subsets are then used to train multiple models. Finally, the predictions of these models are integrated through voting or averaging to produce the ultimate prediction.

Generally, for the regression problem, Bagging's aggregated results often employ either the averaging method or the weighted averaging method to combine the predictions from all base models. Specifically, the averaging method accumulates the predictions of each base model and then divides the sum by the number of base learners to obtain the final prediction value. On the other hand, the weighted averaging method assigns different weights to each base model based on its performance. Then, it calculates the weighted average as the final prediction result [29].

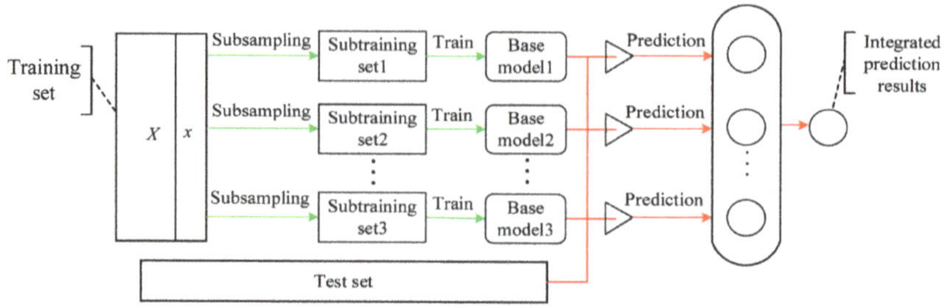

Figure 4. Bagging schematic diagram.

To ensure the quality of the repair while considering the differences and effectiveness of each base learning model, based on the theory of minimizing loss [30], this paper proposes a multi-model prediction result fusion strategy based on an improved weighted averaging method. This method allows multiple models to contribute predictions proportionally based on their trustworthiness or estimated performance, thereby enhancing the performance and stability of the ensemble model and minimizing prediction errors. By analyzing the loss value of each model, weights can be assigned to each model and integrated with the predictions of the base models, ensuring the accuracy of data repair.

Therefore, the weight calculation formulas for the predictions from the modified LSTM algorithm and the WGAN-GP algorithm are as follows:

$$\alpha = \frac{loss\_wgan}{loss\_lstm + loss\_wgan} \tag{7}$$

$$\beta = \frac{loss\_lstm}{loss\_lstm + loss\_wgan} \tag{8}$$

$$Y_{pre} = \omega_1 Y_{pre\_lstm} + \omega_2 Y_{pre\_wgan} \tag{9}$$

where ω_1 denotes the weight of the predicted value of modified LSTM model; ω_2 represents the weight of the predicted value of WGAN-GP model; $loss\_lstm$ indicates the loss function value of the modified LSTM training model; $loss\_wgan$ is the loss function value of the WGAN-GP training model; Y_{pre} is the result value of multi-model fusion prediction; Y_{pre_lstm} is the predicted value of the modified LSTM model; Y_{pre_wgan} is the predicted value of the WGAN-GP model.

3. Bad Data Repair Method for New Energy Station Based on Multi-Model Parallel Ensemble

In order to overcome the potential issues such as overfitting associated with existing single models, in this paper, by combining the modified LSTM model and WGAN-GP model, a novel multi-model parallel ensemble method for the repair of bad data of new energy stations is developed. The proposed method can merge the strengths of multiple models to enhance the repairing performance and robustness. By training multiple models in parallel on different data or feature subsets, it can effectively reduce training time and increase the diversity of models, thereby reducing the risk of overfitting and having better model interpretability.

Specifically, the application of a multi-model parallel ensemble data repair method for the correction process of adverse data in new energy station is illustrated in Figure 5. It mainly consists of two stages: (1) offline training; (2) online repair.

(1) Offline Training

The offline training stage is the key step in building the integrated model, which involves the training and weight determination of the model.

Step 1: Model training. First, the current and voltage historical data of the new energy station are normalized and divided into a training set and a test set, in which the ratio of training set and test set is usually 4:1. Then, self-sampling is utilized to randomly extract multiple subsets from the training data, each containing 70% of the original training data set. Finally, the improved LSTM and WGAN-GP models are trained with each data subset to determine their respective model parameters and structures.

Step 2: Weight calculation. Since the improved LSTM and WGAN-GP each represent different repair strategies, it is necessary to clarify the weight allocation between them. First, loss measurement is used to evaluate the error between the model repair results and the real data, which provides a quantitative evaluation basis for the following steps. Then, the loss values of the improved LSTM and WGAN-GP models in the process of data repair are calculated individually to evaluate the repair efficiency of each model. A lower loss value indicates that the model has a better repair effect, while a higher loss value reflects that the model has a poor repair effect. Finally, the weights of the improved LSTM and WGAN-GP models are determined according to Formulas (7)–(9).

Step 3: Integrated model building. Following the completion of preliminary data preparation, training of individual models, and weight allocation, this stage employs the architecture of parallel ensemble learning. The previously trained, improved LSTM and WGAN-GP models are fused to create an efficient integrated learning model. Initially, the outputs of the two models are fed into a specifically designed fully connected layer, tasked with integrating the outputs of both models to generate a comprehensive repair result. Subsequently, within the fully connected layer, the outputs of the improved LSTM and WGAN-GP models are weighted and fused using the weight strategy determined in Step 2. Finally, for this parallel ensemble learning model, further training and optimization are conducted by adjusting the model's parameters and structure to ensure its effectiveness in handling real-world data repair tasks.

(2) Online Repair

The online stage is the practical application stage, which mainly repairs the real-time measurement data.

Step 1: Real measurement data input. At this stage, the three sets of measured current and voltage data of the new energy station are preprocessed to ensure that the data format is consistent with the model training. Then, the measured data after processing are input into the trained model.

Step 2: Repair bad data online. First, the model loads the parameters and structures learned during the offline training phase. Then, using the knowledge of offline training, the model can repair the actual measurement data.

In summary, establishing an effective bad data repair system for new energy stations requires two key phases: offline training and online repair. The offline phase focuses on training the repair models and determining the weights between different models to ensure the extraction of time-series data characteristics and the generation of high-quality data. The online phase then utilizes the previously trained models to repair bad data.

Remark 1. *In this paper, an innovative approach is developed for the bad data repair of new energy stations in power systems, leveraging a novel integration of modified LSTM and WGAN-GP models in a parallel framework. Compared with a single model, the resulting model significantly improves the robustness and generalization ability of the model, and the accuracy and efficiency of data repair are also significantly improved.*

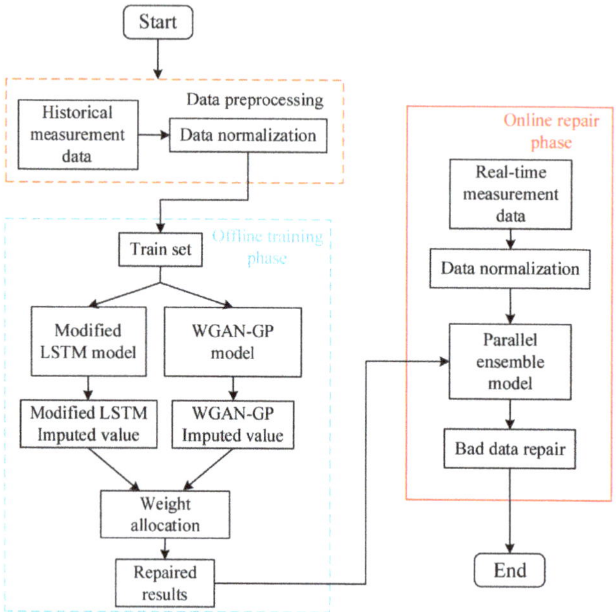

Figure 5. Bad data repair flow chart based on multi-model parallel ensemble learning.

4. Case Analysis

In this paper, a PC configured with AMD Ryzen 7 5800 H 3.20 GHz CPU, 16 GB RAM and a Windows 10 64-bit operating system is used to compile Python programs with Pytorch as the deep learning framework and to build an improved LSTM model and WGAN-GP model. The performance of the proposed algorithm is verified by actual new energy station data.

4.1. Experimental Data Set

In practical applications, the true value of abnormal or corrupted data is often difficult to obtain. This means that the accuracy of data repair cannot be evaluated by directly comparing the true value, so data repair becomes a typical unsupervised learning problem [31]. To more intuitively assess and understand the effects of data repair, a strategy can be employed: simulate the generation of data sets containing bad data from a complete and reliable data set. Then, the proposed method is used to repair these bad data, and the repaired data are compared with the original real data, so as to evaluate the accuracy of the repair effect. In addition, because the data are often affected by random factors and uncertainties, it is difficult to predict the location and amount of bad data. In order to simulate this real scenario and ensure the quantity of bad data in each experiment, this study uses the mask matrix strategy to ensure the randomness and authenticity of the generated bad data, so as to be closer to the abnormal data situation in actual operation.

In order to verify the effectiveness of the proposed method, an actual wind farm in a province was selected as the research object, and the real-time current and voltage measurement data of three PMUs were randomly selected as the test set. Each group contained 200 data points, and the overall data dimension was 200×6. In order to ensure the generalization ability of the model, the complete data set is divided into the training set and the test set in a 4:1 ratio. In the test set, some measurement data in the data set are randomly added with mixed noise to simulate the bad data that may appear in the real scene. In general, various bad data conditions are characterized by increasing or decreasing the normal voltage and current measurement values by 5–10%.

Before training the model, it is necessary to preprocess the historical measurement data to ensure that different data levels or ranges will not adversely affect the training of the model and capture the real characteristics of the data. For data normalization, the calculation formula is as follows

$$x_{Nor} = \frac{x - \min(x)}{\max(x) - \min(x)} \tag{10}$$

where x is the data that need to be normalized; x_{Nor} is the normalized value; $\min(x)$ is the minimum value of the feature x; $\max(x)$ is the minimum value of the feature x.

4.2. Experiment and Result Analysis

4.2.1. Data Repair Evaluation Index

In order to quantitatively test the data repair effect of the proposed model, root-mean-square error (RMSE) and mean absolute percentage error (MAPE) were selected in this paper. MAPE is used as an evaluation index of the data repair effect [31]. Among them, RMSE can measure the deviation between the predicted value and the actual value, while MAPE gives the relative relationship between the predicted error and the actual value, which can more intuitively understand the accuracy of the repair effect.

The performance evaluation indexes are defined as follows:

$$\mu_{RMSE} = \sqrt{\frac{1}{n}\sum_{i=1}^{n} (y_i^{real} - y_i^{pre})^2} \tag{11}$$

$$\mu_{MAPE} = \frac{1}{n}\sum_{i=1}^{n} \left| \frac{y_i^{real} - y_i^{pre}}{y_i^{real}} \right| \tag{12}$$

where n is the number of bad data; y_i^{real} is the actual quantity measurement; y_i^{pre} is the repair value.

4.2.2. Case Setting

In order to accurately evaluate the repair effect of the bad data of the proposed method, two comparative experimental scenarios were set up for verification based on the selected experimental data set.

Scenario 1. *Generate bad data randomly in the original data set, compare the data repair effect of LSTM and the proposed method, and evaluate the bad data repair performance of the proposed method.*

Scenario 2. *Under the condition of different proportions of bad data, the method proposed in this paper is comprehensively compared with several unsupervised data repair methods.*

4.3. Random Bad Data Situation

In order to intuitively demonstrate the superiority in repairing bad data of the parallel integration strategy proposed in this paper over a single model, the single model LSTM and the proposed model are selected in this scenario for comparative experiments. The experiment randomly generates 5% bad data on the original data set, and repaired bad data on the same data set, aiming to reveal the improvement and advantages of the proposed method compared with a single model in processing the same data.

The LSTM method and the method proposed in this paper were used to repair the bad data. The test results are shown in Figure 6. As can be seen from the results shown in Figure 6, the LSTM method has a certain repair effect on a single item of bad data. However, in the face of continuous bad data, the LSTM method has a poor repair performance, and the output is continuous and unchanged data. This is because LSTM over-relies on historical data to make predictions in the face of continuous bad data, fails to effectively capture the characteristics of the current continuous bad data, and so overfitting occurs, resulting in continuous and identical data.

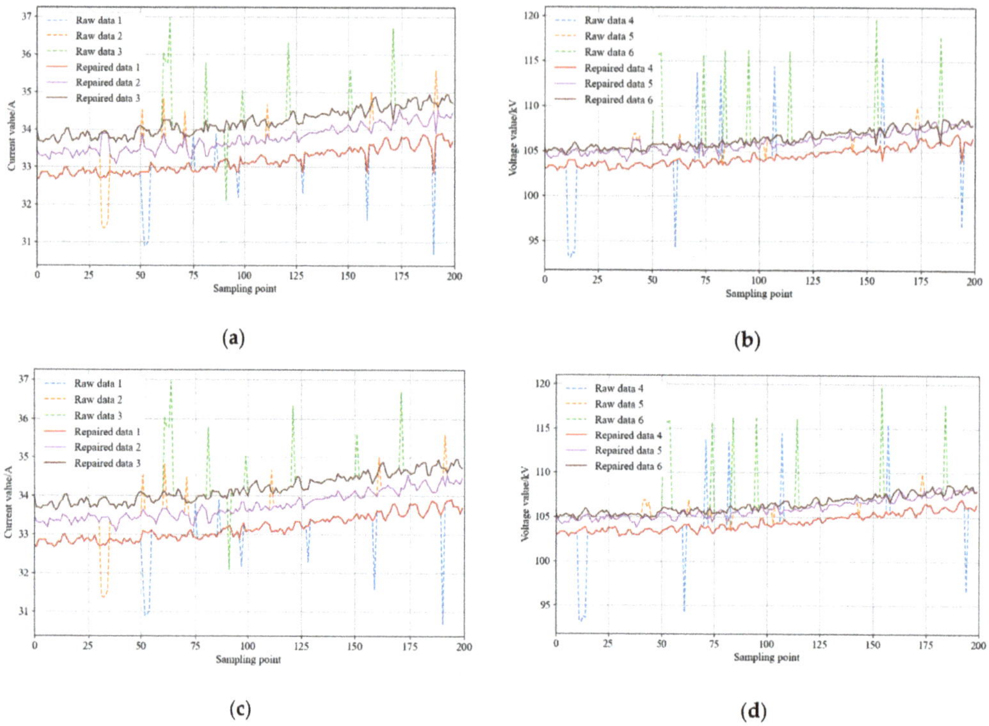

Figure 6. Comparison of repair results between LSTM and the proposed method. (**a**) Current repair results based on LSTM. (**b**) Voltage repair results based on LSTM. (**c**) Current repair results based on proposed method. (**d**) Voltage repair results based on proposed method.

In addition, Figure 6 also shows the repair results of bad data using the method proposed in this paper. It can be seen that when the same continuous bad data are repaired using the proposed method in this paper, the repair result is an item of data with a similar distribution to the original data, and the overfitting phenomenon is effectively alleviated. These results show that the proposed method can better capture data features by integrating the prediction of multiple models, reducing the over-dependence on historical data, avoiding the overfitting problem of a single model, and realizing the high-precision repair of continuous bad data.

In order to evaluate the performance of the proposed method more comprehensively, Table 1 gives the calculation results of the root-mean-square error and mean absolute percentage error of LSTM and the proposed method in this paper. It can be seen from the results in Table 1 that the error of the repair method proposed in this paper is smaller than that of the single LSTM model. This indicates that the proposed method is more accurate and effective in repairing bad data.

Table 1. Case system parameters setting.

Method	RMSE	MAPE
LSTM	0.1859	0.0440%
Proposed method	0.0548	0.0138%

In summary, the method proposed in this paper significantly improves the repair quality of bad data by integrating multiple models. Compared with a single LSTM model, it can effectively alleviate the overfitting phenomenon and show higher accuracy and effect.

4.4. Different Proportions of Bad Data

In order to further systematically and comprehensively verify the performance of the method proposed in this study, in this scenario, three unsupervised bad data repair methods are selected: mean value repair method, KNN repair method, and GAN repair method, and comparative tests are conducted. The experiment randomly generated eight different percentages of bad data on the original data set: 20%, 30%, 40%, 50%, 60%, 70%, and 80%. In each ratio of bad data, four repair methods were tested and compared in detail.

Due to the page limit, this article only displays a comparison of the repair effects of current values under 20% and 80% proportions of bad data. Figures 7 and 8 show the repair results obtained by the mean value method, KNN method, GAN method, and the proposed method under the different proportions of bad data. Clearly, from these figures, it can be observed that under all proportion conditions, the repair effect of the mean value method is significantly inferior to the other three algorithms.

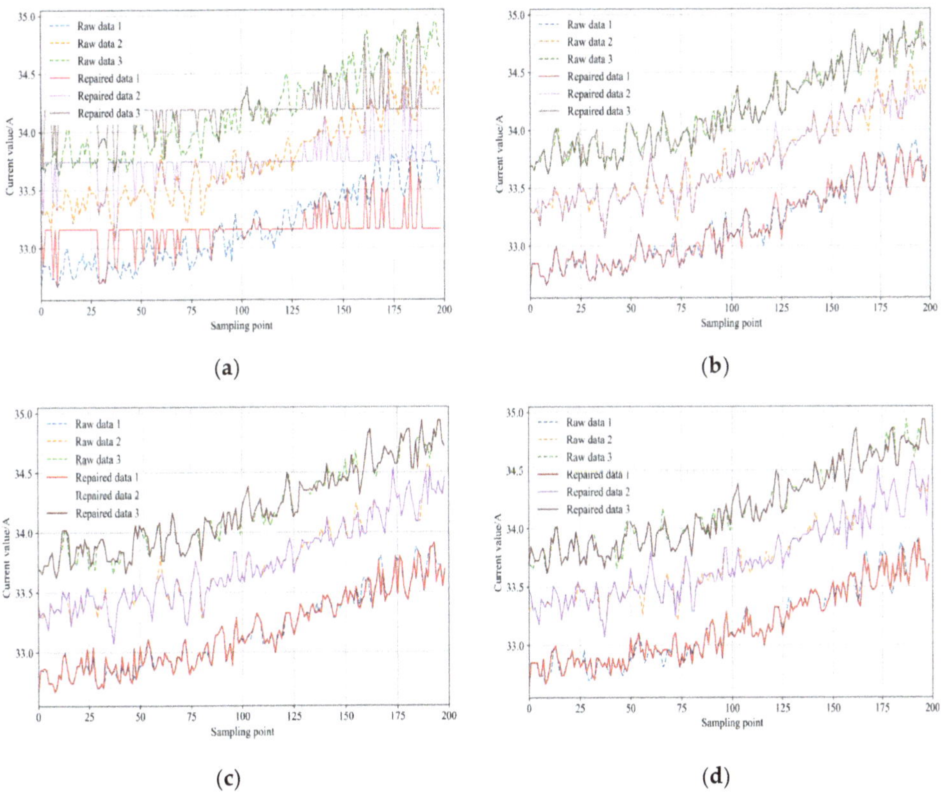

Figure 7. Repair results with different methods under 20% bad data. (**a**) Repair results based on mean value method. (**b**) Repair results based on KNN. (**c**) Repair results based on GAN. (**d**) Repair results based on proposed method.

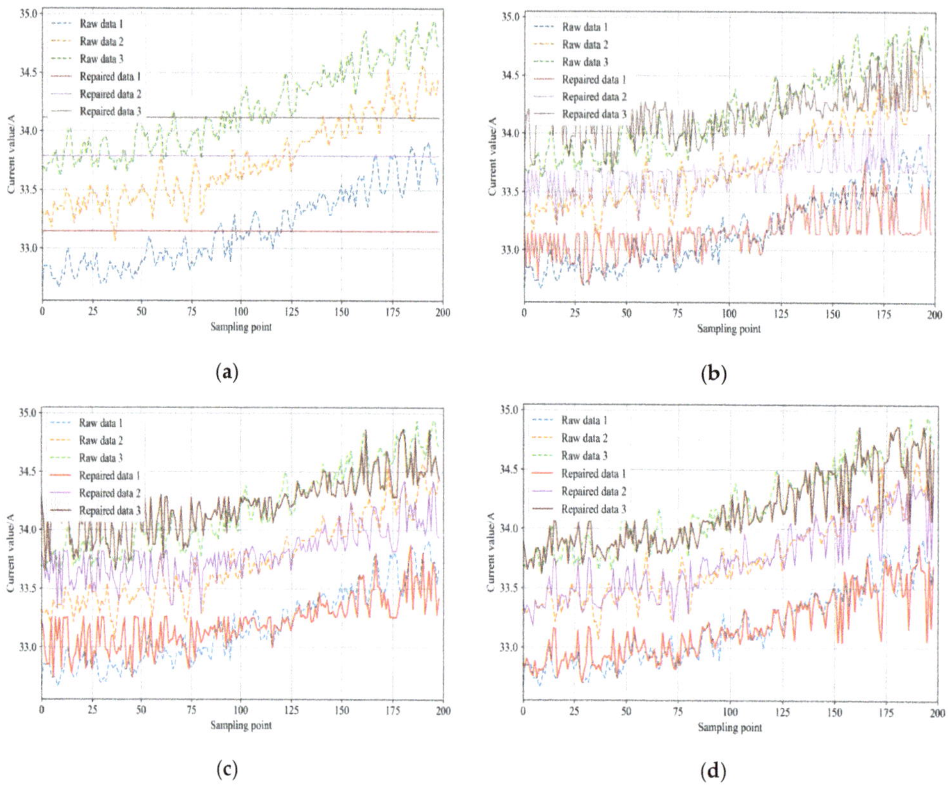

Figure 8. Repair results with different methods under 80% bad data. (**a**) Repair results based on mean value method. (**b**) Repair results based on KNN. (**c**) Repair results based on GAN. (**d**) Repair results based on proposed method.

Under low proportions of bad data conditions, the repair effects of KNN, GAN, and the method proposed in this study are similar. However, with the increase in the proportion of bad data, the repair effects of KNN and GAN gradually deteriorate. In contrast, the method proposed in this paper still has an excellent repair effect. In order to more thoroughly compare the efficiency of these four algorithms in handling different proportions of bad data, this article also provides a comparative analysis of the root-mean-square error and mean absolute percentage error for these methods.

Figure 9 shows the root-mean-square and average percentage errors of each repair method with 20% to 80% bad data. It can be seen that the mean method consistently maintains high RMSE and MAPE values, which indicates that the method is less capable of repairing in the face of bad data. In contrast, the KNN repair method, GAN repair method, and method proposed in this paper can obtain smaller RMSE values and MAPE values when the proportion of bad data is small (<50%), and the error index of the proposed method is significantly smaller than that of other methods. However, in the case of a large proportion of bad data (>50%), as the proportion of bad data increases, the bad data repair effect of each model decreases; however, it can be seen that the performance of the proposed method is still far superior to other methods.

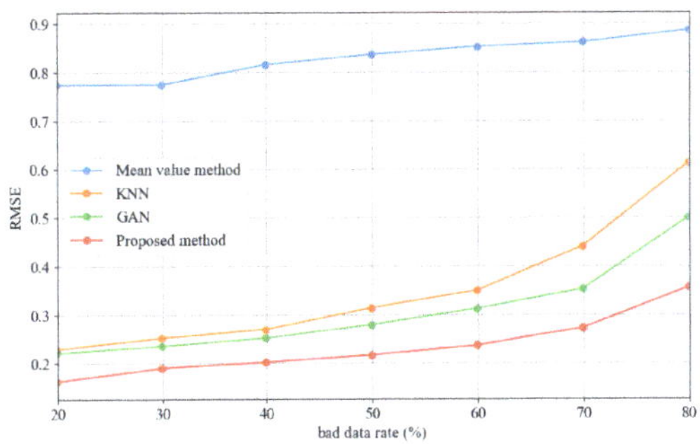

(**a**)

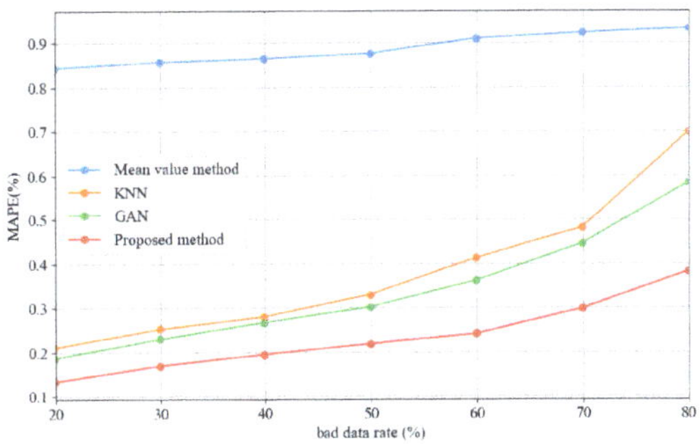

(**b**)

Figure 9. Comparison of each repair algorithm under different bad data rates. (**a**) Root-mean-square error comparison. (**b**) Mean absolute percentage error comparison.

Specifically, from the comparison results of error indicators of each test method under different proportions of bad data in Tables 2 and 3, it can be seen that with a data missing rate of 20%, the data repair accuracy of the proposed method is improved by 36.57% compared with KNN and 28.61% compared with GAN. With a data missing rate of 50%, the proposed method improves the data repair accuracy by 33.71% compared with KNN and 27.63% compared with GAN. Even in the extreme case where the proportion of bad data is as high as 80%, the data repair accuracy of the method proposed in this paper still remains at a high level, with an MAPE of 0.3801%.

Table 2. Root-mean-square error of each method under different proportions of bad data.

Loss Rate/%	RMSE			
	Mean Value Method	LSTM	GAN	Proposed Method
20	0.7742	0.2283	0.2201	0.1621
30	0.7746	0.2516	0.2346	0.1897
40	0.8159	0.2698	0.2512	0.2017
50	0.8363	0.3143	0.2796	0.2162
60	0.8524	0.3496	0.3127	0.2355
70	0.8615	0.4403	0.3528	0.2724
80	0.8862	0.6124	0.5003	0.3564

Table 3. Mean absolute percentage error of each method with different proportions of bad data.

Loss Rate/%	MAPE (%)			
	Mean Value Method	KNN	GAN	Proposed Method
20	0.8435	0.2098	0.1856	0.1325
30	0.8569	0.2517	0.2291	0.1693
40	0.8647	0.2796	0.2667	0.1943
50	0.8762	0.3295	0.3018	0.2184
60	0.9104	0.4121	0.3612	0.2402
70	0.9235	0.4812	0.4454	0.2979
80	0.9331	0.6987	0.5842	0.3801

The above experimental results show that compared with the traditional mean repair method and KNN and GAN single model repair method, the multi-model integration-based bad data repair method proposed in this paper can obtain better bad data repair accuracy and has stronger generalization ability. This is because the proposed method can more comprehensively capture the different distribution characteristics of the new energy station measurement data and mine the potential characteristics and correlation of the data, so that it can better adapt to and deal with bad data, improve the overall performance of the model, and thus provide more accurate repair results under various bad data scenarios.

5. Conclusions

To deal with the issue of low accuracy and poor performance of bad data repair in new energy stations, this paper proposes a novel deep learning approach by combining modified LSTM and WGAN-GP models. Based on the experimental results, the following conclusions can be drawn.

(1) The adoption of the modified LSTM model has effectively mitigated the problems of gradient vanishing or explosion, which is a notable progression in neural network training. This modification can significantly reduce the model's tendency to overfitting of training data, thereby improving both the efficiency of the training and the generalization ability of the model.

(2) This ensemble approach has not only facilitated effective feature extraction from normal measurement data, but has also significantly improved the overall performance of the system. By leveraging the strengths of both models, our approach can achieve a more nuanced and comprehensive analysis of the data, resulting in more accurate and reliable repair data results.

(3) The multi-model parallel ensemble method proposed in this paper has significant advantages over a single model, which can achieve the high-precision repair of bad data under various working conditions and has stronger generalization ability.

Since the approach proposed in this paper for repairing bad data from new energy field stations based on the parallel ensemble of multiple models shows superiority, the conclusions drawn from this study can guide more complex models and integration strategies. Therefore, further research should focus on how to optimize this parallel integration strategy. This may include adjusting the weight allocation between different models, exploring the integration effects of different types of neural network models such as convolutional neural networks (CNNs) or recurrent neural networks (RNNs), as well as improving the existing models to better handle specific types of bad data.

Author Contributions: C.L. and M.L. were responsible for the methodology, simulation, and validation; C.L. conducted the analysis and wrote the paper; conceptualization, C.L. and M.L.; resource management, Z.G. and C.T.; data curation, C.L. and C.T.; original draft preparation, C.L. and M.L.; writing—review and editing, Y.W; visualization, C.L. and M.L.; supervision, Y.W.; project management, M.L.; funding acquisition, Y.W. All authors have read and agreed to the published version of the manuscript.

Funding: This work was supported by the National Natural Science Foundation of China under Grant 62203395, in part by the China Postdoctoral Science Foundation under Grant 2023TQ0306, and in part by the Postdoctoral Research Project of Henan Province under Grant 202101011.

Data Availability Statement: The data presented in this study are available in this article.

Conflicts of Interest: Authors Chenghao Li, Mingyang Liu, Ze Gao, and Chunsun Tian were employed by the company State Grid Henan Electric Power Research Institute. The remaining authors declare that the research was conducted in the absence of any commercial or financial relationships that could be construed as a potential conflict of interest.

References

1. Han, X.; Li, T.; Zhang, D.; Zhou, X. New issues and key technologies of new power system planning under double carbon goals. *High Volt. Eng.* **2021**, *47*, 3036–3046.
2. Zhu, L.; Hill, D.J. Cost-effective bad synchrophasor data detection based on unsupervised time-series data analytic. *IEEE Internet Things J.* **2021**, *8*, 2027–2039. [CrossRef]
3. Peng, Z.; Lu, Y.; Zhang, Y.; Deng, W.; Zeng, Q. The online frequency security assessment of a power system considering the time-varying characteristics of renewable energy inertia. *Electronics* **2023**, *12*, 2205. [CrossRef]
4. Xue, A.; Xu, F.; Martin, K.E.; Xu, J.; You, H.; Bi, T. Linear approximations for the influence of phasor angle difference errors on line parameter calculation. *IEEE Trans. Power Syst.* **2019**, *34*, 3455–3464. [CrossRef]
5. Jiang, X.; Zhang, J.; Harding, B.J.; Makela, J.J.; Domı, A.D. Spoofing GPS receiver clock offset of phasor measurement units. *IEEE Trans. Power Syst.* **2013**, *28*, 3253–3262. [CrossRef]
6. Shepard, D.P.; Humphreys, T.E.; Fansler, A. Evaluation of the vulnerability of phasor measurement units to GPS spoofing attacks. *Int. J. Crit. Infrastruct. Prot.* **2012**, *5*, 146–153. [CrossRef]
7. Jones, K.D.; Pal, A.; Thorp, J.S. Methodology for performing synchrophasor data conditioning and validation. *IEEE Trans. Power Syst.* **2015**, *30*, 1121–1130. [CrossRef]
8. Ghiocel, S.G.; Chow, J.H.; Stefopoulos, G.; Fardanesh, B.; Maragal, D.; Blanchard, B.; Razanousky, M.; Bertagnolli, D.B. Phasor measurement based state estimation for synchronous phasor data quality improvement and power transfer interface monitoring. *IEEE Trans. Power Syst.* **2014**, *29*, 881–888. [CrossRef]
9. Ma, M.; Qin, J.; Yang, D.; Zhou, B.; Pang, Y. A review of the application of artificial intelligence in power systems. *J. Zhengzhou Univ. (Eng. Ed.)* **2019**, *40*, 23–31.
10. Xie, K.; Liu, J.; Liu, Y. A power system timing data recovery method based on improved VMD and attention mechanism Bi-directional CNN-GRU. *Electronics* **2023**, *12*, 1590. [CrossRef]
11. Lattner, S.; Nistal, J. Stochastic restoration of heavily compressed musical audio using generative adversarial networks. *Electronics* **2021**, *10*, 1349. [CrossRef]
12. Ma, J.; Cheng, J.C.P.; Jiang, F.; Chen, W.; Wang, M.; Zhai, C. A bi-directional missing data imputation scheme based on LSTM and transfer learning for building energy data. *Energy Build.* **2020**, *216*, 109941. [CrossRef]
13. Zhang, W.; Jiang, W.; Liu, Q.; Wang, W. AIS data repair model based on generative adversarial network. *Reliab. Eng. Syst. Saf.* **2023**, *240*, 109572. [CrossRef]
14. Lee, G.; Kwon, G.Y.; Kim, S.H.; Kim, D.I.; Kim, Y.; Nam, S.; Ko, B.; Kang, S.; Shin, Y.J. Missing entry estimation of synchrophasor data using artificial neural network. In Proceedings of the 9th International Conference on Power and Energy Systems (ICPES), Perth, WA, Australia, 10–12 December 2019; Institute of Electrical and Electronics Engineers Inc.: Perth, WA, Australia, 2019; pp. 1–6.
15. Wang, S.; Chen, H.; Pan, Z.; Ling, T. A reconstruction method for missing data in power system measurement using an improved generative adversarial network. *Proc. CSEE* **2019**, *39*, 56–64.
16. Wang, Y.; Zhang, Z.; Ma, J.; Jin, Q. KFRNN: An effective false data injection attack detection in smart grid based on Kalman filter and recurrent neural network. *IEEE Internet Things J.* **2022**, *9*, 6893–6904. [CrossRef]
17. Hochreiterand, S.; Schmidhuber, J. Long short-term memory. *Neural Comput.* **1997**, *9*, 1735–1780. [CrossRef] [PubMed]
18. Shen, Q.; Mo, L.; Liu, G.; Zhou, J.; Zhang, Y.; Ren, P. Short-term load forecasting based on multi-scale ensemble deep learning neural network. *IEEE Access* **2023**, *11*, 111963–111975. [CrossRef]
19. Zhao, S.; Dong, X. Research on speech recognition based on improved LSTM deep neural network. *J. Zhengzhou Univ. (Eng. Ed.)* **2018**, *39*, 63–67.
20. Kingma, D.P.; Ba, J.L. Adam: A method for stochastic optimization. In Proceedings of the International Conference on Learning Representations (ICLR), San Diego, CA, USA, 7–9 May 2015; pp. 1–15.
21. Srivastava, N.; Hinton, G.; Krizhevsky, A.; Sutskever, I.; Salakhutdinov, R. Dropout: A simple way to prevent neural networks from overfitting. *J. Mach. Learn. Res.* **2014**, *15*, 1929–1958.
22. Cui, H.; Li, F.; Tomsovic, K. Hybrid symbolic-numeric framework for power system modeling and analysis. *IEEE Trans. Power Syst.* **2021**, *36*, 1373–1384. [CrossRef]
23. Goodfellow, I.; Pouget-Abadie, J.; Mirza, M.; Xu, B.; Warde-Farley, D.; Ozair, S.; Courville, A.; Bengio, Y. Generative adversarial nets. In *Proceedings of Advances in Neural Information Processing Systems*; MIT Press: Toronto, ON, Canada, 2014.

24. Xu, M. Towards generalized implementation of wasserstein distance in GANs. In Proceedings of the AAAI Conference on Artificial Intelligence, Online, 2–9 February 2021; Volume 35, pp. 10514–10522.
25. Dong, X.; Yu, Z.; Cao, W.; Shi, Y.; Ma, Q. A survey on ensemble learning. *Front. Comput. Sci.* **2020**, *14*, 241–258. [CrossRef]
26. Woźniak, M.; Grana, M.; Corchado, E. A survey of multiple classifier systems as hybrid systems. *Inf. Fusion* **2014**, *16*, 3–17. [CrossRef]
27. Odegua, R. An empirical study of ensemble techniques (bagging, boosting and stacking). In Proceedings of the Conference: Deep Learning IndabaXAt, Nigeria, Kenya, 25–31 August 2019.
28. Xu, J.; Yang, Y. Ensemble learning method: Research review. *J. Yunnan Univ. (Nat. Sci. Ed.)* **2018**, *40*, 1082–1092.
29. Ko, H.; Lee, J.; Byun, J.; Son, B.; Park, S. Loss-driven adversarial ensemble deep learning for online time series analysis. *Sustainability* **2019**, *11*, 3489. [CrossRef]
30. Zhao, J.; Wang, J. Repairing adversarial samples based on SmsGAN. *J. Zhengzhou Univ. (Eng. Ed.)* **2021**, *42*, 50–55.
31. Guo, X.; Li, Z.; Liu, H.; Bi, T. PMU loss data recovery method based on enhanced generative adversarial networks. *Power Syst. Technol.* **2022**, *46*, 2114–2121.

electronics

Essay

Convolution Power Ratio Based on Single-Ended Protection Scheme for HVDC Transmission Lines

Guangqiang Peng [1], Lixin Chen [1], Jiyang Wu [1], Huimin Jiang [2], Zhijie Wang [2] and Haifeng Li [2,*]

1 Electric Power Research Institute of EHV, Guangzhou 510633, China; pengguangqiang@ehv.csg.cn (G.P.); chenlixin@ehv.csg.cn (L.C.); wujiyang@ehv.csg.cn (J.W.)

2 School of Electric Power Engineering, South China University of Technology, Guangzhou 510641, China; 202311082961@mail.scut.edu.cn (H.J.); 202311083086@mail.scut.edu.cn (Z.W.)

* Correspondence: lihf@scut.edu.cn

Abstract: In order to solve the problems of insufficient abilities to withstand transition resistance under remote faults and difficulties in identifying internal and external faults for HVDC transmission line protection, a new single-ended protection scheme based on time-domain convolutional power was proposed. In this scheme, the ratio of time-domain convolution power at different frequencies is used to detect internal and external faults, and the long window convolution power is used to form the pole selection criteria. Due to the integration of transient power fault characteristics at high and low frequencies, this scheme amplifies the characteristic differences between internal and external faults caused by DC line boundaries and has a strong ability to withstand transition resistance. Based on PSCAD/EMTDC, simulation verification was conducted on the Yunnan–Guangzhou ±800 kV HVDC project. The results show that the proposed single-ended protection scheme can effectively identify fault poles, as well as internal and external faults. It has strong resistance to transition resistance and certain anti-interference ability and has strong adaptability to DC line boundaries, which meets the protection requirements of HVDC transmission systems for high speed, selectivity and reliability.

Keywords: convolution power; direct current boundary; high voltage direct current transmission; line protection

1. Introduction

Due to remarkable advantages such as lower construction costs, lower losses and bulk and long-distance power transmission capabilities, line-commutated converter high-voltage direct current (LCC-HVDC) systems are widely used [1–4]. The transmission line is the component with the highest fault probability, and line faults affect the whole HVDC system. Therefore, the rapid removal of faults is crucial to the enhancement of the transient stability of power systems [5–8]. Currently, fault identification methods for HVDC transmission lines can be summarized into the following two schemes:

Time domain characteristics based on methods: Traditional time domain schemes typically utilize the transient characteristics of voltage traveling wave attenuation caused by boundaries composed of smoothing reactors at both ends of the line, such as traditional methods of voltage change rate [9,10]. These methods are simple and have low computational complexity and are widely used in practical engineering. However, they will significantly decrease with an increase in transition resistance, thereby affecting the sensitivity of protection. Ref. [11] uses the method of integrating the sum of positive and negative limit current reactance voltages to identify internal and external faults, which has strong anti-noise interference ability and is not affected by AC measurement faults; however, the scheme has a weak ability to withstand transition resistance, and it is easy to misjudge when high-impedance grounding faults occur at the end of the line.

Compared with the above methods based on a certain time domain, utilizing the waveform characteristics of local wave fronts under internal and external faults is also an effective approach. In Refs. [12,13], the Levenberg–Marquardt method was used to fit the initial traveling wave voltage of the measurement point fault, and the failure modes and exponential terms of different fault characteristics were obtained. Reference [14] calculates the concavity and convexity of the current-limiting reactance voltage for fault identification. Reference [15] utilizes a multi-resolution morphological gradient algorithm to extract the polarity information of fault traveling waves, which has low sampling frequency requirements. However, references [12–15] all require correct calibration of the initial fault traveling wave to avoid the influence of reflected traveling waves at the beginning and end of the line. In the event of near-end or terminal faults, the arrival time interval between the first and secondary traveling waves is very short, and there may be significant errors in correctly calibrating the traveling wave within a short time window, resulting in protection malfunction or rejection.

Frequency domain characteristics based on methods: The frequency domain method usually utilizes the high-frequency attenuation of fault traveling waves at the boundaries of DC lines and uses frequency domain extraction tools such as wavelet transforms, and Fourier transforms to extract high-frequency information. Traditional frequency domain extraction methods usually extract transient feature information under a single frequency band. For example, in reference [16], wavelet transforms were used to extract the high-frequency voltage component to identify the faults; reference [17] uses the measured impedance at resonant frequency for fault identification based on the amplitude–frequency characteristics of DC filters. Study [18] extracted the voltage and current data of specific frequency bands on both sides of a reactor, and then calculated the boundary energy to form longitudinal protection. However, the traditional method of extracting a single frequency band has the problem of having poor resistance to transition resistance.

To address the above problems, many methods based on multi-frequency-band transient characteristics are proposed. Reference [19] uses the energy of multiple voltage characteristic frequency bands to construct fault criteria. However, its calculation principle is complex due to voltage energy calculations in multiple frequency bands. The ratios of voltages at different frequency features used in references [20–22] serve as a criterion for fault identification. In [23,24], stationary wavelet transforms and Hilbert–Huang transforms were used to extract the features of multi-frequency-band voltage, which can effectively amplify the characteristic differences of internal and external faults. However, all these methods require higher sampling frequencies and have complex calculation problems.

To improve the above-mentioned problems such as poor resistance and high sampling frequency, a single-ended protection scheme was proposed for high-voltage DC transmission lines based on convolutional power.

1. The differences in the mathematical analytical formulas for transient frequency domain power under internal and external faults in the HVDC transmission line are deduced. Through detailed analysis of the frequency characteristics of faults that are internal and external to the area, it can be concluded that convolutional power is beneficial for amplifying the characteristic differences between internal and external faults to the area caused by the boundary of the DC line.
2. In this study, by introducing the convolutional power in the time domain as the power in the frequency domain, it is proposed that the convolutional power of the short time window is used as the main criterion, and the ratio of the convolutional power of the long and short windows is used as the auxiliary criterion for the identification of high-impedance faults. The two complement each other to realize the fault pole, fault direction and fault identification, internal and external to the area.
3. The proposed scheme can extract high- and low-frequency information without frequency domain extraction tools; the calculation principle is simple; the convolution power ratio at different frequencies is less affected by the transition resistance; and the proposed protection scheme is still adaptable under weak boundary conditions.

2. Analysis of Frequency Domain Characteristics and Influencing Factors

2.1. Frequency Domain Characteristics of Internal and External Faults

Forward faults (FFs) include internal faults and external faults near the inverter side. Taking the line end fault f_1 and the external fault f_2 on the HVDC transmission line in Figure 1 as examples, protection measuring points (M) are installed at the outlet of the rectifier station. The reverse fault (RF) is f_3.

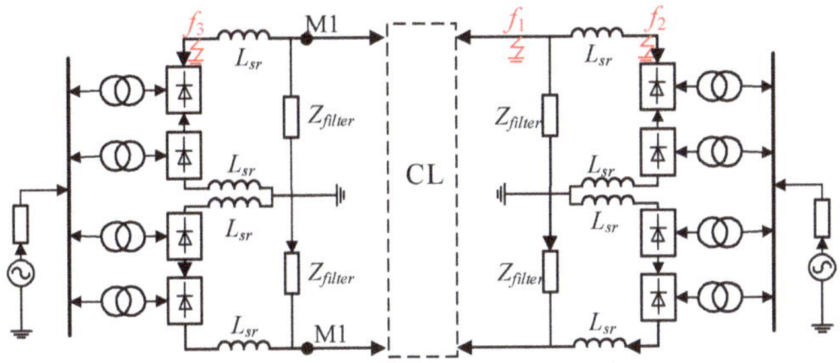

Figure 1. Internal fault and external fault on the HVDC transmission line.

To eliminate the influences of positive and negative pole coupling, the double-pole line is decoupled into a line-model and ground-model network through the Klenberg transform. Moreover, according to Peterson's law and superposition theorem, superimposed circuits for f_1 and f_2 can be obtained for FF, as shown in Figure 2. The black arrow is the current-related direction at M, and the red arrow is the direction of the fault current.

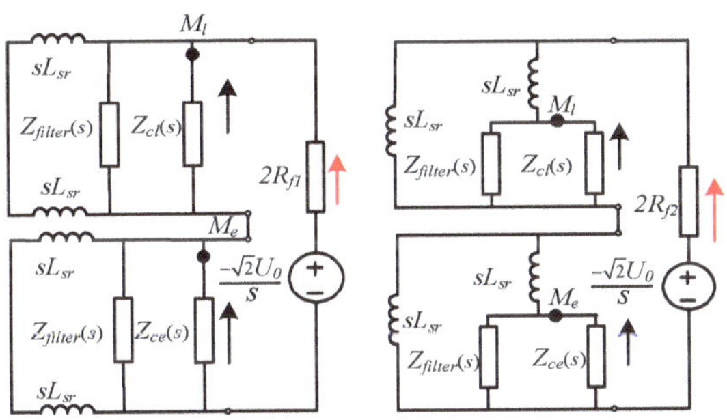

Figure 2. Superimposed circuit of fault point f_1 and f_2.

For Figure 2, the internal equivalent impedance at f_x (x = 1,2) is as follows:

$$Z_{f1k} = Z_{ck}//Z_{filter}//s2L_{sr}(k = l, e) \tag{1}$$

$$Z_{f2k} = (Z_{ck}//Z_{filter} + sL_{sr})//sL_{sr}(k = l, e) \tag{2}$$

where L_{sr} is the current-limiting reactance value on the DC side, $Z_{cx(x=l,e)}$ is the wave impedance of the DC transmission line, $R_{fx(x=1,2)}$ is the transition resistor and U_0 indicates the rated voltage of the line $Z_{filter} = 1/sC_{f1} + sL_{f1} + 1/sC_{f2}//sL_{f2} + 1/sC_{f3}//sL_{f3}$.

Thus, the electrical quantity of the mold network at f_x (x = 1,2) is the following:

$$\begin{cases} U_{fxk}(s) = \frac{-\sqrt{2}U_0 Z_{fxk}(s)}{s(Z_{fxl}(s)+Z_{fxe}(s)+2R_{fx})} \\ I_{fxk}(s) = \frac{\sqrt{2}U_0}{s(Z_{fxl}(s)+Z_{fxe}(s)+2R_{fx})} \end{cases} (k = l, e) \tag{3}$$

where $U_{fxk(k=l,e)}$ and $I_{fxk(k=l,e)}$ are the voltages and currents of the line mode and zero mode components at the fault locations, respectively.

We determined that the TW at f_2 to M, which needs to be multiplied by a partial voltage coefficient λ_u, is as follows:

$$\lambda_u = \frac{Z_{ck}\left(s\right)//Z_{filter}(s)}{Z_{ck}\left(s\right)//Z_{filter}(s)+sL_{sr}}(k=l,e) \tag{4}$$

According to (1)–(4), the general expression of fault TW for f_x (x = 1,2) can be obtained at M as follows:

$$\begin{cases} U_{mf1k}(s) = U_{mf1k}(s); U_{mf2k}(s) = \lambda_u U_{mf2k}(s) \\ I_{mf1k}(s) = U_{mf1k}(s)/Z_{ck}(s); I_{mf2k}(s) = \lambda_u U_{mf2k}(s)/Z_{ck}(s) \end{cases} \tag{5}$$

where $U_{mf1k(k=l,e)}$ and $I_{mf1k(k=l,e)}$ are the voltages and currents of the line mode and zero mode components at the measurement point when a fault occurs at f_1, respectively.

For the RF, the superimposed circuit for f_3 outside the rectifier side is shown in Figure 3.

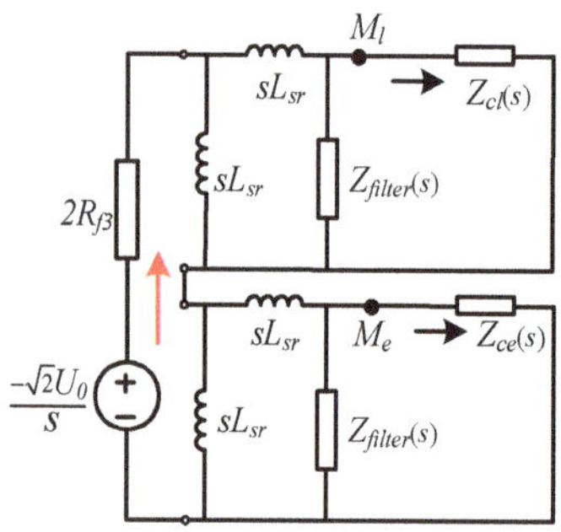

Figure 3. Superimposed circuit of fault point f_3.

Since both ends are on the LCC side, the equivalent impedance at the outlet of the converter station is the same, and the superimposed circuits for f_3 and f_2 are the same, i.e., $Z_{f3k} = Z_{f2k}$. The difference lies in the direction of the fault current. Therefore, the expression of the fault TW at M for f_3 is the following:

$$\begin{cases} U_{mf3k}(s) = \frac{-\lambda_u\sqrt{2}U_0 Z_{f3k}(s)}{s(Z_{f3l}(s)+Z_{f3e}(s)+2R_{f3})}(k=l,e) \\ I_{mf3k}(s) = -U_{mf3k}(s)/Z_{ck}(s) \end{cases} \tag{6}$$

where $U_{mf3k(k=l,e)}$ and $I_{mf3k(k=l,e)}$ are the voltages and currents of the line mode and zero mode components at the measurement point M when the fault occurs at the fault point f_3, respectively.

The convolution in the time domain is equal to the product in the frequency domain [21], the power in the frequency domain corresponding to f_1, f_2, and f_3 can be obtained as follows:

$$P_{mf1k}(s) = U_{mf1k}(s)I_{mf1k}(s) = -\frac{2U_0{}^2(Z_{f1k}(s))^2}{s^2(Z_{f1l}(s)+Z_{f1e}(s)+2R_{f1})^2 Z_{ck}(s)} \tag{7}$$

$$P_{mf2k}(s) = U_{mf2k}(s)I_{mf2k}(s) = -\frac{2U_0{}^2(Z_{f2k}(s)\lambda_u)^2}{s^2(Z_{f2l}(s)+Z_{f2e}(s)+2R_{f2})^2 Z_{ck}(s)} \tag{8}$$

$$P_{mf3k}(s) = U_{mf3k}(s)I_{mf3k}(s) = \frac{2U_0{}^2(Z_{f3k}(s)\lambda_u)^2}{s^2(Z_{f3l}(s) + Z_{f3e}(s)+2R_{f3})^2 Z_{ck}(s)} \tag{9}$$

where $P_{mf1k(k=l,e)}$, $P_{mf2k(k=l,e)}$ and $P_{mf3k(k=l,e)}$ are the power values of the line mode and zero mode components at the measurement point M when faults occur at f_1, f_2 and f_3, respectively.

To visually evaluate the characteristic differences of different electrical quantities at f_1 and f_2, for $G_U(s) = U_{f2}(s)/U_{f1}(s)$, $G_I(s) = I_{f2}(s)/I_{f1}(s)$ and $G_P(s) = P_{f2}(s)/P_{f1}(s)$, the amplitude–frequency curve is shown in Figure 4.

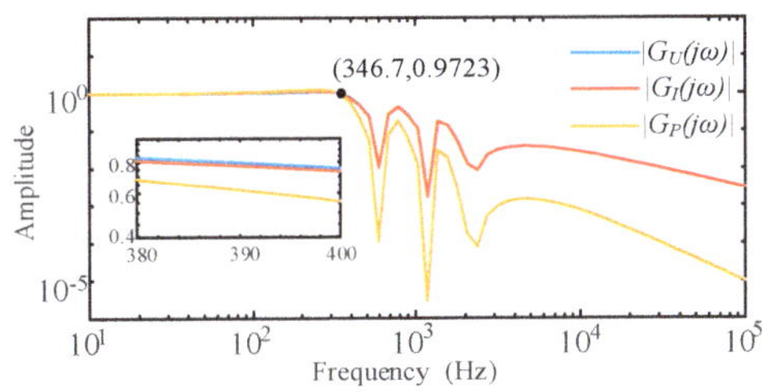

Figure 4. Amplitude-frequency curves of $|G_U(j\omega)|$, $|G_I(j\omega)|$ and $|G_P(j\omega)|$ ($R_{f1} = R_{f2} = 0$).

As a result of the influence of the DC boundary, the above ratios are each less than 1. As shown in Figure 4, the amplitude–frequency curves of $|G_U(j\omega)|$ and $|G_I(j\omega)|$ are very similar, and $|G_P(j\omega)|$ is not as different from these values in the frequency range of 0 to 340 Hz. However, when the frequency is greater than 340 Hz, the $|G_P(j\omega)|$ curve becomes lower than these curves. The gap between $|G_P(j\omega)|$ and either $|G_U(j\omega)|$ or $|G_I(j\omega)|$ increases significantly as the frequency increases. Therefore, in the high-frequency band, the attenuation of the frequency domain power by the DC boundary is more serious than that of a single voltage or current, and the degree of attenuation intensifies as the frequency increases. Hence, compared to using voltage or current alone, $P_{fi}(s)$ can be used to more effectively identify internal and external faults.

2.2. Influencing Factors

The generation of fault TWs is mainly affected by three factors: dispersion, refraction and reflection and fault resistance.

2.2.1. The Dispersion of Traveling Waves

As traveling waves propagate, fault decay due to the influence of long-distance transmission lines occurs, as shown in Figure 5.

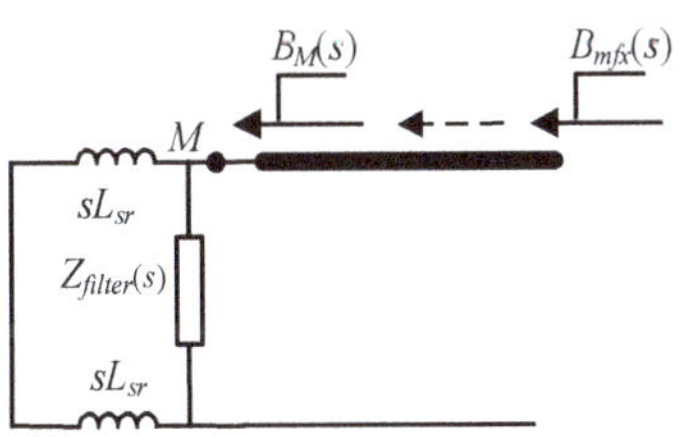

Figure 5. Diagram of TW propagation.

The process satisfies the following frequency domain relationship:

$$\begin{array}{c} B_{mx}(s) = e^{-\gamma_k L} B_{mfx}(s) \\ \gamma_k(s) = \sqrt{(R(\omega) + j\omega L(\omega))(G(\omega) + j\omega C(\omega))} \end{array} \quad (x = 1, 2) \tag{10}$$

where $R(\omega)$, $L(\omega)$, $C(\omega)$ and $G(\omega)$ are the resistance, inductance, capacitance and conductivity of the transmission line per unit length, respectively, and L is the length of line propagation. B_{mfx} is the initial traveling wave at the fault point, B_{mx} is the reverse traveling wave transmitted from the fault point to the measurement point M, and $e^{-\gamma kL}$ represents the transmission formula of the fault traveling wave.

The amplitude–frequency characteristics of the line transfer function are shown in Figure 6.

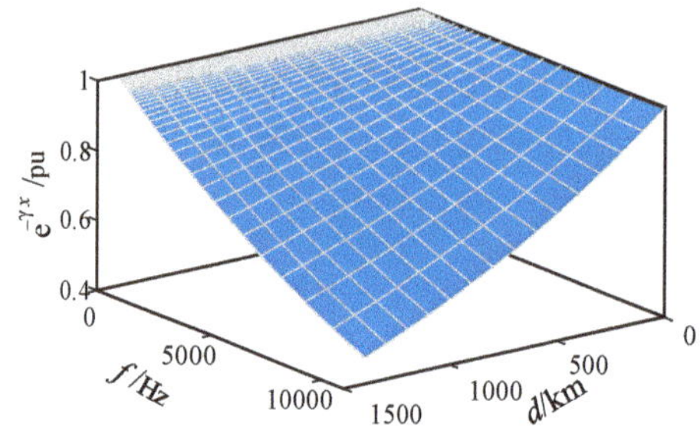

Figure 6. Amplitude–frequency characteristics of the line transmission function.

It can be seen from Figure 6 that the transfer function $e^{-\gamma L}$ of the line increases with the increase in frequency and distance; the high-frequency component and the length of the line will cause serious attenuation of the voltage and current traveling waves at the measuring point. Therefore, when the fault occurs in a long-distance line, it will affect the protection sensitivity that uses only high-frequency components.

2.2.2. The Reflection and Refraction of Traveling Waves

Refraction and reflection occur when the fault TW reaches the node with discontinuous wave impedance, as shown in Figure 7. Considering only the forward TW, the discontinuity of surge impedance is mainly at the outlet of each converter station.

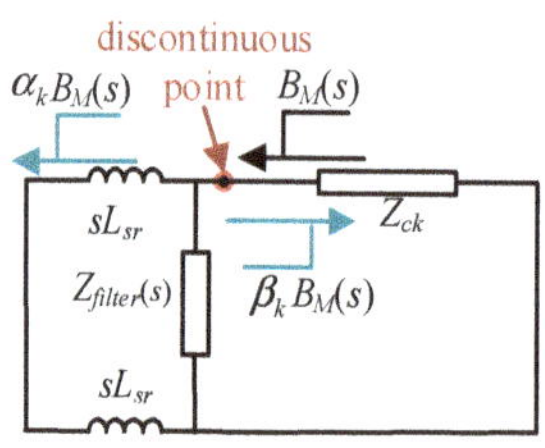

Figure 7. Refraction and reflection of the TW.

According to the Peterson principle, α_k and β_k are the refractive coefficient and the reflection coefficient, respectively:

$$\begin{cases} \alpha_k(s) = \frac{2(2sL_{sr}//Z_{filter}(s))}{2sL_{sr}//Z_{filter}(s)+Z_{ck}(s)} \\ \beta_k(s) = \frac{2sL_{sr}//Z_{filter}(s)-Z_{ck}(s)}{2sL_{sr}//Z_{filter}(s)+Z_{ck}(s)} \end{cases} (k = l, e). \tag{11}$$

According to the refraction and reflection principles of TWs, the first TW at M is the superposition of the incident wave and the reflected wave. Therefore, the first TW of the voltage and current for f_x (x = 1,2) at M are shown in (12).

$$\begin{cases} U_{Mfxk}(s) = (1+\beta_k)e^{-\gamma L}U_{mfxk}(s) \\ I_{Mfxk}(s) = U_{Mfxk}(s)/Z_{ck}(s) \end{cases} \tag{12}$$

Taken together, the frequency domain power for f_x (x = 1,2,3) at M near the rectifier station are as follows:

$$P_{Mf1k}(s) = U_{Mf1k}(s)I_{Mf1k}(s) = -\frac{2((1+\beta_k)e^{-\gamma_k L})^2 U_0{}^2(Z_{f1k}(s))^2}{s^2(Z_{f1l}(s) + Z_{f1e}(s)+2R_{f1})^2 Z_{ck}(s)} \quad (13)$$

$$P_{Mf2k}(s) = U_{Mf2k}(s)I_{Mf2k}(s) = -\frac{2((1+\beta_k)e^{-\gamma_k L})^2 U_0{}^2(Z_{f2k}(s)\lambda_u)^2}{s^2(Z_{f2l}(s) + Z_{f2e}(s)+2R_{f2})^2 Z_{ck}(s)} \quad (14)$$

$$P_{Mf3}(s) = U_{Mf3}(s)I_{Mf3}(s) = \frac{2U_0{}^2(Z_{f3k}(s)\lambda_u)^2}{s^2(Z_{f3l}(s) + Z_{f3e}(s)+2R_{f3})^2 Z_{ck}(s)} \quad (15)$$

Comparing (8)–(9) and (14)–(15) shows that the frequency domain power retains the polarity of the current. For the FF, the power value is negative, and for the RF, it is positive. Hence, the power polarity can be used to effectively identify external faults on the rectifier side.

2.2.3. Influence of Fault Resistance

According to (13)–(14), the frequency domain is inversely proportional to the square of the fault resistance, meaning that it is more affected by fault resistance than by voltage and current. In the worst-case scenario, R_{f2} = 0 Ω. By changing the value of the fault resistance R_{f1}, the $G_P(j\omega)$ curves at different frequencies and different R_{f1} values can be obtained, as shown in Figure 8.

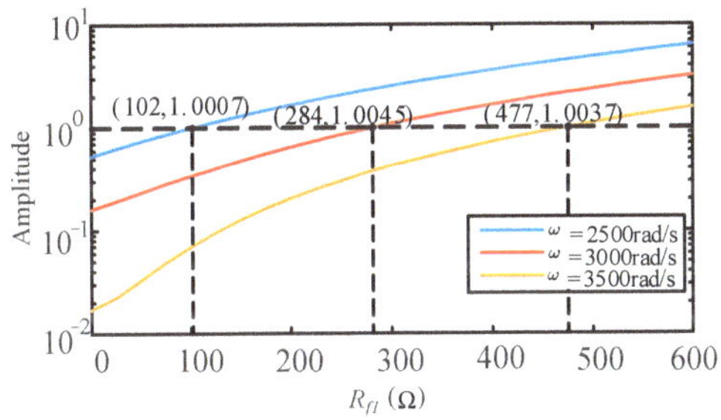

Figure 8. $|G_P(j\omega)|$ at different frequencies and different R_{f1} values (R_{f2} = 0).

$G_P(j\omega) = P_{f2}(j\omega)/P_{f1}(j\omega)$, which indicates the ratio of the transient frequency of faults outside the zone to the transient frequency of faults inside the zone. It can be seen from Figure 8:

(1) In the same frequency band, as R_{f1} increases, $P_{f1}(j\omega)$ decreases, resulting in $|G_P(j\omega)|$ rapidly increasing; that is, the $G_P(j\omega)$ characteristics are greatly affected by the transition resistance.

(2) At the same transition resistor, as the frequency increases, $|G_P(j\omega)|$ decreases, indicating that in the high-frequency band, the fault protection method based on frequency extraction can better detect internal and external faults.

Therefore, the protection scheme based on frequency extraction can identify low-resistance faults, but when a high-impedance grounding fault occurs in the DC line, this method is no longer applicable, and theoretically it is impossible to distinguish between internal and external faults based on the power in the frequency domain.

3. Ratio Characteristics of the Frequency Domain Power at Different Frequencies

Based on the above analysis, only the frequency domain power of a certain frequency (frequency band) is greatly affected by the fault resistance. To weaken this influence, a quantitative index P_{ratio} can be constructed, which is defined as the ratio of the frequency domain power values at different frequencies. We suppose that two different frequencies are ω_1 and ω_2, and that $\omega_1 > \omega_2$. $U_f(j\omega)$ is the step signal. The ratios P_{ratio1} and P_{ratio2} can be obtained under internal and external faults, respectively, as follows:

$$P_{ratio1} = P_{f1}(j\omega_1)/P_{f1}(j\omega_2); P_{ratio2} = P_{f2}(j\omega_1)/P_{f2}(j\omega_2) \quad (16)$$

The line wave impedance also changes with the frequency; therefore, it is necessary to consider the relationship between the line wave impedance and the frequency and draw the waveform of the line wave impedance with frequency according to Equation (10), as shown in Figure 9.

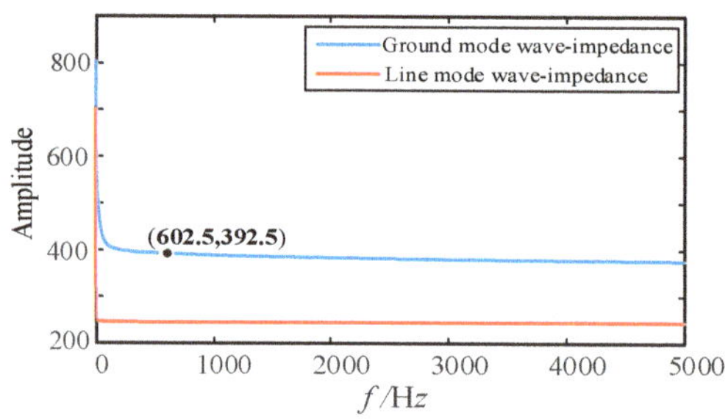

Figure 9. Wave impedance of DC line.

As can be seen from Figure 9, the line wave impedance changes very little with frequency in the high-frequency band and can be approximated into a constant in the high-frequency band. In the same way, it can be seen from Figure 9 that the linear mode wave impedance changes little with frequency compared with the ground mode wave impedance, so the linear mode component is considered for analysis and discussion.

Therefore, the expressions of P_{ratio1} and P_{ratio2} are as follows:

$$\begin{aligned} P_{ratio1} &= U_{Mf1}(j\omega_1)I_{Mf1}(j\omega_1)/U_{Mf1}(j\omega_2)I_{Mf1}(j\omega_2) \\ &\approx \omega_2{}^2(1+\beta(j\omega_1))^2/\omega_1{}^2(1+\beta(j\omega_2))^2 \\ &= \frac{2Z_{filter}(j\omega_2)j\omega_2 L_{sr} + \left(2j\omega_2 L_{sr} + Z_{filter}(j\omega_2)\right)Z_{c1}}{2Z_{filter}(j\omega_1)j\omega_1 L_{sr} + \left(2j\omega_1 L_{sr} + Z_{filter}(j\omega_1)\right)Z_{c1}} \end{aligned} \quad (17)$$

$$\begin{aligned} P_{ratio2} &= U_{Mf2}(j\omega_1)I_{Mf2}(j\omega_1)/U_{Mf2}(j\omega_2)I_{Mf2}(j\omega_2) \\ &\approx \omega_2{}^2\lambda_u(\omega_1)^2(1+\beta(\omega_1))^2/\omega_1{}^2\lambda_u(\omega_2)^2(1+\beta(\omega_2))^2 \\ &= P_{ratio1}(j\omega_2 L_{sr} + Z_{c1})^2/(j\omega_1 L_{sr} + Z_{c1})^2 \end{aligned} \quad (18)$$

As shown in (17) and (18), P_{ratio1} and P_{ratio2} are only related to the frequency, current-limiting reactance and wave impedance, all of which are basically not affected by the fault resistance.

Letting $|G_{Pratio}| = P_{ratio1}/P_{ratio2}$, we can obtain the following:

$$|G_{Pratio}| = \frac{P_{ratio2}}{P_{ratio1}} \approx \frac{(j\omega_2 L_{sr} + Z_{c1})^2}{(j\omega_1 L_{sr} + Z_{c1})^2} \quad (19)$$

Substituting the model project parameters of the Yun–Guang UHVDC and taking ω_2 = 600 rad/s, the $|G_{Pratio}|$ curves at different frequencies ω_1 and fault resistances R_{f1} can be obtained, as shown in Figure 10.

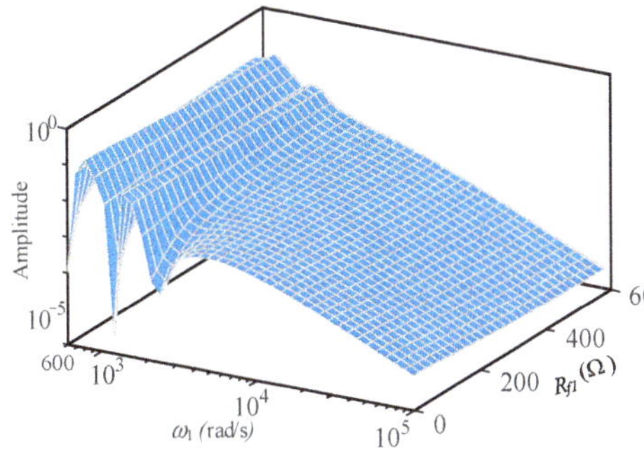

Figure 10. $|G_{Pratio}|$ at different R_{f1} and ω_1.

As can be seen from Figure 10:

(1) In the case of $\omega_1 > \omega_2 = 600$ rad/s, regardless of the transition resistance and frequency ω_1, $G_{Pratio} < 1$, $P_{ratio1} < P_{ratio2}$.

(2) As ω_1 gradually increases, $|G_{Pratio}|$ also further decreases, while $|G_{Pratio}|$ denotes P_{ratio2}/P_{ratio1}, so the difference between the quantitative indicators P_{ratio1} and P_{ratio2} is more obvious, which further amplifies the difference between internal and external faults.

(3) In the same way, when ω_1 is fixed, as the transition resistance R_{f1} gradually increases, the value of $|G_{Pratio}|$ changes slightly, indicating that the ability of $|G_{Pratio}|$ to identify internal and external faults is less affected by the transition resistance ability R_{f1} at different frequencies.

4. Single-Ended Protection Scheme of HVDC Transmission Lines

First, the CP $P_{fi}(t)$ is defined as the convolution of two functions, $U_{fi}(t)$ and $I_{fi}(t)$, in the time domain [24]:

$$P_{fi}(t) = (u_{fi} \times i_{fi})(t) = \int_{t_0}^{t_0+T_s} u_{fi}(t-\tau)i_{fi}(\tau)d\tau \tag{20}$$

where t_0 and T_s are the starting time of the data window and the length of the time window, respectively.

From the convolution theorem, the Laplace transform of function convolution is the product of the function Laplace transform [25]; that is, $U(t)$ and $I(t)$ undergo the Laplace transform to obtain $U(j\omega)$ and $I(j\omega)$ in the frequency domain, and the expression for this relationship is the following:

$$(U \times I)(t) \overset{\text{Laplace transform}}{\underset{\text{Equivalent}}{\rightarrow}} U(j\omega)I(j\omega) = P_{fi}(s) \tag{21}$$

Therefore, $P_{fi}(t)$ is the CP and $P_{fi}(s)$ is the CP in the frequency domain.

Specifically, the CP represents the application of voltage as an input quantity to all current points within T_s, and is generated by varying the voltage quantity on the current. The convolution operation convolves the input data and can be regarded as a weighted summation, whereas the result of the convolution is the feature extraction and the amplification of the fault characteristics. The convolution calculation process is shown in Figure 11. The input signal u(t) is regarded as a discrete sequence u(n) = {u_1, u_2, u_3, . . ., u_n}; i(t) is regarded as a discrete sequence i(n) = {i_1, i_2, i_3, . . ., i_n}.

It can be seen from Figure 11 that the convolution power result under the sampling point can be obtained by adding each column in Figure 11, and the convolution result presents the arrangement results from small to large and then large to small, and the discrete signal convolution power solution is obtained through the convolution calculation process of Figure 11:

$$P_a(X) = \sum_{i=1}^{N} U_a(i) I_a(X - i + 1) \ (a = p, n) \tag{22}$$

where P_a is the CP; X is the Xth sampling point; and N is the total number of sampling points, which reflects the length of the time window. N is calculated as follows, where f_s is the sampling frequency:

$$N = T_s f_s \tag{23}$$

u(X) \ X	1	2	3	4	n	2n−2	n+n−1
u_1	$u_1 i_1$	$u_1 i_2$	$u_1 i_3$	$u_1 i_4$	$u_1 i_n$	0	0
u_2	0	$u_2 i_1$	$u_2 i_2$	$u_2 i_3$	$u_2 i_{n-1}$	0	0
u_3	0	0	$u_3 i_1$	$u_3 i_2$	$u_3 i_{n-2}$	0	0
u_4	0	0	0	$u_4 i_1$	$u_4 i_{n-3}$	0	0
⋮	0	0	0	0	⋮	⋮	⋮
u_n	0	0	0	0	$u_n i_1$	$u_{n-1} i_1$	$u_n i_1$

Figure 11. Specific convolution calculation process.

In fact, because the TW is a nonstationary time signal, its spectrum distribution is related to the time window. For different time windows, the spectrum distribution is not consistent. Therefore, the spectrum distribution of the CP is also affected by the time window. From the perspective of signal processing, when the sampling frequency is fixed, the length of the time window reflects different frequency resolutions. According to the Shannon sampling theorem, the fault information also contains many harmonic component frequencies, except the fundamental frequency signal in the long and short windows, but the frequency resolution of each time window is inversely proportional to the length of the time window. Taking the sampling frequency of 10 kHz as an example, when the short window is 1 ms, the minimum frequency interval on the frequency axis is 1000 Hz, whereas when the long window is 3 ms, the minimum frequency interval is 333.33 Hz. Therefore, it is clear that compared with that of the short window, the frequency component of the long window is greater [26]. In detail, the fault frequency information in the short window can be regarded as high-frequency information. Similarly, the long window can be regarded as the wide band of low-frequency and high-frequency information.

The single-ended protection scheme for HVDC transmission lines is described in the following sections.

4.1. *Starting Criterion*

When a fault occurs in the line, the pole line voltage suddenly changes. Therefore, the variation in polar line voltage ΔU_a can be used as the starting criterion of the protection:

$$\sum_{j=0,1,2} |\Delta U_a(j+i) - \Delta U_a(j+i-1)| > k_v U_{ref} (a = p, n) \tag{24}$$

where U_{ref} is the rated voltage and a represents the fault pole. To avoid the influence of steady-state voltage fluctuations, k_v is the voltage fluctuation coefficient and is taken as 0.08.

4.2. *Fault Identification Criteria of the Internal and External Faults*

The maximum value of the CP sequence under long and short windows is selected as the characteristic quantity.

4.2.1. Main Criterion 1

Main criterion 1 adopts the CP with a short window of T_1. The main function of the criterion is to quickly identify serious faults, including the metallic and low-resistance faults. Main criterion 1 is given as follows:

$$P_{aT1max} > P_{aT1.set} = k_{rel.1} P_{f2T1max} (a = p, n) \quad (25)$$

where $P_{aT1max} = max(|P_{aT1}|)$, and P_{aT1} is the discrete CP value obtained by convolution of the voltage and current within T_1, taken as 1 ms. $P_{f2T1max}$ represents the metal external fault inverter side within T_1; $k_{rel.1}$ is the reliability coefficient, taken as 1.5.

4.2.2. Main Criterion 2

Main criterion 1 is based on the CP with a short window, and has the advantages of simplification and rapid tripping, but its sensitivity is reduced with increasing fault resistance. To address this issue, a criterion composed of the CP ratio (R_a) at different frequencies is added to assist in identifying high-resistance faults. Main criterion 2 is the following:

$$R_a > R_{a.set} = k_{rel.2} R_{f2a} (a = p, n) \quad (26)$$

where $\begin{cases} R_a = P_{aT1\max} / P_{aT2\max} \\ P_{aT2\max} = \max(|P_{aT2}|) \end{cases}$, $R_{a.set}$ represents the setting value, R_{f2a} is the ratio of long- and short-window CPs of the metal external fault inverter side, P_{aT2} is the discrete CP value obtained by convolution of the voltage and current within T_2, taken as 3 ms, and $k_{rel.\ 2}$ is the reliability coefficient, taken as 2.

Compared to main criterion 1, the above criterion requires waiting for a long window and increasing the number of calculations. However, when a high-resistance fault occurs, the rising speed of the fault current is relatively slow, and the system allows a longer response time for the protection action.

4.3. Fault Direction Identification Criterion

As mentioned in Section 2, the polarity of the CP can be used to distinguish the RF and FF, so the fault direction can be identified by the polarity of the short-window CP, including Δ_{th1} as the threshold value:

$$\begin{cases} A_1 = \min(P_{aT1}) > \Delta_{th1} & \text{RF} \\ A_1 = \min(P_{aT1}) < -\Delta_{th1} & \text{FF} \end{cases} \quad (27)$$

4.4. Fault Pole Selection Identification Criterion

The coupling strength between DC lines is related to frequency. In the high-frequency band, the coupling strength is relatively large, so the high-frequency components of the sound pole and fault pole are very similar. However, in the low-frequency band, the coupling strength is weak, so the low-frequency component of the fault pole is more significant than that of the sound pole. Therefore, a pole selection criterion based on a low-frequency CP under a long window can be constructed as follows:

$$\begin{cases} \frac{P_{pT2\max}}{P_{nT2\max}} > \Delta_{th2} & \text{PGF} \\ \frac{P_{pT2\max}}{P_{nT2\max}} < \frac{1}{\Delta_{th2}} & \text{NGF} \\ \Delta_{th2} > \frac{P_{pT2\max}}{P_{nT2\max}} > \frac{1}{\Delta_{th2}} & \text{PPF} \end{cases} \quad (28)$$

where Δ_{th2} is the threshold value and $k_{rel.3}$ is the reliability coefficient, which is 1.5. The rectifier station acts according to the protection result after fault identification, and then the DC system enters the fault isolation and recovery phase. PGF is a positive ground fault, NGF is a negative ground fault, and PPF is a fault between poles.

In summary, this study designed a single-ended protection scheme that has both advantages with the CP ratios of short and long windows. The specific flow chart is shown in Figure 12, which is composed of main criterion 1 and main criterion 2.

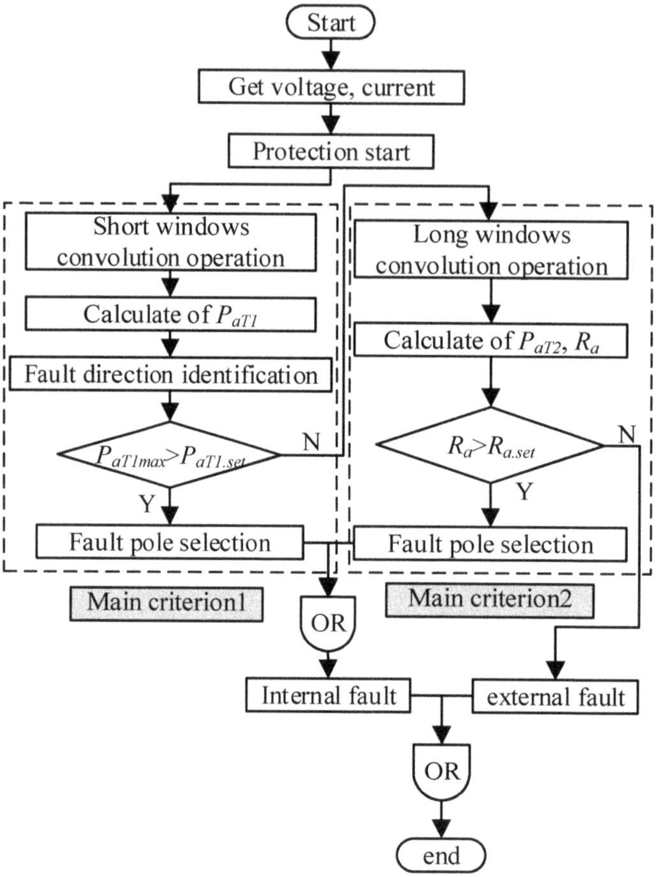

Figure 12. Flowchart of a novel single-ended protection scheme based on the CP ratio.

5. Case Analysis

A ±800 kV UHVDC transmission system model was established in PSCAD/EMTDC with respect to the parameters of the Yunnan to Guangdong UHVDC transmission project. The specific model parameters are shown in Table 1. The structures of the DC line are depicted in Figure 13. The sampling frequency is 10 kHz. The time windows of the CPs are 1 ms and 3 ms, so the numbers of sampling points N are 10 and 30, respectively. The protection algorithm was written in MATLAB software. According to the setting principle and simulation results, the criterion-setting value of the protection scheme is shown in Table 2. The steps in the following sections are used to verify the action of protection under different faults.

Table 1. Model parameters.

Parameter	Rectifier Side	Inverter Side
Rated power	5000 MW	5000 MW
Rated DC current(I_0)	3125 A	3125 A
Rated DC voltage(U_0)	±800 kV	±800 kV
Boundary element (L_{sr})	0.15 H	0.15 H
Filter	C_{f1} = 1.2 μF C_{f2} = 2.824 μF C_{f3} = 2.647 μF L_{f1} = 9.345 mH L_{f2} = 15.919 mH L_{f3} = 4.6556 mH	

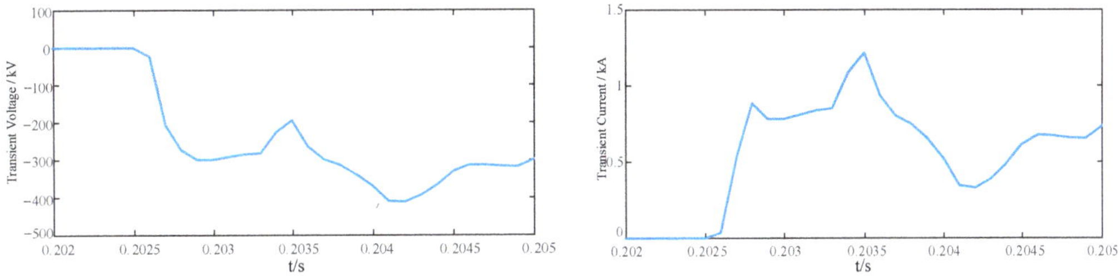

(**a**) (**b**)

Figure 13. Transient current under internal faults. (**a**) Transient voltage under internal faults. (**b**) Transient current under internal faults.

Table 2. Setting value of protection criterion.

Criterion	Δ_{th1}	Δ_{th2}	$P_{pT1.set}$	R_P
Setting value	20	1	900	0.08

5.1. Identification of Internal and External Faults

The fault occurs at t = 0.2 s, and when a PGF occurs at point f_1 (internal fault), the fault voltage and fault current at the boundary measurement point after the fault are shown in Figure 13.

The transient convolution power graphs of the long and short windows calculated with Equations (20)–(22) in the case of internal faults are shown in Figure 14.

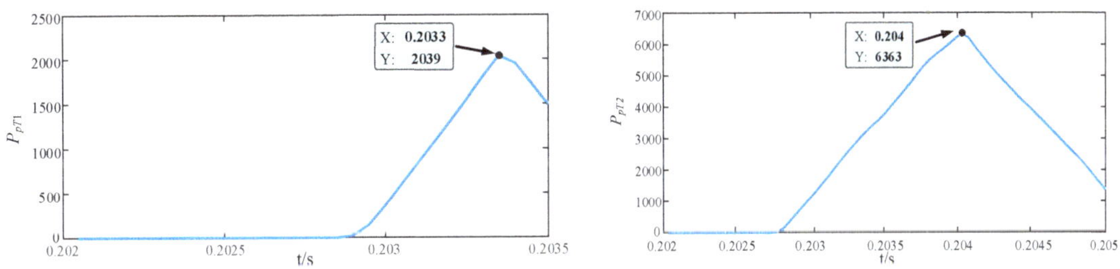

(a) (b)

Figure 14. Long and short window transient convolutional power. (**a**) Short time window transient convolutional power. (**b**) Long time window transient convolutional power.

As shown in Figure 14, $P_{\text{pT1\_MAX}}$ = 2039 > $P_{\text{pT1\_set}}$ = 900 in main criterion I; similarly, the ratio of the convolution power calculated with main criterion II according to Formulas (25) and (26), is P_{ratio} = 0.32 > R_p = 0.08. Therefore, the protection is determined as an internal fault.

When a PGF occurs at point f_2 (external fault), the fault voltage and fault current at the boundary measurement point after the fault are shown in Figure 15.

The transient convolution power graphs of the fault for long and short windows calculated using Equations (20)–(22) during external faults are shown in Figure 16.

From Figure 16, it can be seen that P_{pT1_max} = 681 < P_{pT1_set} = 900 for main criterion I; similarly, it can be seen that the ratio of convolution power calculated via main criterion II according to Formulas (25) and (26), is P_{ratio} = 0.045 < R_p = 0.08. Therefore, the protection is determined as an external fault.

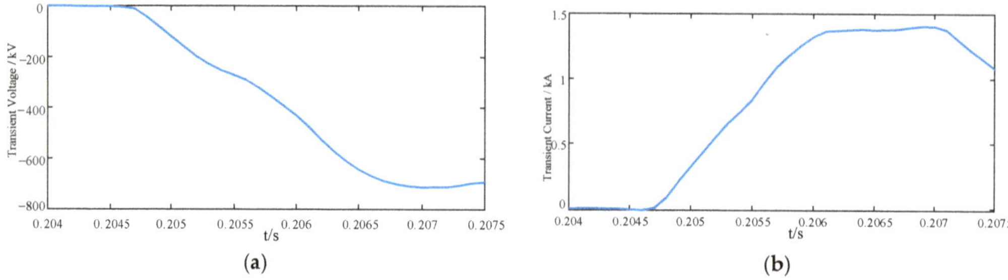

Figure 15. Transient current under internal faults. (**a**) Transient voltage under internal faults. (**b**) Transient current under internal faults.

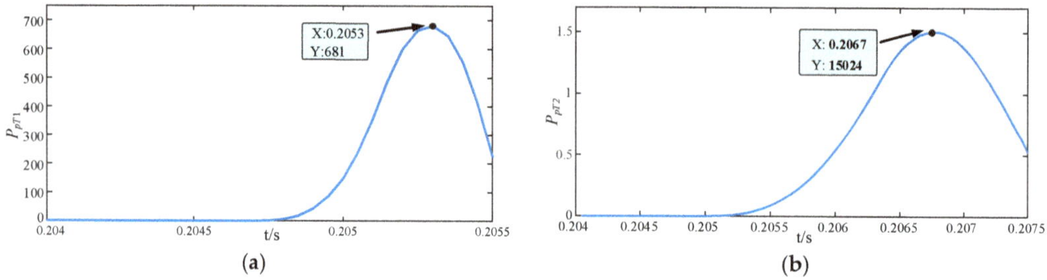

Figure 16. Long- and short-window transient convolutional power. (**a**) Short-time-window transient convolutional power. (**b**) Long-time-window transient convolutional power.

To verify the accuracy and applicability of the protection scheme, the simulation was set to have high-resistance faults at every 200 km interval of the lines, with their resistances varying from 0 to 600 Ω. Considering the most serious situation, f_2 and f_3 were set as metallic ground faults in the later analysis. Since the positive and negative poles are symmetrical, only the positive fault is discussed here. In the table below, √ represents the protection action; conversely, × represents no action. Limited by the article space, only some fault point simulation results are shown here. The simulation test results are shown in Table 3.

Table 3 shows the following:

(1) In the case of low-impedance grounding faults (internal faults), the maximum value of the high-frequency convolution power of each fault type, $P_{pT1\max}$, is much greater than that of P_{pT1set}, and the A1 value is obtained by the calculation using Equation (27) in order to determine the fault direction; therefore, main criterion 1 in the proposed protection scheme can quickly and reliably identify the low-impedance grounding faults in the region.

(2) As shown in the shaded font in Table 3, when the transition resistance is set at a distance of 1414 km from the protection installation to a fault of 600 Ω, $P_{pT1\max}$ is equal to 882, which is less than the setting value P_{pT1set}, which may cause the effect of rejection.

(3) In order to prevent the rejection phenomenon when the high-impedance grounding fault is made, this study proposes to use main criterion 2 to identify the high-impedance grounding fault inside and outside the region. From the observations in the table, it can be seen that the P_{ratio} under the internal fault (f_1) is greater than that under the external fault (f_2), and the value of the P_{ratio} is still greater than that of the metal fault outside the area on the inverter side, even if the transition resistance is 600 Ω, which ensures the identification of the high-impedance fault in the area via main criterion 2.

Table 3. Fault identification results for internal and external faults.

Fault Distance/km	Fault Type	Transition Resistance/Ω	Fault Direction	A_1	P_{pT2max}	P_{pT1max}	R_P	Action of Main Criteria 1 and 2
0	PGF	0	FF	−5916	18,743.6	5915.6	0.316	(1√, 2√)
		160	FF	−1354.8	4235.4	1354.8	0.320	(1√, 2√)
		200	FF	−1051	3301.7	1050.7	0.318	(1√, 2×)
		400	FF	−396	1258.2	395.9	0.315	(1×, 2√)
		600	FF	−196	623.4	196.1	0.314	(1×, 2√)
	PPF	0	FF	−30,961	88,281.1	30,960.9	0.350	(1√, 2√)
		600	FF	−2899	9027.5	2898.8	0.321	(1√, 2√)
400	PGF	0	FF	−3871	15,410.5	3870.6	0.251	(1√, 2√)
		160	FF	−1111.5	3933.8	1111.5	0.283	(1√, 2√)
		200	FF	−876	3087.5	875.6	0.284	(1×, 2√)
		400	FF	−353	1212.4	352.7	0.291	(1×, 2√)
		600	FF	−186	615.9	186.3	0.303	(1×, 2√)
	PPF	0	FF	−27,464	116,074.2	27,464.2	0.237	(1√, 2√)
		600	FF	−2482	9400.4	2481.7	0.264	(1√, 2√)
800	PGF	0	FF	−3802	14,702.4	3801.7	0.259	(1√, 2√)
		160	FF	−952.9	3736.3	952.9	0.255	(1√, 2√)
		200	FF	−752	2936.1	751.9	0.256	(1×, 2√)
		400	FF	−358	1194.2	357.8	0.300	(1×, 2√)
		600	FF	−190	616.0	190.0	0.308	(1×, 2√)
	PPF	0	FF	−22,507	105,594.7	22,506.7	0.213	(1√, 2√)
		600	FF	−2227	9245.6	2227.2	0.241	(1√, 2√)
1418 (f_1)	PGF	0	FF	−3795	17,621.5	3795.1	0.215	(1√, 2√)
		160	FF	−882.0	3985.7	882.0	0.221	(1×, 2√)
		200	FF	−719	3691.0	718.8	0.195	(1×, 2√)
		400	FF	−283	1520.2	283.1	0.186	(1×, 2√)
		600	FF	−167	808.6	166.5	0.206	(1×, 2√)
	PPF	0	FF	−16,061	15,331.7	16,061.0	0.179	(1√, 2√)
		600	FF	−1918	614.0	1918.3	0.178	(1√, 2√)
f_2	PGF	0	FF	−601	15,429.1	600.9	0.039	(1×, 2×)
f_3	PGF	0	RF	0.00345	/	2235.11	/	/

5.2. Faulty Pole Selection

When the protection method in this research determines that the fault is an internal fault, it is necessary to further select the fault pole selection criterion for the fault in the region. The waveforms of the long window convolution power P_{pT2} for a PGF, NGF and PPF occurring at point f_1 are shown in Figure 17.

As shown in Figure 17:

(1) When a positive pole fault occurs in the system, the ratio of transient convolution power between the positive and negative poles on the rectifier side is much greater than 1.5.
(2) When a negative pole fault occurs in the system, the ratio of transient convolution power between the positive and negative poles on the rectifier side is less than 0.67.
(3) When an interpole fault occurs in the system, the ratio of transient convolution power between the positive and negative poles on the rectifier side is between 0.67 and 1.5.

At the 400 km location in the DC line, the NGF and PGF are applied under different resistances, and the corresponding waveform of P_{pT2} with a long window is shown in Figure 18.

Figure 18 shows that although the CP with a long window decreases with increasing fault resistance, the waveform of P_{pT2} of the PGF is always above that of the NGF, within 600 Ω. Therefore, the CP with a long window can be used to identify the faulty pole, which is consistent with the description in Section 4. In the actual criterion, the fault pole is judged by calculating the ratio (P_{pT2max}/P_{nT2max}). The simulation test results of specific fault pole selection are shown in Table 4.

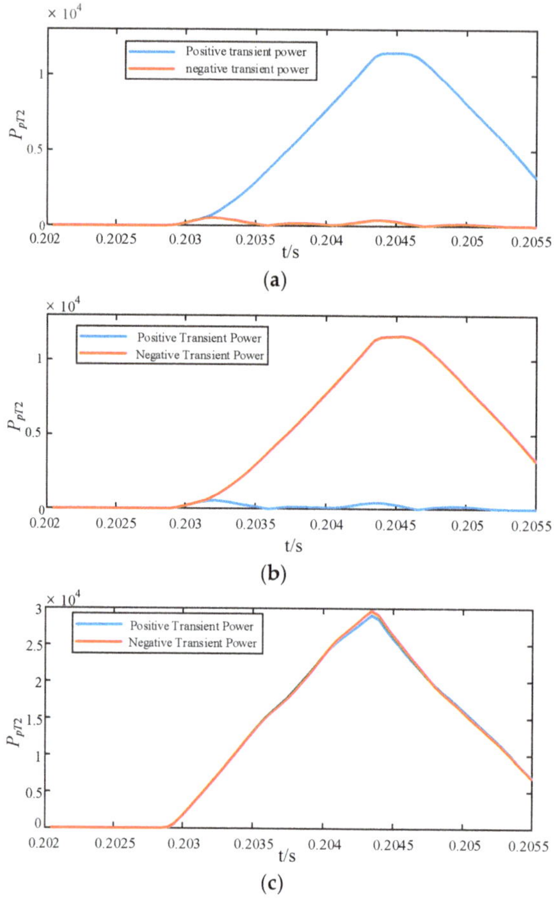

Figure 17. Convolutional power under different fault types. (**a**) Convolutional power during positive pole fault. (**b**) Convolutional power during negative pole fault. (**c**) Convolutional power during interpole faults.

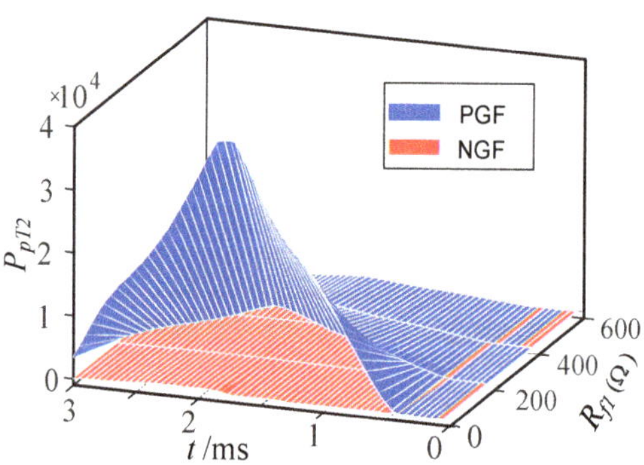

Figure 18. P_{pT2} in case of a PGF and NGF.

Table 4. Faulty pole selection results.

Fault Type	Resistance/Ω	P_{pT2max}/P_{nT2max}
400 km PGF	0	64.710
	200	65.547
	400	52.141
	600	43.037
400 km NGF	0	0.0181
	200	0.0220
	400	0.0208
	600	0.0228
400 km PPG	0	0.9989
	200	0.9988
	400	0.9990
	600	0.9991

Table 4 shows that in various fault conditions, the proposed criterion can reliably distinguish the fault pole.

5.3. Adaptability Analysis of the Noise Interference

In practical engineering, the line may encounter noise interference, which affects the reliability of the line protection action. After sampling interference signals such as noise, the obtained digital signal contains more high-frequency components, which affect the action performance of the criterion based on transient variation. To verify the anti-interference ability of the protection scheme, Gaussian white noise based on the actual signal power value was added to the voltage and current measured at M_1, and the signal-to-noise ratio (SNR) was 40 dB. Taking the middle of the line as an example, the specific simulation results of the scheme proposed in [26] and the scheme proposed in this research are shown under severe Gaussian white noise interference in Table 5.

Table 5. Action performance of main criteria 1 and 2 after adding noise.

Fault Distance	Fault Resistance	Noisy P_{pT1max}	Noisy R_P	Criterion Action	No-Noise *TTV*	Noisy *TTV*
600 km	0 Ω	3694.5	0.254	(1√, 2√)	0.413	0.396
	200 Ω	768.5	0.267	(1×, 2√)	0.357	0.365
	400 Ω	280.9	0.253	(1×, 2√)	0.383	0.435
	600 Ω	158.2	0.272	(1×, 2√)	0.318	0.350
f_2	0 Ω	851.7	0.053	(1×, 2×)	0.168	0.265

The simulation results in Table 5 show that the protection criteria proposed in this study can withstand 40 dB of Gaussian white noise and maintain the original performance. However, the protection scheme based on the time-domain transient voltage in [22] is vulnerable to Gaussian white noise, especially in the case of external faults. It is necessary to recalculate the setting value to avoid protection misoperation. Therefore, the proposed protection scheme has a better ability to resist noise interference.

5.4. Adaptability Analysis of the Boundary Element

When the high-frequency component passes through the boundary element, such as a smoothing reactor, it is greatly attenuated, and the attenuation degree of the high-frequency component is different for smoothing reactors with different values, which affects the adaptability of the protection principle. Therefore, the simulation verifies the adaptability of the method to smoothing reactors with different values when a fault occurs. The specific simulation results are shown in Table 6.

Table 6. Results under different values of smoothing reactor.

Fault Distance	Fault Resistance	L_{sr} /H	P_{pT2max}	P_{pT1max}	R_P	Protection Margin ($P_{pT1max}/P_{pT1.set}$, $R_P/R_{P.set}$)
0 km	0 Ω	0.15	18,743	5915.6	0.316	(6.572, 3.950)
		0.10	2215	6682.4	0.302	(4.330, 2.649)
		0.05	26,628	7505.4	0.282	(2.414, 1.270)
	600 Ω	0.15	623.4	196.1	0.314	(0.218, 3.925)
		0.10	587.0	185.3	0.316	(0.120, 2.772)
		0.05	567.2	176.0	0.310	(0.056, 1.396)
1418 km (f_1)	0 Ω	0.15	17,621	3795.1	0.286	(4.217, 2.687)
		0.10	17,145	3175.6	0.185	(2.058, 1.614)
		0.05	18,172	4062.3	0.224	(1.307, 1.010)
	600 Ω	0.15	808.6	166.5	0.290	(0.185, 2.575)
		0.10	698.6	132.2	0.292	(0.128, 2.561)
		0.05	663.7	138.1	0.295	(0.044, 1.329)
f_2	0 Ω	0.15	15,429	600.9	0.039	/
		0.10	18,008	1028.8	0.057	/
		0.05	18,684	2072.7	0.111	/

As shown in Table 6, with the reduction in the smoothing reactor value, the CP value of the short window of an external fault continues to increase, resulting in a reduction in the discrimination degree of the out-of-area faults. Thus, protection criteria based on high frequencies, such as main criterion 1, are greatly affected by boundary elements. However, different smoothing reactor values have little effect on the action performance of main criterion 2 proposed in this study. When L_{sr} > 50 mH and the fault resistance is 600 Ω, main criterion 2 still has a sufficient margin for identifying faults. Therefore, the proposed protection scheme is also suitable for HVDC transmission systems with weak boundary characteristics, which are easy to implement and have good universality.

5.5. Comparison Studies

(1) Time-domain method:

Using the rate of change of DC voltage (ROCOV)-based fault protection scheme proposed in reference [6], the superiority of the proposed scheme is verified through simulation comparison.

Set a positive grounding fault with a transition resistance of 400 Ω at f_1 (internal fault), and a metallic grounding fault (external fault) at f_2. The ROCOV waveform is shown in Figure 19.

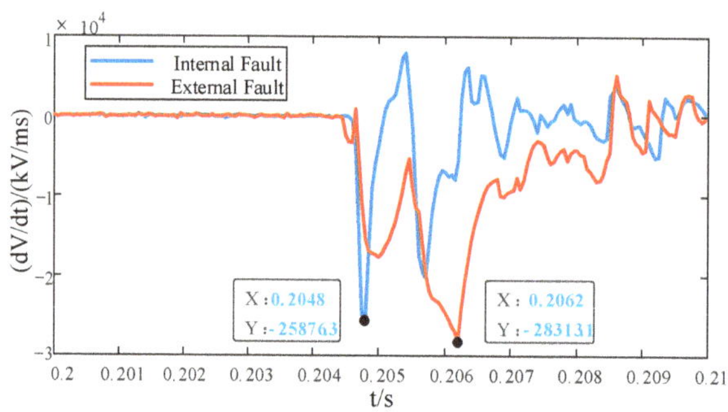

Figure 19. ROCOV under internal and external faults.

As shown in Figure 18, the ROCOV at f_2 is greater than that at f_1, and the ROCOV's ability to withstand transition resistance is less than 400 Ω. However, the protection scheme proposed in this research has a resistance to transition resistance of 600 Ω. Therefore, using ROCOV alone cannot identify high-resistance faults.

(2) Frequency domain method

A protection scheme based on wavelet transforms for measuring high-frequency component values of point voltages was proposed in reference [12]. Similarly, a positive grounding fault with a transition resistance of 500 Ω was set at f_1 under a weak boundary (L_{dc} = 50 mH) (internal fault), and a metallic grounding fault was set at f_2 (external fault). The transient high-frequency voltage components obtained through wavelet transforms are shown in Figure 20.

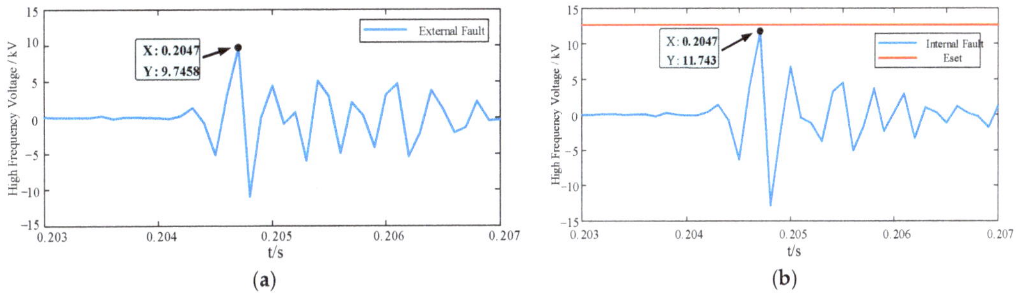

Figure 20. High-Frequency Voltage under internal and external faults. (**a**) High-Frequency Voltage under internal faults. (**b**) High-Frequency Voltage under external faults.

By setting the metal grounding fault outside the zone, the maximum transient high-frequency voltage of the most severe external fault is 9.7485. Setting the reliability coefficient K_{rel} = 1.3 can obtain the setting value for internal and external faults in the zone. From Figure 20, the grounding fault with a transition resistance of 500 Ω in the area is less than the set value, so the protection will refuse to operate. As the smoothing reactor of this scheme adopts 50 mH, it is not suitable for weak boundary systems. Meanwhile, as a time–frequency analysis tool, wavelet transform has complex computational principles. And this article uses time-domain convolution power to extract high-frequency signals with a simple principle, and the proposed protection scheme has a resistance to transition resistance of 600 Ω.

6. Conclusions

To resolve the problem of the insufficiency of HVDC transmission systems to withstand fault resistance, a fault characteristic quantity based on the CP was constructed, and a novel single-ended protection scheme based on the CP ratio for HVDC transmission lines was proposed. The results of the mathematical analysis and numerical examples show the following achievements:

(1) By analyzing the characteristic difference in frequency domain power after DC line fault in detail, a new protection criterion was proposed using the power ratio of different frequencies under long and short windows. The scheme is less affected by transition resistance and fault distance, and the principle of the proposed scheme is simple. It can obtain the frequency domain component of a fault traveling wave without relying on other methods, such as extracting frequency domain information.

(2) The Yunnan–Guangzhou ±800 kV HVDC project was built to verify the performance of the proposed fault rapid identification scheme. In all cases, it can accurately identify faults, has the identification ability of up to 600 Ω transition resistance, and has low requirements for the sampling frequency of the protection device. Only a 10 kHz sampling frequency is required to identify faults inside and outside the

area. Moreover, when the SNR is greater than 40 dB, noise has little effect on the proposed method.
(3) Compared with other time-domain methods based on measuring point voltage change and frequency domain methods based on wavelet transform extraction, the proposed protection scheme can identify different types of short-circuit faults and has a strong ability to withstand excessive resistance. The proposed protection scheme can also be applied to HVDC transmission systems with weak boundary characteristics.

Future work plans to apply this method to flexible HVDC transmission systems under weak boundaries.

Author Contributions: Conceptualization, G.P., L.C. and J.W.; methodology, G.P., L.C., J.W. and H.J.; validation, H.J., Z.W. and H.L.; formal analysis, G.P., L.C. and J.W.; investigation, G.P., L.C. and J.W.; resources, G.P., L.C. and J.W.; data curation, H.J., Z.W. and H.L.; writing—original draft preparation, G.P., L.C., J.W., H.J., Z.W. and H.L.; writing—review and editing, G.P., L.C., J.W., H.J., Z.W. and H.L. All authors have read and agreed to the published version of the manuscript.

Funding: This research was funded by the China Southern Power Grid Extra High Voltage Power Transmission Company Electric Power Research Institute. (CGYKJXM20220107).

Data Availability Statement: All data generated or analyzed during this study are included in this published article.

Conflicts of Interest: The authors declare no conflict of interest.

References

1. Ma, J.; Liu, C.; Wu, Y. A Pilot Directional Protection Scheme for LCC-HVDC Lines Based on Grounding Resistance. *IEEE Trans. Power Deliv.* **2022**, *37*, 4460–4473. [CrossRef]
2. Chen, Y.; Wen, M.; Wang, Z. A pilot directional protection based on low frequency voltage variable quantity for LCC-HVDC transmission lines. *Int. J. Electr. Power Energy Syst.* **2023**, *145*, 108512. [CrossRef]
3. Zhang, Y.; Li, Y.; Song, J. A New Protection Scheme for HVDC Transmission Lines Based on the Specific Frequency Current of DC Filter. *IEEE Trans. Power Deliv.* **2019**, *34*, 420–429. [CrossRef]
4. Yu, S.; Wang, X.; Zhang, X. A sensitive single-end DC line fault detection method for MMC-HVDC grids using reactor voltage ratio. *Int. J. Electr. Power Energy Syst.* **2023**, *148*, 108953. [CrossRef]
5. Ma, J.; Kang, J.; Zhou, Y.; Cheng, P. A Novel Pilot Protection Scheme for AC Line Connected to LCC-HVDC Inverter Based on Composite-Mode Model. *IEEE Trans. Power Deliv.* **2022**, *37*, 1627–1639. [CrossRef]
6. Liu, Z.; Gao, H.; Luo, S. Non-unit protection for long-distance LCC-HVDC transmission lines based on waveform characteristics of voltage travelling waves. *Int. J. Electr. Power Energy Syst.* **2023**, *150*, 109079. [CrossRef]
7. Rafael, L. One-terminal traveling wave-based transmission line protection for LCC-HVDC systems. *Electr. Power Syst. Res.* **2023**, *223*, 109650.
8. Liu, J.; Lin, S.; Zhong, W.; Liu, L. Improved Identification Method and Fault Current Limiting Strategy for Commutation Failure in LCC-HVDC. *J. Mod. Power Syst. Clean Energy* **2022**, *10*, 1761–1772. [CrossRef]
9. Li, R.; Xu, L.; Yao, L. DC fault detection and location in meshed multi-terminal HVDC systems based on DC reactor voltage change rate. *IEEE Trans. Power Deliv.* **2016**, *32*, 1516–1526. [CrossRef]
10. Sneath, J.; Rajapakse, A. Fault detection and interruption in an earthed HVDC grid using ROCOV and hybrid DC breakers. *IEEE Trans. Power Deliv.* **2016**, *31*, 973–981. [CrossRef]
11. Yang, S.; Xiang, W.; Li, R.; Lu, X.; Zuo, W.; Wen, J. An improved DC fault protection algorithm for MMC HVDC grids based on modal domain analysis. *IEEE J. Emerg. Sel. Top. Power Electron.* **2020**, *8*, 4086–4099. [CrossRef]
12. Zhang, C.; Song, G.; Dong, X. A Novel Traveling Wave Protection Method for DC Transmission Lines Using Current Fitting. *IEEE Trans. Power Deliv.* **2020**, *35*, 2980–2991. [CrossRef]
13. Zhang, C.; Li, Y.; Song, G.; Dong, X. Fast and Sensitive Nonunit Protection Method for HVDC Grids Using Levenberg–Marquardt Algorithm. *IEEE Trans. Ind. Electron.* **2022**, *69*, 9064–9074. [CrossRef]
14. Jia, K.; Chen, M.; Shi, Z.; Chen, C.; Zhao, G.; Bi, T. Single-ended protection scheme of flexible DC transmission line based on concavity and convexity of the current limiting reactor voltage integration. *Power Syst. Technol.* **2021**, *45*, 4498–4505.
15. Yu, X.; Gu, J.; Zhang, X. A non-unit transmission line protection scheme for MMC-HVDC grids based on a novel distance criterion. *Int. J. Electr. Power Energy Syst.* **2023**, *151*, 109151. [CrossRef]
16. Li, B.; Li, Y.; He, J.; Wen, W. A Novel Single-Ended Transient-Voltage-Based Protection Strategy for Flexible DC Grid. *IEEE Trans. Power Deliv.* **2019**, *34*, 1925–1937. [CrossRef]
17. Liu, N.; Li, Y.; Li, S.; Li, T.; Yu, L. A pilot protection for LCC-HVDC transmission lines based on measured surge impedance at tuning frequency. *IEEE Trans. Power Deliv.* **2022**, *37*, 2090–2103. [CrossRef]

18. Dai, Z.; Liu, N.; Zhang, C.; Pan, X.; Wang, J. A pilot protection for HVDC transmission lines based on transient energy ratio of DC filter link. *IEEE Trans. Power Deliv.* **2020**, *35*, 1695–1706. [CrossRef]
19. Shu, H.; Wang, S.; Lei, S. Single-ended protection method for hybrid HVDC transmission line based on transient voltage characteristic frequency band. *Prot. Control. Mod. Power Syst.* **2023**, *26*, 8. [CrossRef]
20. Duan, J.; Li, H.; Lei, Y.; Tuo, L. A novel non-unit transient based boundary protection for HVDC transmission lines using synchrosqueezing wavelet transform. *Int. J. Electr. Power Energy Syst.* **2020**, *115*, 105478. [CrossRef]
21. Zhang, D.; Wu, C.; He, J.; Luo, G. Research on single-ended protection principle of LCC-VSC three-terminal DC transmission line. *Int. J. Electr. Power Energy Syst.* **2022**, *135*, 107512. [CrossRef]
22. Zhang, Y.; Cong, W. An improved single-ended frequency-domain-based fault detection scheme for MMC-HVDC transmission lines. *Int. J. Electr. Power Energy Syst.* **2021**, *125*, 106463. [CrossRef]
23. Psaras, V.; Tzelepis, D.; Vozikis, D.; Adam, G.; Burt, G. Non-unit protection for HVDC grids: An analytical approach for wavelet transform-based schemes. *IEEE Trans. Power Deliv.* **2021**, *36*, 2634–2645. [CrossRef]
24. Ye, D.; Xie, F.; Hao, Z. A novel identification scheme of lightning disturbance in HVDC transmission lines based on CEEMD-HHT. *CPSS Trans. Power Electron.* **2021**, *6*, 145–154. [CrossRef]
25. Wei, D. Novel convolution and correlation theorems for the fractional fourier transform. *Optik* **2016**, *127*, 3669–3675. [CrossRef]
26. Chen, X.; Li, H.; Liang, Y.; Wang, G. A protection scheme for hybrid multi-terminal HVDC networks utilizing a time-domain transient voltage based on fault-blocking converters. *Int. J. Electr. Power Energy Syst.* **2020**, *118*, 105825. [CrossRef]

electronics

MDPI

Article

A Fusion Adaptive Cubature Kalman Filter Approach for False Data Injection Attack Detection of DC Microgrids

Po Wu [1], Jiangnan Zhang [1], Shengyao Luo [2], Yanlou Song [1], Jiawei Zhang [2] and Yi Wang [2,*]

[1] State Grid Henan Electric Power Research Institute, Zhengzhou 450052, China; wupo@ha.sgcc.com.cn (P.W.); zhangjiangnan@ha.sgcc.com.cn (J.Z.); songyanlou@ha.sgcc.com.cn (Y.S.)
[2] School of Electrical and Information Engineering, Zhengzhou University, Zhengzhou 450001, China; luosy@gs.zzu.edu.cn (S.L.); zjw20221820@gs.zzu.edu.cn (J.Z.)
* Correspondence: yiwang@zzu.edu.cn

Abstract: With the widespread application of information technology in microgrids, microgrids are evolving into a class of power cyber–physical systems (CPSs) that are deeply integrated with physical and information systems. Due to the high dependence of microgrids' distributed cooperative control on real-time communication and system state information, they are increasingly susceptible to false data injection attacks (FDIAs). To deal with this issue, in this paper, a novel false data injection attack detection method for direct-current microgrids (DC MGs) was proposed, based on fusion adaptive cubature Kalman filter (FACKF) approach. Firstly, a DC MG model with false data injection attack is established, and the system under attack is analyzed. Subsequently, an FACKF approach is proposed to detect attacks, capable of accurately identifying the attacks on the DC MG and determining the measurement units injected with false data. Finally, simulation validations were conducted under various DC MG model conditions. The extensive simulation results demonstrate that the proposed method surpasses traditional CKF detection methods in accuracy and effectiveness across different conditions.

Keywords: DC MG; FDIA; fusion adaptive CKF algorithm; attack detection

1. Introduction

With new energy technologies rapidly advancing and energy policies gradually adapting, the integration of distributed energy resources and the construction of the energy internet have become significant trends in global energy development. Direct-current microgrids (DC MGs), with their efficiency, flexibility, and reliability, play a crucial role in facilitating the access to distributed energy and accelerating the development of the energy internet [1–5]. However, as typical examples of cyber–physical systems (CPSs), the open, distributed, and networked characteristics of nonlinear DC-MG systems require extensive data exchange and information processing in communication and collaboration processes, leading to increased security risks such as data tampering and attacks [6–9]. These attacks can compromise the accurate estimation of the DC MG's state, and inaccurate state assessments can lead to incorrect monitoring decisions. Therefore, in-depth research into the secure operation of DC MGs to ensure their stability and reliability is of significant importance for promoting the development of distributed energy systems and the construction of the energy internet.

In general, a false data injection attack (FDIA) is a prevalent attack that disrupts the normal operation of a system by tampering with its data. This can cause the system to crash or malfunction. In the case of DC MGs, FDIAs can result in voltage fluctuations, load imbalances, and other issues that significantly impact the system's stability and reliability. Therefore, detecting and preventing FDIAs effectively is one of the crucial issues in the field of DC MG research [10–14]. Currently, FDIA detection methods are mainly classified into two categories: model-based and data-based detection methods [15–17]. Model-based detection methods determine whether it contains false data attack injection by calculating the residual difference between the actual output of the process and the output predicted based on the mathematical model. References [18,19] have received attention for their robustness to model uncertainties and disturbances due to the unknown input observer (UIO) approach to decouple FDIA and perturbation for attack detection. In addition, reference [20] proposed a false attack detection algorithm based on the cubature Kalman filter for detecting attack injections in nonlinear systems. Reference [21] proposed a square root maximum correlation entropy volume Kalman filter with adaptive kernel width for estimating continuous discrete nonlinear dynamic systems with non-Gaussian non-zero mean noise. In references [22,23], the detection problem was treated as an augmented and generalized state space system. The cubature Kalman filter (CKF) and unscented Kalman filter (UKF) were employed to estimate the sensor states. The main drawback of detection schemes based on system models is the requirement for system parameters and models. These parameters are not constant and can be subject to slight uncertainties and fluctuations, which can impact the effectiveness of the detection.

In comparison, the advantage of data-driven detection schemes is their independence from system parameters. These schemes only require the design of a model and training the model using historical data, thereby avoiding the negative impact caused by minor fluctuations in system parameters. Reference [24] employed multilayer perceptron (MLP) neural networks for fault detection and learning vector quantization (LVQ) classification for fault diagnosis. References [25,26] used long short-term memory (LSTM) neural networks and deep convolutional neural network (DCNN) deep learning techniques to improve detection efficiency. Reference [27] proposed a data-driven state estimation method based on data fusion techniques to obtain the optimal estimate, which significantly improved detection accuracy. The drawback of these data-driven fault detection methods is the requirement for a large amount of data, and during the training process, overfitting may occur, making them unsuitable for online implementation.

Although the aforementioned literature provides some effective attack detection strategies, there are still certain limitations. The literature [18–27] primarily focuses on studying whether the system contains false data injection, rather than considering the specific unit of attack injection [28]. This lack of consideration makes it difficult to achieve accurate detection of false data attacks. To address this issue, a false data injection attack detection method based on a fusion adaptive cubature Kalman filter is proposed. The method consists of multiple local units and an information fusion module, which detects attacks by evaluating the estimation results of each local unit. It not only achieves accurate detection of false data injection attacks but also identifies the specific units targeted by the attack through the magnitude of local residuals, enabling the isolation of faulty units. During the information fusion process, after detecting and isolating the corresponding measurements and estimates, the method is able to obtain the correct system estimation. The proposed method is more accurate and effective than the traditional cubature Kalman filter detection method under different working conditions.

The rest of this paper is organized as follows: Section 2 introduces the DC MG models. Section 3 discusses the FACKF algorithm in detail. Section 4 presents simulations to evaluate the proposed algorithm. Finally, Section 5 provides concluding remarks and future perspectives.

2. DC MG Model

2.1. DC MG Structure

Microgrids are classified into three main categories based on the distribution method: DC, AC, and AC-DC hybrid. Among them, the DC microgrid is a small power system that uses DC current for energy transmission and distribution. It mainly consists of DC sources, DC loads, DC centralized controllers, DC grids, and energy storage devices. Figure 1 shows a detailed and simplified diagram of the DC microgrid system structure.

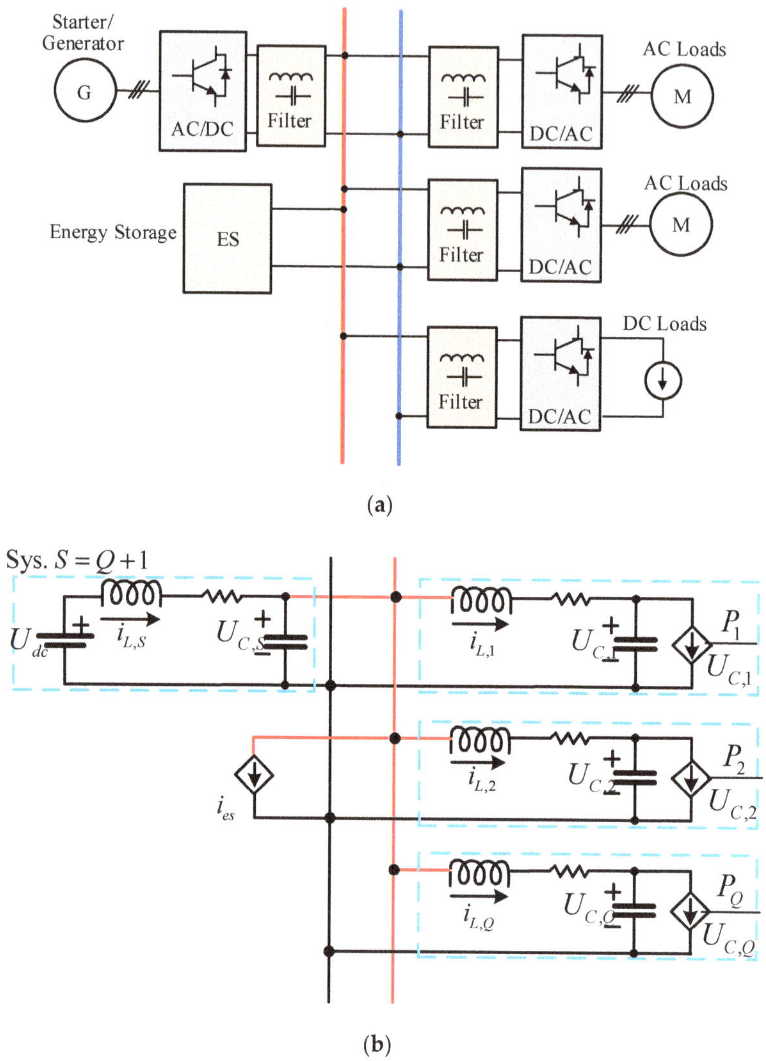

Figure 1. DC MG system structure diagram. (**a**) Detailed system structure of DC MG. (**b**) Simplified system structure of DC MG.

Figure 1a shows an islanded DC microgrid comprising an energy storage system (ESS), a DC source (consisting of a generator and power electronics), and nonlinear loads. Typically, DC and AC loads are connected to the microgrid via tightly regulated converters and inverters. If the loads are connected to the microgrid using multiple power converters and inverters to maintain constant power, they are considered constant power loads (CPLs).

2.2. DC MG Model

To simplify the analysis, Figure 1b presents a simplified model of the DC MG. It is evident that the structure can be decoupled into multiple CPLs and an ESS. Assuming a coordinated variation near the operating point and letting the energy storage current serve as the control input, the nonlinear equation of state of the whole DC MG is shown below:

$$\dot{x}(t) = \mathrm{A}x(t) + \mathrm{DH}(x(t)) + B_{es}\tilde{i}_{es}(t) + B_s U_{dc} \tag{1}$$

where $x(t)$ indicates the state quantity of the system. $x(t) = \left[x_1^T(t) \cdots x_n^T(t) x_s^T(t)\right]^T$; $x_n(t) = \left[i_{L,n} \quad U_{C,n}\right]^T$. $i_{L,n}$ and $U_{C,n}$ denote the inductor current and capacitor voltage in the n-th CPL, respectively. $x_s(t) = \left[i_{L,s}\ U_{C,s}\right]^T$; $i_{L,s}$ and $U_{C,s}$ are the inductor current and capacitor voltage in the ESS. $\dot{x}(t)$ represents the integral of $x(t)$ with respect to the time; DH is the transfer matrix of the perturbation caused by the current fluctuation of the n-th CPL; U_{dc} means the voltage of the storage power supply; B_s is the coupling matrix between U_{dc} and the system; $\tilde{i}_{es}(t)$ is the injected power, which can stabilize the DC MG and improve the robustness of the closed-loop system to disturbances and temporary faults; B_{es} is the coupling matrix. In general, the DC MG model can be expressed as

$$\begin{cases} x_{k+1} = f(x_k, u_k) + w_k \\ y_k = h(x_k, u_k) + v_k \end{cases} \tag{2}$$

where x_{k+1} represents the state vector at moment $k+1$; y_k indicates the measurement output vector at moment k; $f(x_k, u_k)$ and $h(x_k, u_k)$ are the nonlinear functions; u_k means the control input matrix; w_k and v_k are the system and measurement noise, respectively, both of which are Gaussian white noise.

2.3. DC MG Model with FDIA

FDIA is a type of cyber attack that involves injecting false data into real information. This attack typically occurs when the attacker lacks knowledge of system parameters or previous data. A sophisticated FDIA can cause system instability, posing a serious threat to network security. Therefore, real-time and accurate detection of false data is essential to maintain system stability.

The DC MG model in Equation (2) states that the injection of FDIA into the system will cause interference and alter the measurement information. Therefore, the DC MG measurement equation model with false data injection can be expressed by

$$y_k = h(x_k, u_k) + v_k + f_{k,\varphi}\gamma \tag{3}$$

$$f_{k,\varphi} = \begin{cases} 1, & k \geq \varphi \\ 0, & k < \varphi \end{cases} \tag{4}$$

where γ is the type and extent of the false data attack; k denotes the moment of system operation; $f_{k,\varphi}$ and φ are the location of the fault and the moment of the fault, respectively.

3. FDIA Detection Based on an FACKF

3.1. Fusion Adaptive CKF Approach

To implement an effective detection of false data injection in sensors, this section proposes a nonlinear detection method based on the FACKF algorithm. The method comprises local filters and an information fusion module. The local filters perform adaptive cubature estimation on individual parts of the overall unit, and the information fusion module combines the state estimates of all the local filters to obtain a global estimate. Attacks are detected during the information fusion process based on the residuals of the local units, and the sensors injected by the attacks are localized. The computational burden is distributed across various local filters in this structure, rather than the main filter. This approach can enhance the accuracy and reliability of the decision while also reducing the impact of noise by combining the local filters and sharing information. Figure 2 illustrates the FACKF framework.

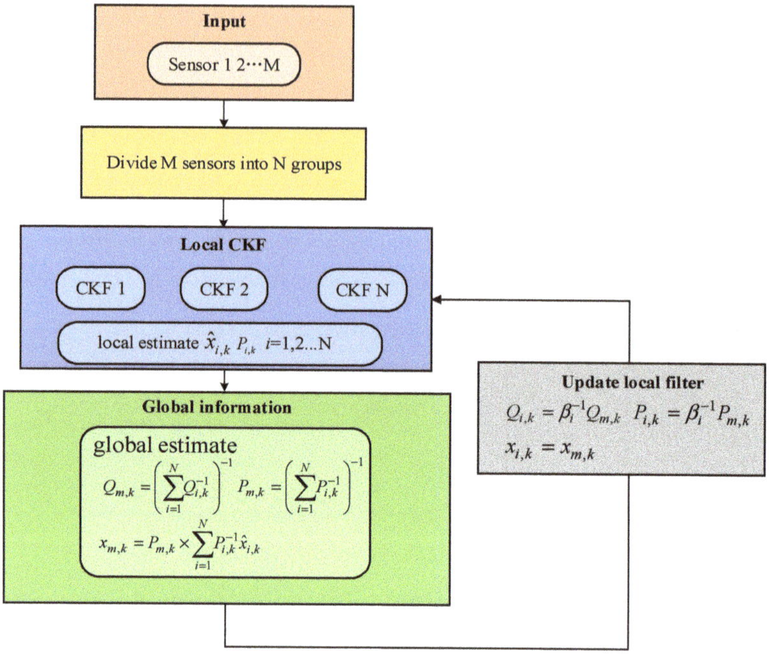

Figure 2. The framework of the FACKF.

Figure 2 illustrates the three main stages of the FACKF. Firstly, M sensors are grouped into N local units, each containing M − 1 sensor values. Secondly, each local unit performs adaptive CKF estimation to obtain a local estimate, $\hat{x}_{i,k}$, and performs information fusion to obtain the global estimate, $x_{m,k}$. Finally, the global estimate is sent back to the local filters and each local filter is updated. The global state estimates, $x_{m,k}$ and $P_{m,k}$, are determined by information fusion and sent to the local units for updating. The information distribution factor (β_i) of the i-th local filter is used to weight the localization. If all local filters have the same weight, 1/N is the number of localization. The nonlinear equations in fusion adaptive CKF describe the localized sensor system.

$$\begin{aligned} x_{k+1} &= f(x_k, u_k) + w_k \\ y_{i,k} &= h_i(x_k, u_k) + v_{i,k} \end{aligned} \tag{5}$$

$$\begin{cases} Q_k = E\left[w_k w_k^T\right] \\ R_{i,k} = E\left[v_{i,k} v_{i,k}^T\right] \end{cases} \tag{6}$$

where x_{k+1} represents the state vector and $y_{i,k}$ indicates the measurement output vector of the i-th local sensor; $f(x_k, u_k)$ and $h_i(x_k, u_k)$ are the system function and measurement function, respectively; w_k is the system noise at the time instant, k; $v_{i,k}$ represents the measurement noise at moment k, and both are Gaussian white noise; Q_k and $R_{i,k}$ are the corresponding covariance matrices, respectively.

The implementation steps for the FACKF method are as follows:

Step 1: Firstly, initialize the initial values of the parameter states and the initial covariance matrix.

Step 2: Generate $2n + 2$ cubature points:

$$x_{i,j,k-1} = \sqrt{P_{i,k-1}}\zeta_i + \hat{x}_{i,k-1},\ i = 1, \cdots, 2n+2 \tag{7}$$

where $x_{i,j,k-1}$ is the i-th cubature point at moment $k-1$ calculated for the j-th local unit, $\hat{x}_{i,k-1}$ is the estimated value at moment $k-1$ for the local unit, ζ_i is the matrix of cubature points, and $P_{i,k-1}$ is the estimated covariance at moment $k-1$ for the i-th local unit.

Step 3: The local CKF is updated in time and measurements, and the estimated state of the CKF for the first local unit is

$$\hat{x}_{i,k|k-1} = \frac{1}{2n}\sum_{j=0}^{2n+2} f\left(x_{i,j,k-1}, u_{i,k-1}\right) \tag{8}$$

$$P_{i,x_{ik|k-1}} = \frac{1}{2n}\sum_{j=0}^{2n+2}\left[x_{i,j,k|k-1} - \hat{x}_{i,k|k-1}\right] \times \left[x_{i,j,k|k-1} - \hat{x}_{i,k|k-1}\right]^T + Q_{i,k-1} \tag{9}$$

$$\hat{y}_{i,j,k|k-1} = \frac{1}{2n}\sum_{j=0}^{2n+2} h\left(x_{i,j,k-1}, u_{i,k-1}\right) \tag{10}$$

where $x_{i,j,k-1}, u_{i,k-1}$ are, respectively, the cubature points of the generated state and control variables; $\hat{x}_{i,k|k-1}$ and $P_{i,x_{ik|k-1}}$ are the a priori estimates of the state quantities and the a priori covariance matrices of the state quantities at the k-th moment in the i-th local unit, which are obtained by averaging over the individual $2n+2$ cubature points; $\hat{y}_{i,j,k|k-1}$ is the estimate of the a priori quantities at the k-th moment of the i-th local unit.

Step 4: Construct adaptive factors and perform correction of the covariance matrix.

To improve the adaptivity and robustness of the CKF algorithm, its filtering performance is enhanced by correcting the covariance matrix online through the construction of an adaptive factor. The main steps can be expressed as

$$e_k = y_{i,k} - \hat{y}_{i,j,k|k-1} \tag{11}$$

$$\Delta e = e_k e_k^T \tag{12}$$

$$\eta_k = \begin{cases} 1 & tr(\Delta e) \le tr(P_{i,y_{i,k}}) \\ \dfrac{tr(P_{i,y_{i,k}})}{tr(\Delta e)} & tr(\Delta e) > tr(P_{i,y_{i,k}}) \end{cases} \tag{13}$$

where e_k represents the innovation vector, Δe denotes the innovation vector matrix, and η_k is the adaptive factor.

Based on the constructed adaptive factor, the self-covariance matrix and cross-covariance matrix are dynamically adjusted through online correction.

Self-covariance matrix correction:

$$P_{i,y_i,k} = \eta_k[\frac{1}{2n}\sum_{j=0}^{2n}(y_{i,j,k|k-1} - \hat{y}_{i,j,k|k-1}) \times (y_{i,j,k|k-1} - \hat{y}_{i,j,k|k-1})^T] + R_{i,k} \tag{14}$$

Cross-covariance matrix correction:

$$P_{i,x_{i,k},y_{i,k}} = \eta_k[\frac{1}{2n}\sum_{j=0}^{2n}(x_{i,j,k|k-1} - \hat{x}_{i,k|k-1}) \times (y_{i,j,k|k-1} - \hat{y}_{i,j,k|k-1})^T] \tag{15}$$

The local cubature Kalman filter gain ($K_{i,k}$) at the current moment is then calculated to obtain the a posteriori estimate, $\hat{x}_{i,k}$:

$$K_{i,k} = P_{i,x_{i,k},y_{i,k}} P_{i,y_{i,k}}{}^{-1} \tag{16}$$

$$\hat{x}_{i,k} = \hat{x}_{i,k|k-1} + K_{i,k}\left(y_{i,k} - \hat{y}_{i,j,k|k-1}\right) \tag{17}$$

$$P_{i,k} = P_{i,x_{ik|k-1}} - K_{i,k} P_{i,y_{i,k}} K_{i,k}^T \tag{18}$$

where $P_{i,x_{i,k},y_{i,k}}$ and $P_{i,y_{i,k}}$ are, respectively, the self-covariance matrix and the mutual covariance matrix at moment k of the i-th local unit; $K_{i,k}$ and $\hat{x}_{i,k}$ are the Kalman gain and a posteriori estimates of the i-th local unit, respectively; and $P_{i,x_{i,k}}$ is the updated estimated covariance matrix.

Step 5: The estimates of all local state vectors ($\hat{x}_{i,k}$), the state error covariance ($P_{i,k}$), and the process noise covariance ($Q_{i,k}$) of the local CKF are integrated into the information fusion to obtain the global estimate:

$$\begin{cases} P_{m,k} = \left(\sum_{i=1}^{N} P_{i,k}^{-1}\right)^{-1} \\ Q_{m,k} = \left(\sum_{i=1}^{N} Q_{i,k}^{-1}\right)^{-1} \\ x_{m,k} = P_{m,k} \times \sum_{i=1}^{N} P_{i,k}^{-1}\hat{x}_{i,k} \end{cases} \tag{19}$$

where $P_{m,k}$ and $Q_{m,k}$ are the overall state error covariance and process noise covariance obtained by fusing the N local unit estimates, respectively; $x_{m,k}$ is the overall estimate.

Step 6: The global estimates from Step 4 are assigned to each local unit, and the local units use them as previous information in the next step, returning state estimates, state error covariance, and process noise for the first local filter CKF.

$$\begin{cases} Q_{i,k} = \beta_i^{-1} Q_{m,k} \\ P_{i,k} = \beta_i^{-1} P_{m,k} \\ x_{i,k} = x_{m,k} \end{cases} \tag{20}$$

where $Q_{i,k}$ and $P_{i,k}$ are the process noise and state error covariances returned to each local estimator; $x_{i,k}$ is the state estimate returned as the current estimate of the local unit; β_i is the information distribution factor of the i-th local filter and $\sum_{i=1}^{N} \beta_i^{-1} = 1$.

3.2. False Data Injection Attack Detection Based on FACKF

Traditional false data injection detection is based on a single residual size judgement, which is unable to effectively detect the unit of false data injection. When false data are injected, they will quickly spread to other locations and affect the overall system, and the estimated values of other parameters will be affected and deviate from the true values over time, resulting in the final estimation results deviating from the true values. In order to accurately detect the attack and determine the attack injection unit to achieve accurate and reliable state estimation, the state variance (V_k) and local state residual $(r_{i,k})$ are defined:

$$V_k = \left[(y_k - h(x_{m,k}, u_k))^{\mathrm{T}}(y_k - h(x_{m,k}, u_k))\right]^{\frac{1}{2}} \tag{21}$$

$$r_{i,k} = \left[(\hat{x}_{i,k} - x_{m,k})^{\mathrm{T}}(\hat{x}_{i,k} - x_{m,k})\right]^{\frac{1}{2}} \tag{22}$$

The state variance, V_k, determines whether a sensor experiences FDIA, while the residual, $r_{i,k}$, identifies the specific measurement unit in which attack injection occurs in FDIA detection. In the absence of FDIA, the locally estimated states of the local adaptive CKF are close to their true values, with V_k and $r_{i,k}$ approaching 0. When false attack injection occurs in a measurement unit, only one local unit is correct. In this case, the state estimate, $x_{m,k}$, is inaccurate and close to the fault estimate. Based on the maximum value of $r_{i,k}$, it is possible to diagnose which local unit is correct, while the measurements from the excluded sensors originating from that local unit are considered incorrect. The steps of the FACKF are summarized in Algorithm 1.

Algorithm 1: Fusion Adaptive Cubature Kalman Filter

1: Generate local measurements.
2: Initialize the parameters X_0, P_0 of each local.
3: Calculate cubature points:

$$x_{i,j,k-1} = \sqrt{P_{i,k-1}}\zeta_i + \hat{x}_{i,k-1};$$

4: Perform time and measurement updates for each local CKF:

$$P_{i,y_i,k} = \eta_k [\frac{1}{2n} \sum_{j=0}^{2n} \left(y_{i,j,k|k-1} - \hat{y}_{i,j,k|k-1}\right) \times \left(y_{i,j,k|k-1} - \hat{y}_{i,j,k|k-1}\right)^T] + R_{i,k}$$

$$P_{i,x_{i,k},y_{i,k}} = \eta_k [\frac{1}{2n} \sum_{j=0}^{2n} \left(x_{i,j,k|k-1} - \hat{x}_{i,k|k-1}\right) \times \left(y_{i,j,k|k-1} - \hat{y}_{i,j,k|k-1}\right)^T]$$

$$\hat{x}_{i,k} = \hat{x}_{i,k|k-1} + K_{i,k}\left(y_{i,k} - \hat{y}_{i,j,k|k-1}\right)$$

5: Calculate global estimation:

$$Q_{m,k} = \left(\sum_{i=1}^{N} Q_{i,k}^{-1}\right)^{-1}, P_{m,k} = \left(\sum_{i=1}^{N} P_{i,k}^{-1}\right)^{-1};$$

$$x_{m,k} = P_{m,k} \times \sum_{i=1}^{N} P_{i,k}^{-1}\hat{x}_{i,k};$$

6: Calculate state variance and residual of locals:

$$V_{j,k} = \frac{1}{N}\sum_{i=1}^{N}\left(\hat{x}_{i,k}^{(j)} - \frac{1}{N}\sum_{i=1}^{N}\hat{x}_{i,k}^{(j)}\right);$$

$$r_{i,k} = \left[\left(\hat{x}_{i,k} - x_{m,k}\right)^T\left(\hat{x}_{i,k} - x_{m,k}\right)\right]^{\frac{1}{2}};$$

7: Compare state variance with threshold. If it is larger than the threshold, go to 9 and alarm a faulty situation. Otherwise, go to 11.
8: Evaluate the residual value and determine the faulty sensor based on correct locals.
9: Eliminate faulty locals and calculate the global estimates based on healthy locals.
10: Assign global estimates to locals, update them, and come back to 3.

$$Q_{i,k} = \beta_i^{-1}Q_{m,k}, P_{i,k} = \beta_i^{-1}P_{m,k}, x_{i,k} = x_{m,k}$$

4. Simulation Results

In order to prove the validity of the proposed FACKF method, a DC microgrid system is set up under different working conditions, and the estimation performance of a DC microgrid under normal operating conditions and the detection and robustness estimation ability of a fake data attack injection scenario are, respectively, compared and analyzed. Through this series of comparative experiments, not only can the FACKF method demonstrate high estimation accuracy under normal operating conditions, but it can also validate its excellent detection efficiency and robustness when faced with security threats. This fully proves the practical value and effectiveness of the method in ensuring the stable operation and security protection of DC microgrids.

4.1. Test Systems

Case 1: A CPL source and a DC source are set up in the DC microgrid. In the absence of network attacks, the CKF and FACKF are, respectively, used for state estimation to evaluate the estimation accuracy of the two algorithms in the presence of noise.

Case 2: On the basis of Case 1, the case of false data injection attack is considered; the traditional residual detection method and the FACKF method proposed in this paper are used to detect the false data injection attack.

Case 3: The DC grid unit is extended to 2 CPL, and the false data injection attack is set to test the validity of the detection algorithm considering multiple CPL.

Case 4: The computational efficiency of the conventional CKF and the proposed FACKF is investigated.

In order to accurately simulate the field measurements measured by the PMUs, additional noise is added to the simulated data. It is worth pointing out that all the discussed methods were executed in MATLAB R2021b on a PC with Intel Core CPU E5-11400H, 4.5 GHz, and 16 GB memory.

Furthermore, for the purpose of conducting a quantitative analysis and comparison of various methods' performance, the performance metric known as root mean square deviation (RMSD) is used to comprehensively assess the effectiveness of the CKF and FACKF. It is defined as follows:

$$RMSD = \sqrt{\frac{\sum_{k=1}^{n} (\hat{x}_k - x_k)^2}{n}} \tag{23}$$

where k represents the time instant, x denotes the state variable, n represents the number of time steps, $\hat{x}_k$ represents the estimated value of the state variable, and x_k signifies the true value.

4.2. Case 1: Comparison of Estimation Accuracy

To compare the estimated performance of the two in the absence of an attack, consider a CPL and a source. The system variables are $\boldsymbol{X} = [i_{L1}, v_{c1}, i_{Ls}, v_{cs}]$, the system parameters are given in Table 1, and the initial conditions of the DC microgrid are chosen as follows: $\boldsymbol{X}_0 = [1.5, 200, 1.5, 200]$. The initial conditions of the Kalman filter are as follows: $\hat{x}_{0|0} = [1, 200, 1, 200]$. The initial covariance is $10^{-6}\boldsymbol{I}_{2\times2}$; the standard deviation of the system and measurement noise is $10^{-8}\boldsymbol{I}_{2\times2}$.

Table 1. The parameters of DC MG system with a CPL.

$r_{L,1} = 1.1\ \Omega$	$L_1 = 39.5$ mH	$C_1 = 500\ \mu$F	$P_1 = 300$ W
$r_s = 1\ \Omega$	$L_s = 17$ mH	$C_S = 550\ \mu$F	$V_{dc} = 200$ V

In order to study the performance of the proposed FACKF algorithm for the attack-free injection case, we assume that all state measurements are available. Therefore, we consider four local variables (M = 4), each consisting of measurements from three sensors, as follows:

$$y_1 = [v_{c1}, i_{Ls}, v_{cs}],\ y_2 = [i_{L1}, i_{Ls}, v_{cs}]\ y_3 = [v_{c1}, i_{L1}, v_{cs}]\ y_4 = [v_{c1}, i_{Ls}, i_{L1}] \tag{24}$$

The CKF and the FACKF proposed in this paper are applied to estimate the system state for comparison, respectively, and the estimation results are shown in Figure 3.

From the test results, it can be seen that without FDIA, both the traditional CKF method and the FACKF method proposed in this paper can accurately track the operating state of the DC microgrid system with good dynamic performance, and the state estimation accuracy of the two methods can satisfy the system monitoring requirements.

The results of the root mean square error (RMSE), root mean square deviation (RMSD), and constant norm-2 error (E_{n2}) calculated for both the traditional CKF method and the FACKF method are shown in Table 2. From the results in Table 2, it can be seen that the RMSEs of both the traditional CKF method and the FACKF method are smaller, which is consistent with the results presented in Figure 3; i.e., both the CKF method and the proposed method have higher tracking accuracy without spurious data injection. It can also be found that the RMSE, RMSD, and E_{n2} of each state component of the proposed partitioned FACKF method are slightly smaller than that of the traditional CKF method, compared to the traditional CKF method, and this result indicates that the proposed FACKF method has a higher state estimation accuracy, which is attributed to its use of the distributed estimation strategy.

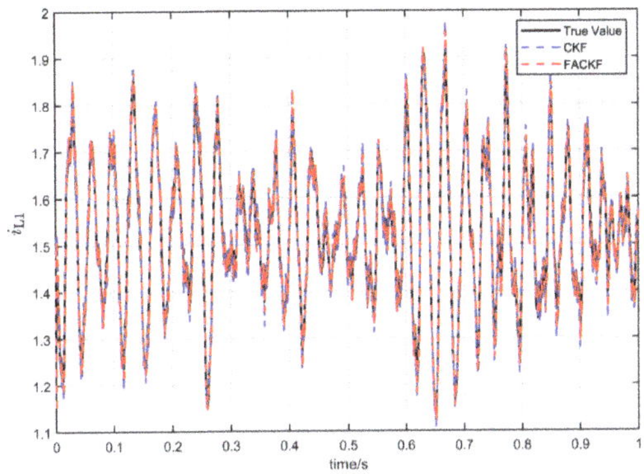

(**a**)

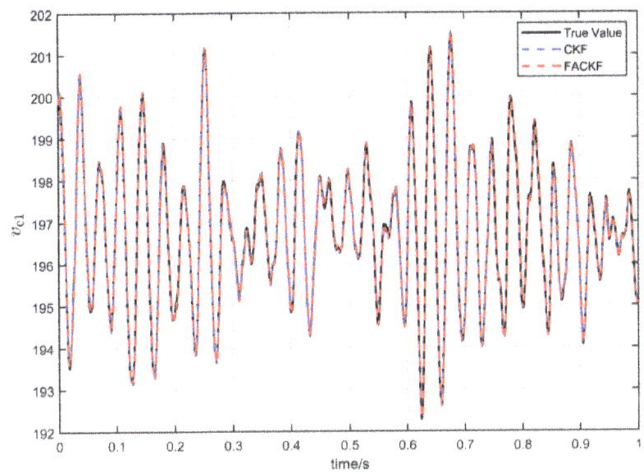

(**b**)

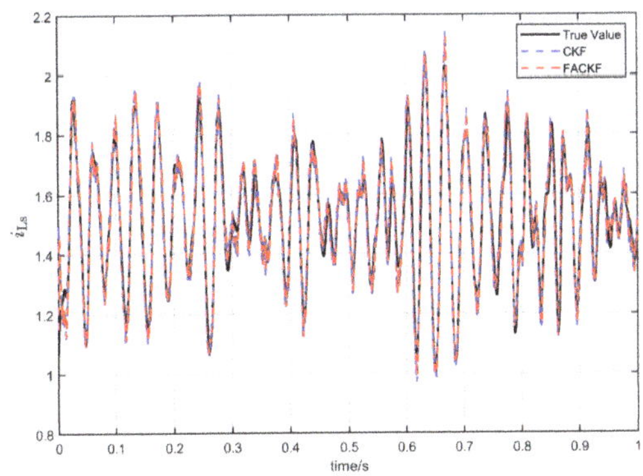

(**c**)

Figure 3. *Cont.*

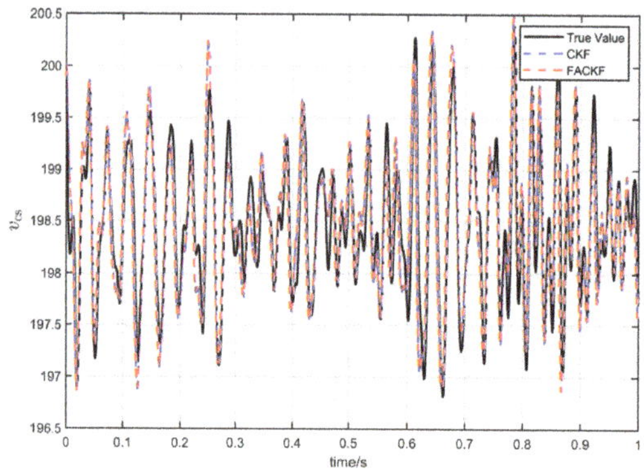

(d)

Figure 3. Estimated results of the two algorithms in the no-attack case. (**a**) Estimation results of two methods for the state variable i_{L1}. (**b**) Estimation results of two methods for the state variable v_{c1}. (**c**) Estimation results of two methods for the state variable $i_{L\ 1}$. (**d**) Estimation results of two methods for the state variable v_{cs}.

Table 2. Comparison between different methods.

State	Method	RMSE	RMSD	E_{n2}
i_{L1}	CKF Proposed method	0.0099 0.0108	0.0741 0.0712	0.0357 0.0302
v_{c1}	CKF Proposed method	0.0142 0.0120	0.0624 0.0584	0.0421 0.0367
i_{Ls}	CKF Proposed method	0.0142 0.0097	0.0725 0.0676	0.0483 0.0321
v_{cs}	CKF Proposed method	0.0129 0.0099	0.0863 0.0825	0.0546 0.0474

4.3. Case 2: Single CPL DC MG System

In the case of a single attack, in order to verify the detection and localization ability of the fusion CKF method proposed in this paper for false data attacks, an attack is set up on v_{c1} based on case 1. The traditional residual detection method and the partitioned fusion CKF method proposed in this paper are used to detect the system false data injection attack, respectively.

As can be seen from Figure 4, under the false attack injection, the residual difference increases significantly. Although the system can be detected by FDIA at this time, the specific location of FDIA injection cannot be determined. The results of the system using the conventional residual method are shown in Figure 5, where the residual values are the normalized data with no units.

It can be seen from Figure 5 that the state variance amplitude of the proposed FACKF method increases significantly when 0.3 s $\leq$ t $\leq$ 0.5 s, and FDIA can be judged to exist in the DC microgrid. When the voltage state is attacked, it has a larger value than other local residuals, because there is no measurement information containing attacks in the second local unit. In conclusion, the FACKF method proposed in this paper can not only achieve accurate detection of false data injection, but also locate the injected specific measurement unit.

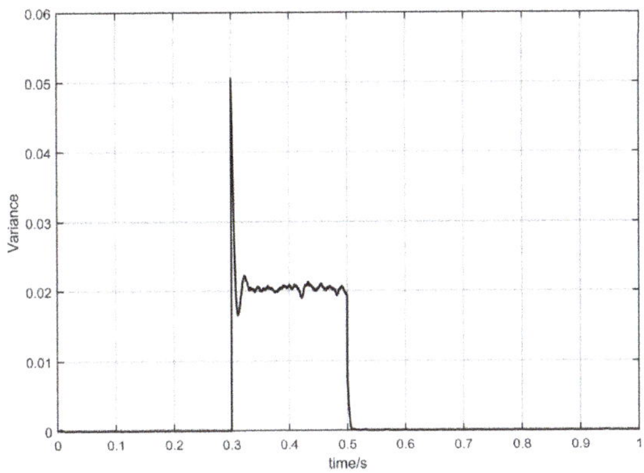

Figure 4. Traditional residual method detects the system.

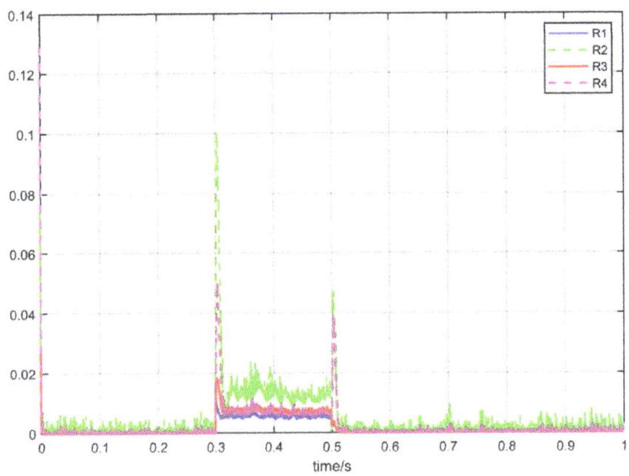

Figure 5. Based on fusion CKF residual detection.

4.4. Case 3: Effectiveness of the FACKF under Multiple CPLs

To validate the effectiveness of the FACKF method under multiple CPL scenarios, the model is extended to include 2 CPLs, and false data injection is employed to detect the system. Setting $\boldsymbol{P}_{i,0}$ as $10^{-6}\boldsymbol{I}_{6\times 6}$, $\boldsymbol{Q}_{i,k}$ as $10^{-6}\boldsymbol{I}_{5\times 5}$, and $\boldsymbol{R}_{i,k}$ as $10^{-6}\boldsymbol{I}_{5\times 5}$, an attack is initiated by injecting false data into v_{c1}, starting at 0.4. In the attack scenario, we compare the state estimation results of the proposed method in this paper and the CKF.

From Figure 6, it can be seen that under attack conditions, the system state estimation of the traditional CKF starts to deviate and gradually moves away from the true values. This is primarily due to erroneous data progressively influencing the system's state. However, the method proposed in this paper is capable of accurately detecting FDIA and identifying the affected measurement units. Upon detecting FDIA, this method isolates the injected unit and uses correct estimates from unaffected local units for overall estimation. As a result, this approach not only precisely tracks the operating state of the DC microgrid system but also demonstrates excellent dynamic performance.

Table 3 shows the comparison of various estimation metrics in the case of an attack. It can be clearly seen that once a network attack occurs, the RMSE, RAMD, and E_{n2} of the CKF are significantly larger than those of the FACKF, which is due to the fact that the traditional CKF design framework does not take into account the effect of attacks. Therefore, once an attack occurs, the estimation results of the CKF will be disturbed and seriously deviate from the real value or even regress. Compared with the traditional CKF, the proposed FACKF method can isolate the attack-induced bias and thus obtain better estimation performance. These results validate the effectiveness and excellent performance of the proposed FACKF method.

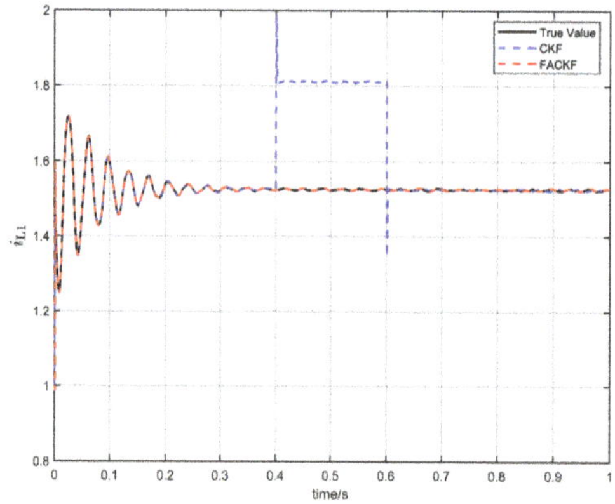

(a)

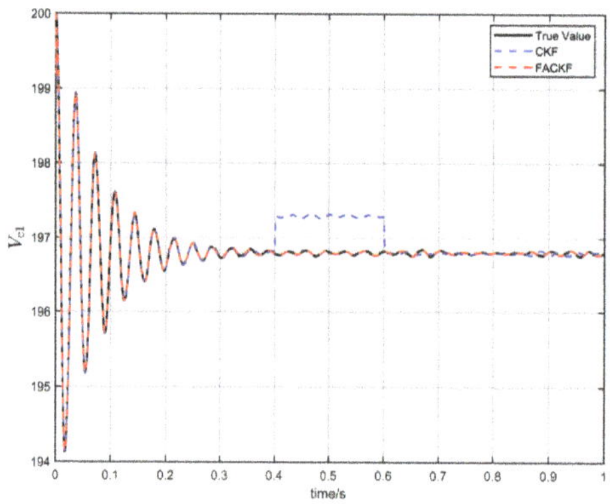

(b)

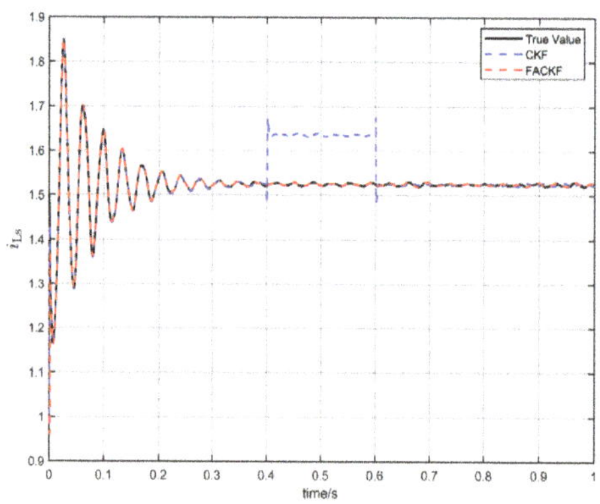

(c)

Figure 6. *Cont.*

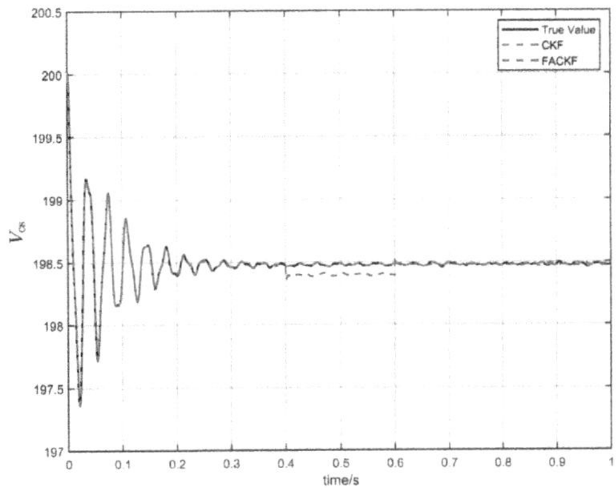

(d)

Figure 6. Comparison of state estimation results for FACKF and CKF under attack. (**a**) Comparison of estimation results for i_{L1}. (**b**) Comparison of estimation results for V_{c1}. (**c**) Comparison of estimation results for i_{Ls}. (**d**) Comparison of estimation results for V_{cs}.

Table 3. Comparison between different methods under attack.

State	Method	RMSE	RMSD	E_{n2}
i_{L1}	CKF	0.1437	0.3524	0.2417
	Proposed method	0.0112	0.0693	0.0321
V_{c1}	CKF	0.1120	0.4351	0.3442
	Proposed method	0.0105	0.0542	0.0402
i_{Ls}	CKF	0.1406	0.3346	0.3157
	Proposed method	0.0120	0.0443	0.0297
V_{cs}	CKF	0.2598	0.4887	0.4021
	Proposed method	0.0139	0.0632	0.0214

4.5. Case 4: Computational Efficiency Test

In order to meet the requirements of various real-time applications in energy management systems (EMSs), dynamic estimation in DC microgrids needs to be not only accurate but also computationally efficient. This section aims to validate the feasibility of the proposed FACKF method for adapting to the sampling rate of 30–60 samples per second from PMUs, ensuring its suitability for real-time applications. By conducting tests on the computation time of the FACKF method and the traditional CKF method under different conditions, the computation times of both algorithms are shown in Figure 7.

The results demonstrate that although the FACKF method has a slightly longer computation time compared to the traditional CKF method, this difference is primarily due to the introduction of more complex formulas in the state estimation process of the FACKF. However, it is worth noting that the average computation time for both methods is significantly lower than the PMU sampling rate. This indicates that the FACKF method can meet the real-time requirements while maintaining accurate synchronous estimation capabilities. Furthermore, despite involving more complex computational steps, the additional computation time of the FACKF method has minimal impact on performance in practical applications, effectively meeting the real-time processing requirements of the EMS in DC microgrids. Therefore, the proposed FACKF method not only demonstrates superior estimation accuracy but also confirms its practicality and efficiency in fast, dynamic environments. It provides a reliable technical solution for energy management in DC microgrids.

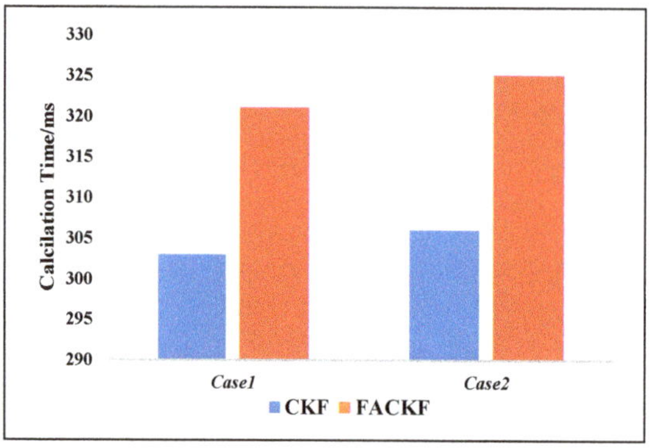

Figure 7. The comparison of computation times.

5. Conclusions

In the safe operation of a DC microgrid, real-time and accurate state estimation is crucial. To address this requirement, this paper introduces a novel method using fusion adaptive cubature Kalman filter (FACKF) technology, aiming to effectively detect and precisely locate the source of false data injection attacks. Through the analysis of experimental results, several conclusions can be drawn:

(1) The FACKF method proposed in this study can effectively detect false data injection attacks and accurately identify the attacked injection units, enhancing the system's protection against malicious attacks.
(2) After applying this research method, once an attack is identified, isolating the attacked unit and replacing the overall estimate with the correct local estimate can significantly enhance the robustness of the estimation process, thereby improving the stability and reliability of the system.
(3) The method proposed in this study exhibits low computational complexity. It not only effectively detects and locates attacks but also accurately estimates the system state under attack conditions, further ensuring the safe operation of the DC microgrid.

Although the FACKF method significantly enhances the system's protection against fake data injection attacks, its performance will vary with different attack modes and intensification. For the future work, we put forward the following suggestions: (1) adjust the algorithm to improve the adaptability to various attack scenarios and system configurations and (2) consider introducing advanced data analysis and machine learning techniques to improve the sensitivity of attack detection.

Author Contributions: P.W. and J.Z. (Jiangnan Zhang) were responsible for methodology, simulation, and validation. P.W. conducted the analysis and wrote the paper. Conceptualization was handled by S.L. and J.Z. (Jiangnan Zhang); resource management by S.L. and Y.S.; data curation by P.W. and Y.S.; original draft preparation by S.L. and J.Z. (Jiawei Zhang); review and editing by Y.W.; visualization by J.Z. (Jiangnan Zhang) and J.Z. (Jiawei Zhang); supervision by Y.W.; project management by J.Z. (Jiangnan Zhang); and funding acquisition by Y.W. All authors have read and agreed to the published version of the manuscript.

Funding: This work was supported by the Science and Technology Project of State Grid Henan Electric Power Company under Grant 521702230004, the National Natural Science Foundation of China under Grant 62203395, in part by the China Postdoctoral Science Foundation under Grant 2023TQ0306, and in part by the Postdoctoral Research Project of Henan Province under Grant 202101011.

Data Availability Statement: Data are contained within the article.

Conflicts of Interest: The authors declare no conflicts of interest.

References

1. Ansari, S.; Zhang, J.; Singh, R.E. A review of stabilization methods for DCMG with CPL, the role of bandwidth limits and droop control. *Prot. Control Mod. Power Syst.* **2022**, *7*, 12–19. [CrossRef]
2. Liu, B.; Li, H.; Zhang, H.; Han, M. A reactive power injection algorithm for improving the microgrid operational reliability. *Electronics* **2023**, *12*, 2932. [CrossRef]
3. Jiang, X.; Liu, W.; Yan, G.; Shao, H.; Zhang, L.; Wen, Y. Research on load carrying capacity evaluation of main grid with multiple DC feeds. *J. Electr. Power Sci. Technol.* **2023**, *38*, 216–225.
4. Liu, J.; Zhang, L.; Zhao, B.; Li, X. Online dynamic black start strategy for multi-power microgrid with energy storage. *Distrib. Util.* **2023**, *40*, 13–20.
5. Jasim, A.M.; Jasim, B.H.; Neagu, B.-C.; Alhasnawi, B.N. Coordination vontrol of a hybrid AC/DC smart microgrid with online fault detection, diagnostics, and localization using artificial neural networks. *Electronics* **2023**, *12*, 187. [CrossRef]
6. Fu, Y.; Zhang, Y.; Tian, S.; Shen, J.; Li, H.; Geng, F. State estimation method of active distribution network resisting multi-point false data attack. *Smart Power* **2023**, *51*, 69–76+83.
7. Xie, W.; Han, M.; Cao, W.; Guerrero, J.M.; Vasquez, J.C. Virtual resistance trade off design for DCMG grid-forming converters considering static-and large-signal dynamic constraints. *IEEE Trans. Power Electron.* **2020**, *36*, 5582–5593. [CrossRef]
8. Zhang, J.; Wu, Z.; Lu, X.; Wang, C. Research on coordinated control strategy of "optical storage direct and flexible" system considering DC power spring. *Distrib. Util.* **2023**, *40*, 82–92.
9. Yi, R. Research on the defense mechanism based on FDIA in smart grid. *Front. Comput. Intell. Syst.* **2022**, *2*, 84–88. [CrossRef]
10. Madichetty, S.; Mishra, S. Cyber attack detection and correction mechanisms in a distributed DC microgrid. *IEEE Trans. Power Electron.* **2021**, *37*, 1476–1485.
11. Dehghani, M.; Niknam, T.; Ghiasi, M.; Bayati, N.; Savaghebi, M. Cyber-attack detection in DC microgrids based on deep machine learning and wavelet singular values approach. *Electronics* **2021**, *10*, 1914. [CrossRef]
12. Huang, C.; Hong, M.; Fu, S. Distributed state estimation of active distribution network considering false data injection attack. *Electr. Power Eng. Technol.* **2022**, *41*, 22–31.
13. Xie, Y.; Yan, X.; Sang, Z.; Yang, X.; Ying, M.; Zhou, Y. Fake data injection attack method for hybrid AC-DC systems. *Electr. Power Eng. Technol.* **2022**, *41*, 165–172.
14. Xia, Y.S.; Wang, Y.; Zhou, L.; Fan, R.S. False data injection attack detection method based on improved generative adversarial network. *Electr. Power Constr.* **2022**, *43*, 58–65.
15. Wang, B.; Peng, X.; Jiang, M.; Liu, D. Real-time fault detection for UAV based on model acceleration engine. *IEEE Trans. Instrum. Meas.* **2020**, *69*, 9505–9516. [CrossRef]
16. Bhandari, G.; Lyth, A.; Shalaginov, A.; Grønli, T.-M. Distributed deep neural-network-based middleware for cyber-attacks detection in smart IoT ecosystem: A novel framework and performance evaluation approach. *Electronics* **2023**, *12*, 298. [CrossRef]
17. Musleh, A.S.; Chen, G.; Dong, Z.Y. A survey on the detection algorithms for false data injection attacks in smart grids. *IEEE Trans. Smart Grid* **2019**, *11*, 2218–2234. [CrossRef]
18. Yang, J.; Zhu, F.; Wang, X.; Bu, X. Robust sliding-mode observer-based sensor fault estimation, actuator fault detection and isolation for uncertain nonlinear systems. *Int. J. Control Autom. Syst.* **2015**, *13*, 1037–1046. [CrossRef]
19. Liu, M.; Zhao, C.; Deng, R.; Cheng, P.; Chen, J. False data injection attacks and the distributed countermeasure in DC microgrids. *IEEE Trans. Control Netw. Syst.* **2022**, *9*, 1962–1974. [CrossRef]
20. Yan, L.; Zhang, Y.; Xiao, B.; Xia, Y.; Fu, M. Fault detection for nonlinear systems with unreliable measurements based on hierarchy cubature Kalman filter. *Can. J. Chem. Eng.* **2018**, *96*, 497–506. [CrossRef]
21. Wang, Y.; Liu, D. Maximum correntropy cubature Kalman filter and smoother for continuous-discrete nonlinear systems with non-Gaussian noises. *ISA Trans.* **2023**, *137*, 436–445. [CrossRef] [PubMed]
22. El Sayed, W.; Abd El Geliel, M.; Lotfy, A. Fault diagnosis of PMSG stator inter-turn fault using extended Kalman filter and unscented Kalman filter. *Energies* **2020**, *13*, 2972. [CrossRef]
23. Wang, S.; Zhang, W.; Sun, Y.; Trivedi, A.; Chung, C.; Srinivasan, D. Wind Power Forecasting in the presence of data scarcity: A very short-term conditional probabilistic modeling framework. *Energy* **2024**, *291*, 130305. [CrossRef]
24. Tang, H.; Tang, Y.; Su, Y.; Feng, W.; Wang, B.; Chen, P.; Zuo, D. Feature extraction of multi-sensors for early bearing fault diagnosis using deep learning based on minimum unscented kalman filter. *Eng. Appl. Artif. Intell.* **2024**, *127*, 107138. [CrossRef]
25. Yang, J.; Guo, Y.; Zhao, W. Long short-term memory neural network based fault detection and isolation for electro-mechanical actuators. *Neurocomputing* **2019**, *360*, 85–96. [CrossRef]
26. Ma, Y.; Li, C.; Cao, Z.; Pan, L.; Yang, C.; Yang, Z. Robust dynamic state estimation method of power system based on data fusion technology. *Smart Power* **2023**, *51*, 78–84.

27. Vafamand, A.; Moshiri, B.; Vafamand, N. Fusing unscented Kalman filter to detect and isolate sensor faults in DC microgrids with CPLs. *IEEE Trans. Instrum. Meas.* **2021**, *71*, 1–8. [CrossRef]
28. Shao, S.; Yan, R.; Lu, Y.; Wang, P.; Gao, R.X. DCNN-based multi-signal induction motor fault diagnosis. *IEEE Trans. Instrum. Meas.* **2019**, *69*, 2658–2669. [CrossRef]

electronics

MDPI

Article

Study on Instability Mechanism and Compensation Strategy for Distributed Energy Storage Systems

Yuda Ning [1], Hu Lin [1], Xiaoteng Wan [1], Yumeng Huang [1], Zhiqin Yang [1], Qing Wang [2,*], Tangchong Wang [2] and Yongqing Deng [2]

[1] Power China Jiangxi Hydropower Engineering Burea Co., Ltd., Nanchang 330096, China; ningyuda-jxsd@powerchina.cn (Y.N.); linhu1-jxsd@powerchina.cn (H.L.); wanxiaoteng-jxsd@powerchina.cn (X.W.); huangyuming-jxsd@powerchina.cn (Y.H.); yangzhiqin@email.ncu.edu.cn (Z.Y.)

[2] School of Information Engineering, Nanchang University, Nanchang 330031, China; 416100230060@email.ncu.edu.cn (T.W.); dengyq@ncu.edu.cn (Y.D.)

* Correspondence: wangq@ncu.edu.cn

Abstract: Distributed energy storage systems (DESSs), which would become key components in a new power system, can flexibly deliver peak load shaving and demand management. With the popularization of distributed renewable energy generation in a distribution network, the grid impedance varies and DESSs thus have to face stability issues. In order to enhance the system's stability, a compensation strategy is proposed for the inverter in a DESS. First, a stability analysis model is developed to show the main factors that affect system stability. Then, an improved compensation strategy is proposed for the phase-locked loop (PLL) in a DESS, in which control parameters are adaptively tuned on-line according to real-time conditions to improve the stability of a grid-tied DESS. Simulation and hardware-in-the-loop (HIL) experimental results are given to validate the effectiveness of the proposed strategy. Simulation and experimental results show that the proposed strategy significantly increases the system's tolerance to grid impedance variations, maintains total harmonic distortion (THD) below 5% during normal operation, and effectively reduces low-order harmonic content caused by impedance fluctuations. Moreover, the strategy is demonstrated to enhance system stability under low state-of-charge (SOC) conditions, showcasing its robustness and adaptability across various operating scenarios.

Keywords: distributed energy storage system; grid-tied inverter; impedance variation; phase-locked loop; compensation strategy; system stability

1. Introduction

Thanks to the characteristics of a decentralized layout and flexible charge–discharge operations, distributed energy storage systems (DESSs), which can serve a wide range of applications at different levels in a power system, are considered key components in a new power system [1,2]. In distribution networks and microgrids, DESSs can serve as backup bidirectional power supplies to achieve dynamic power balance [3]. For end users, DESSs can shift loads to price valleys and thus save on electricity bills [4,5]. Until now, DESSs have attracted great research attention and have been applied to many commercial applications.

However, grid-connected DESSs have to face stability issues [6,7]. On the one hand, the distribution network is generally controlled by feeder line switching operations, which leads to the impedance variation in distribution networks. On the other hand, according to the electrochemical characteristics of batteries in DESSs, the internal resistance, storage capacity, and terminal voltage would be changed due to its aging and charging/discharging operations [8–12]. Impedance variation from both the DESS side and the distribution network side bring complex high nonlinearities to this grid-tied system and even fail the system. Since most commonly used DESSs are controlled by a current controller, the stability issue becomes a serious problem for further commercial use. Thus, it is necessary to study the instability mechanism of a grid-tied DESS and develop a compensation strategy accordingly to enhance the system's stability.

Harmonics lead to deteriorated power quality and even damage equipment. Thus, harmonics are chosen as a major factor that can be used to evaluate the system's stability [13–15]. Since the power quality might deteriorate as the battery impedance fluctuates, some efforts have been published to study the possibility of using DESSs to improve the power quality considering impedance fluctuation in DESSs [16–18]. Meanwhile, some scholars have conducted in-depth investigations on the instability mechanisms of grid-tied systems with regards to grid impedance variations. Although some engineering solutions have been proposed in [19–23] to address the instability issues by using typical stability criteria, impedance fluctuation in DESSs and impedance fluctuation in a distribution network are analyzed separately, which is inconsistent for practical applications.

Parameter identification is usually used to establish an analysis model of a DESS [24–27]. In [28], the impedance of the battery is interpreted by an equivalent circuit model, by which the input and output impedance of the inverter are designed to ensure stable operation. In [29], the battery is numerically modelled and the stability operating region is determined by a function of the internal resistance and internal voltage of the battery, both of which would be influenced by either aging or variation in the state of charge (SOC). Referring to the distribution network, in [30], a Nyquist theorem-based stabilization criterion is proposed for a grid-tied inverter, according to which the control loop is optimized. In [31,32], the research results show that interaction between a DESS inverter and the weak grid brings significant threats to the stability of the power system. Thus, it is necessary to study the stability of a grid-tied DESS under different conditions.

A phase-locked loop (PLL) has proven to be a key factor that affects the system's stability in a weak grid. In [33], a novel feed-forward control method is introduced to the controller, by which low-order harmonics of a grid current induced by PLL perturbation can be suppressed. In [34], the control parameters of a proportional–integral (PI) controller in a PLL are optimized to ensure the steady state performance and dynamic response of the inverter. However, the proposed method cannot achieve adaptive adjustment when impedance varies. In [35], an adaptive impedance-phased compensation control strategy is proposed in consideration of a PLL. However, the main purpose is to improve the insufficient phase margin caused by limited grid voltage feedback control.

The above literature shows that the system should overcome two issues to enhance the stability of grid-tied DESSs: (i) The system's stability is affected by both impedance variation in the distribution network and impedance fluctuation in the DESS. Thus, impedance variations on both sides should be considered in the controller. (ii) Neither impedance fluctuation in the DESS nor impedance variation in the distribution network are predictable, making it necessary to design an adaptive control strategy to address different situations. However, few published studies satisfy both issues simultaneously, limiting their adaptability in dynamic grid conditions.

To address these challenges, this paper proposes an improved compensation strategy with adaptive parameter regulation for the PLL in a DESS, which dynamically compensates impedance variations caused by both the DESS and the distribution network. Compared with existing approaches, this strategy introduces a novel adaptive compensation link that simultaneously considers real-time impedance variations on both sides, significantly enhancing the system's stability and adaptability under varying conditions. By providing a comprehensive stability analysis and proposing a dynamic compensation approach, this study aims to enhance the performance and reliability of grid-tied DESSs. The proposed strategy is designed to handle the complex interaction between the DESS and the distribution network, offering a practical solution to mitigate stability issues in systems with fluctuating impedance. This approach not only contributes to the theoretical understanding of stability in distributed systems but also provides insights that can guide future development of robust energy management strategies.

In this paper, the system's stability is firstly checked by the mathematical model of the grid-connected DESS. Then, considering the real-time grid impedance information as well as the output voltage ripple of the DESS, a novel dynamic compensation link is introduced in the PLL. The impact of the proposed compensation strategy on the system's stability is analyzed. In the end, simulation and experimental results are given to verify the proposed strategy.

2. Stability Analysis of DESS

The typical structure of a grid-tied DESS is shown in Figure 1 and the batteries in a DESS can be equivalent to the Thevenin model introduced by [36]. Corresponding parameters are explained in the appendix.

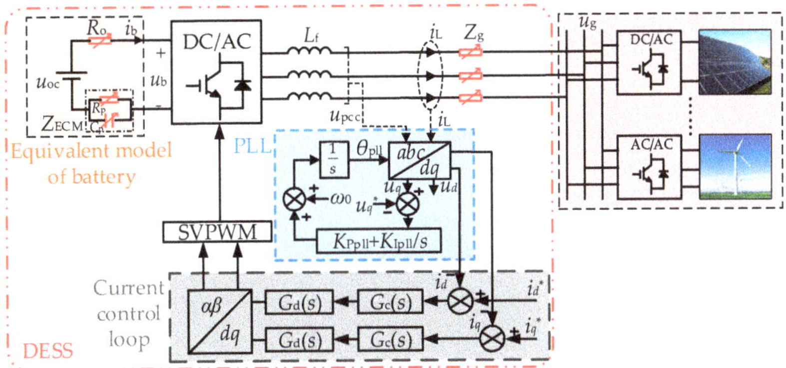

Figure 1. Topology and control structure of energy storage-based grid-tied inverter system.

According to the PLL structure shown in Figure 1, the transfer function of the PLL can be written as a typical second-order structure as follows

$$G_{\text{pll\_cl}}(s) = \frac{\Delta\theta_{pll}}{\Delta v_{gt}} = \frac{K_{\text{Ppll}}s + K_{\text{Ipll}}}{s^2 + U_{\text{q}}K_{\text{Ppll}}s + U_{\text{q}}K_{\text{Ipll}}} = \frac{2\zeta\omega_n s + \omega_n{}^2}{U_{\text{q}}(s^2 + 2\zeta\omega_n s + \omega_n{}^2)} \tag{1}$$

The natural frequency and damping ratio coefficient of the PLL can be expressed as

$$\begin{cases} \omega_n = \sqrt{U_{\text{q}}K_{\text{Ipll}}} \\ \zeta = \frac{U_{\text{gd}}K_{\text{Ppll}}}{2\sqrt{U_{\text{q}}K_{\text{Ipll}}}} \end{cases} \tag{2}$$

Accordingly, the bandwidth of the PLL can be configured by $\omega_{\text{bpll}} \approx \sqrt{2+\sqrt{2}}\omega_{\text{n}} \approx 2\omega_{\text{n}}$, $f_{\text{bpll}} = \omega_{\text{bpll}}/2\pi$.

2.1. Output Impedance of Three-Phase Grid-Tied Inverter

According to [37], there is a transfer relationship between the actual phase of the grid voltage and the calculated phase obtained from the PLL, which affects the q-axis current control loop shown in Figure 1. In this case, the interaction between the DESS inverter and grid impedance might lead to instable oscillations on the q-axis. Thus, this paper mainly focuses on the q-axis impedance and its stability criterion. The coupling effect between the d-axis and the q-axis can be ignored for ease of analysis.

Taking the influence of the PLL and the inductance variation in the DESS into consideration, the small-signal model of the q-axis current loop can be expressed as in Figure 2, where the q-axis current i_q is influenced by the PLL.

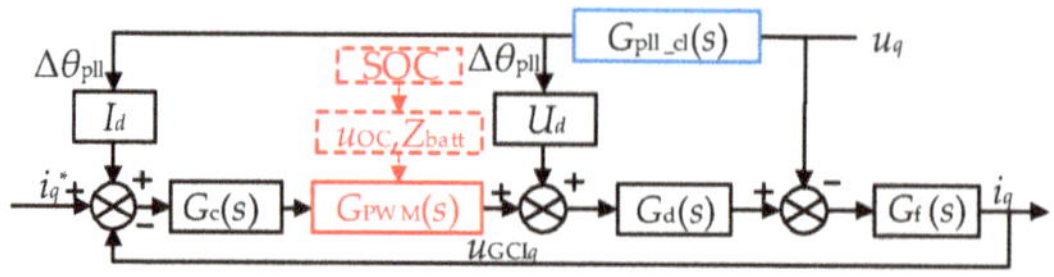

Figure 2. Small-signal model of q-axis current loop considering the influence of DESS and PLL.

According to the structure shown in Figure 2, the transfer function $G_{cl}(s)$ of Figure 2 can be expressed by

$$G_{\mathrm{cl}}(s) = \frac{i_q(s)}{i_q{}^*(s)} = \frac{G_{\mathrm{c}}(s)G_{\mathrm{PWM}}G_{\mathrm{d}}(s)G_{\mathrm{f}}(s)}{1 + G_{\mathrm{c}}(s)G_{\mathrm{PWM}}G_{\mathrm{d}}(s)G_{\mathrm{f}}(s)} \tag{3}$$

However, the terminal voltage and internal impedance of the DESS would be changed by the variation in the SOC, and the transfer function of the PWM modulation system $G_{\mathrm{PWM}}(s)$ can thus be expressed by

$$G_{\mathrm{PWM}}(s) = \frac{u_b(SOC)}{U_{tri}} = \frac{u_{OC}(SOC) - i_b Z_{batt}(SOC)}{U_{tri}} \tag{4}$$

where u_{OC}, i_b, and Z_{batt} stand for battery voltage, battery current, and impedance of battery, respectively, all of which could be affected by the variation in the SOC. Since the DESS is controlled by the current loop, in order to simplify the calculation, G_{PWM} is simplified as in the following

$$G_{\mathrm{PWM}}(s) = \frac{u_b(SOC)}{U_{tri}} = \frac{u_{oc}(SOC) + u_{bx}(SOC)}{U_{tri}} = \frac{u_{OC} - i_b(\frac{R_p}{1+R_pC_ps} + R_0)}{U_{tri}} \tag{5}$$

It is clear that u_{bx} is caused by the impedance fluctuation of the DESS, and hence can be adopted for reflecting the variation in DESS impedance in terms of the SOC. Moreover, combining (1)~(5), the output conductance $Y_{\mathrm{GCI}}(s)$ and output impedance Z_{GCI} of the DESS can be derived by

$$Y_{\mathrm{GCI}}(s) = -\frac{i_q(s)}{u_q(s)} = \frac{G_{\mathrm{f}}(s)}{1 + G_{\mathrm{c}}(s)G_{\mathrm{PWM}}(s)G_{\mathrm{d}}(s)G_{\mathrm{f}}(s)} - G_{\mathrm{pll\_cl}}(s)G_{\mathrm{cl}}(s)[I_q + \frac{U_q}{G_{\mathrm{c}}(s)G_{\mathrm{PWM}}(s)}] \tag{6}$$

$$Z_{\mathrm{GCI}}(s) = \frac{1}{Y_{\mathrm{GCI}}(s)} \tag{7}$$

2.2. *Stability Analysis of Grid-Connected DESS*

According to the impedance stability criterion given in [38], the ratio between grid impedance Z_g and DESS impedance Z_{GCI} can be applied to judge the stability requirement of the phase margin (PM) and magnitude margin (GM) at the same time. The phase margin and magnitude margin can be expressed as

$$\begin{cases} \mathrm{PM} = \angle \frac{Z_{\mathrm{g}}(j\omega_{\mathrm{c}})}{Z_{\mathrm{GCI}}(j\omega_{\mathrm{c}})} \\ \mathrm{GM} = -20\lg\left|\frac{Z_{\mathrm{g}}(j\omega_{\mathrm{x}})}{Z_{\mathrm{GCI}}(j\omega_{\mathrm{x}})}\right| \end{cases} \tag{8}$$

where ω_x is the angle frequency when the phase difference between Z_{g} and Z_{GCI} is 180°, and ω_{c} is the angle frequency when the magnitude of Z_{g} equals the magnitude of Z_{GCI}. Therefore, the impedance ratio reflects the stability problem of a grid-tied DESS. Since the increase in grid impedance will influence $Z_{\mathrm{g}}(s)/Z_{\mathrm{GCI}}(s)$, considering the worst case, Z_{g} is considered as a pure inductive impedance L_{g}, which can be represented by

$$Z_{\mathrm{g}}(s) = sL_{\mathrm{g}} \tag{9}$$

The Bode plots of sensitivity analysis between $Z_g(s)/Z_{GCI}(s)$ and L_g are shown in Figure 3, where the increasing direction of L_g is indicated by the red solid arrow. In this figure, different coloured circles and triangles represent the magnitude margin points and phase margin points for different values of grid impedance L_g, respectively.

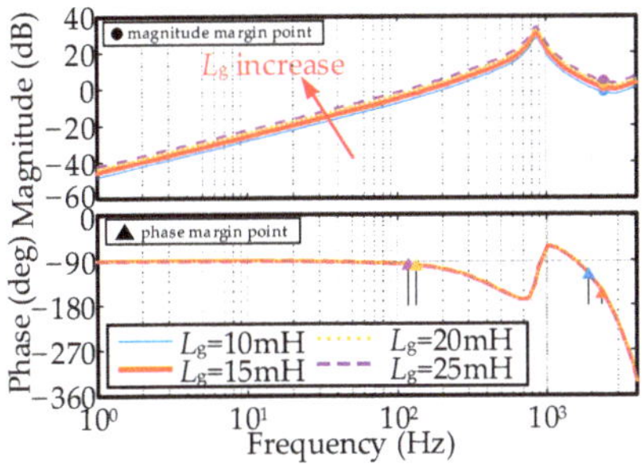

Figure 3. Bode plots of impedance ratios when L_g increases from 10 mH to 25 mH.

According to the analysis result in Figure 3, the magnitude–frequency response curve of $Z_g(s)/Z_{GCI}(s)$ shifts upward with an increment in L_g, and the phase–frequency response curve remains unchanged. When the value of L_g is small, the gain margin is greater than 0 in the frequency domain, which satisfies the stability criterion and the system remains in stable operation. With the increment in grid impedance L_g, the magnitude–frequency response curve gradually shifts upward, and the gain margin gradually decreases to 0. At this time, the system operates under a critical stable state. As the grid impedance L_g keeps increasing until the gain margin is smaller than 0, the system will not satisfy the stability criterion and the system becomes unstable.

Meanwhile, the output characteristic of the DESS is influenced by the variation in the SOC during the charging and discharging process of the battery. Key parameters (such as U_{OC}, R_0, R_p, and C_p), explanations of which are listed in the appendix, cannot be ignored in such systems. The equivalent Thevenin model is adopted for the batteries in the DESS due to its simplicity and ability to accurately represent key dynamic characteristics, such as open-circuit voltage and internal impedance, under varying SOCs. Compared to more complex models, the Thevenin model provides sufficient accuracy for analyzing the system's behavior while maintaining computational efficiency, making it ideal for real-time control and implementation in the proposed PLL compensation strategy. Combining the equivalent Thevenin model of batteries in the DESS shown in Figure 1 with the key parameters' fitting function introduced in [39], the variation in key parameters versus the SOC can be found in Figure 4.

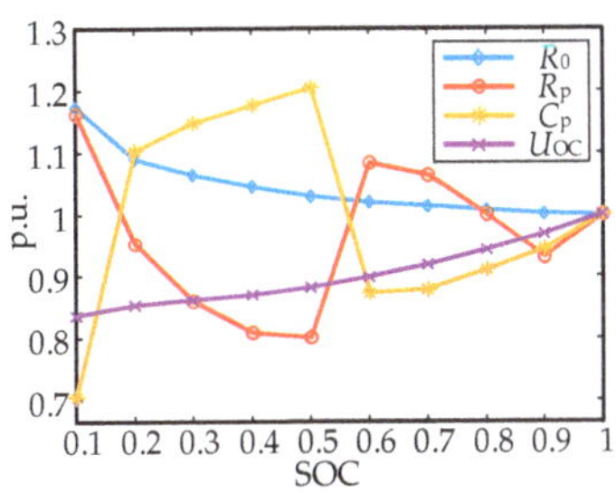

Figure 4. Variations in key parameters in DESS batteries concerning the change in SOC.

As shown in Figure 4, during the discharging process, the SOC drops from 1 to 0.1, u_{OC} would be decreased, and R_0 would be increased. The Bode plot of impedance ratios under different SOCs is given in Figure 5. In this figure, different colored circles and triangles represent the magnitude margin points and phase margin points at different SOC of the battery, respectively. The diamond symbols represent the peak points of amplitude characteristics at different SOC. As shown in the figure, as the SOC decreases, the peak value of the amplitude–frequency characteristic curve would be significantly changed and the local minimum value of the phase–frequency response curve would be decreased. When the SOC is lower than 0.3 p.u., the frequency that the phase–frequency response curve rides through $-180°$ comes ahead and the gain margin of the system is below 0. In this case, the DESS brings negative efforts to the system's stability.

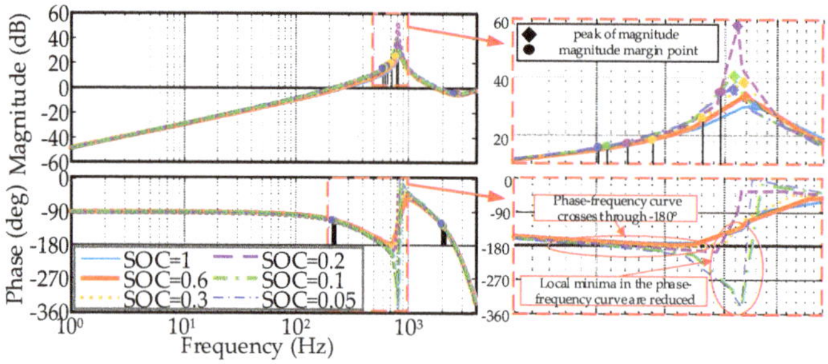

Figure 5. Bode plots of impedance ratios under different SOCs.

The above analysis shows that the impedance variation might result in high risk to system stability. Thus, it is necessary to improve the robustness by comprehensively considering the impedance fluctuation brought by both the DESS and the distribution network.

3. Proposed Strategy with Adaptive PLL Compensation

In order to enhance the stability of a grid-connected DESS, an improved PLL strategy is proposed and is shown in Figure 6. As shown in the figure, compared with a traditional PLL, a compensation link H_c(s) is employed in the proposed structure, by which the impedance on the q-axis of the DESS can be compensated to improve the system's stability. Moreover, in order to deal with global impedance changes for the concerned grid-connected system, H_c(s) is adaptively adjusted according to both the impedance of the distribution network and the impedance of the DESS. Since the impedance of the DESS relates to the RMS value output voltage ripple U_{bx}, U_{bx} is employed to reflect the impedance of the DESS and can be obtained by a high-pass filter G_{HPF}(s) with the measured transient voltage of the DESS inverter u_b.

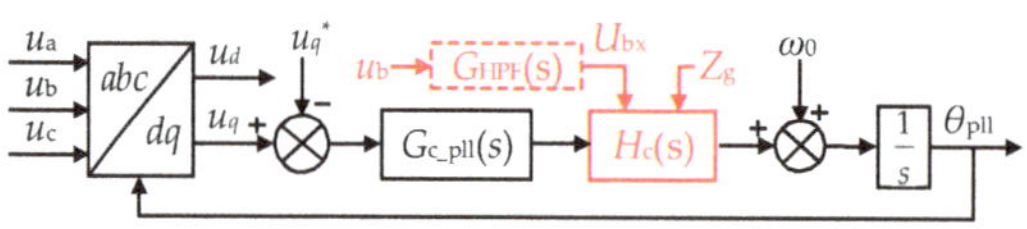

Figure 6. Block diagram of PLL compensation control.

The transfer function of the improved PLL can be expressed as

$$G_{\text{pll\_cl}}'(s) = \frac{(K_{\text{Ppll}}s + K_{\text{Ipll}})H_c}{s^2 + H_c U_q K_{\text{Ppll}}s + H_c U_q K_{\text{Ipll}}} \tag{10}$$

3.1. Structure of Compensation Link in PLL

A general function expression for the proposed compensation link H_c(s), which is added for the PLL as shown in Figure 6, can be expressed as

$$H_c(s) = \frac{\sum_{k=0}^{m} b_k s^k}{\sum_{h=0}^{n} a_h s^h} = \frac{b_m s^m + b_{m-1} s^{m-1} + \cdots + b_0}{a_n s^n + a_{n-1} s^{n-1} + \cdots + a_0} = \frac{b_m + b_{m-1} s^{-1} + \cdots + b_0 s^{-m}}{s^{n-m} + a_{n-1} s^{n-m-1} + \cdots + a_0 s^{-m}} \tag{11}$$

where n and m represent the highest order terms in the numerator and denominator, respectively; a_h and b_k are the coefficients in the polynomials of the numerator and denominator, respectively. To simplify the analysis and calculation, the highest order term of the denominator, which is marked as a_n, is set to 1. It is clear that the structure of H_c(s) would be changed by setting n, m, a_h, and b_k to different values. For example, by setting n = 2 and m = 2, the compensation link becomes a second-order compensation structure, which is particularly suitable for adjusting PLL parameters dynamically to stabilize the system under varying impedance conditions. Such flexibility makes $H_c(s)$ a versatile tool for enhancing PLL performance in grid-tied DESS applications.

Referring to H_c(s) shown in (11), compared with the highest order term of the denominator, other terms with high order are relatively small and can thus be ignored for instability analysis. Thus, the effort of H_c(s) can be categorized as three situations: n − m < 0, n − m = 0, and n − m > 0, the curves of which are shown in Figure 7. Different colored circles and triangles represent the magnitude margin points and phase margin points with different n-m, respectively.

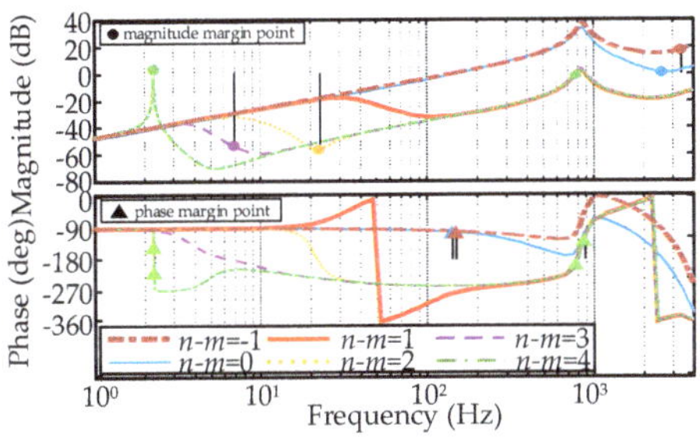

Figure 7. The Bode diagram of impedance ratio when $m < n$.

According to the analysis result in Figure 7, the following conclusions can be obtained.

- If $n - m = 0$, the compensation link can be treated as a proportional link, whose transmission gain is the ratio of the highest order coefficient of the numerator and denominator. In this condition, the PLL is still a typical second-order system, the bandwidth of which would be changed by H_c(s).
- If $n - m < 0$, taking $n - m = -1$ as an example, the compensation is barely detectable in the low-frequency domain. However, the decline rate of the phase–frequency response curve slows down by nearly 200 Hz and the magnitude of the impedance ratio is not suppressed. After achieving the peak value, there is a magnitude increment to the reference and the gain margin of the system is still below 0, which would even enlarge the instability margin of the system. A similar phenomenon can be obtained with different values of $n - m < 0$ and the system stability will not be improved in these conditions.
- If $n - m > 0$, the compensation link with different structures can consistently suppress the magnitude of the impedance ratio in the high-frequency band. However, a local minimum point shows up in the low-frequency band and the local minimum point moves left as $n - m$ increases. It should be noted that if $n - m = 4$, a significant increment would be obtained on the left of the local minimum point, which might lead to instability in the grid-tied DESS.

Thus, to ensure the reliability of the PLL, coefficients in the proposed compensation link should satisfy $4 > n - m > 0$.

Meanwhile, the phase–frequency characteristics are also changed by the proposed compensation link. As shown in Figure 7, if n − m = 1, the phase maintains at −90° in the low-frequency band and the DESS is negatively resistive. As the frequency increases, the phase–frequency curve crosses 0° and the DESS becomes approximately resistive. Increasing $n - m$ from 2 to 4, the phase curve moves to the opposite direction and crosses −180° in the low-frequency band. As shown in Figure 7, the mentioned crossing frequency in the low-frequency band would be decreased by the increment in $n - m$. In the high-frequency band, after the phase–frequency curve achieves −270°, the compensation link with a different $n - m$ shows similar effects on the phase–frequency characteristics. In this case, where GM > 0, the phase margin locates in the interval of [30°,60°] and the DESS is stable.

According to the above analysis, the newly added compensation link in (11) can effectively improve the stability margin by setting $n - m = 1, 2, 3$. Moreover, the structure is reliable and less likely to bring a negative effect to the system's dynamic response.

3.2. Parameters' Design for Compensation Link in PLL

According to the analysis in Section 3.1, a typical second-order compensation structure, which is expressed by (12), can be employed to improve the system's stability.

$$H_c(s) = \frac{b}{s^2 + a_1 s + a_0} \tag{12}$$

It is clear that a_1, a_0, and b are key parameters that affect the system's stability. By taking one of the parameters as variable, the effects of every parameter on the impedance ratio of the DESS can be obtained as in Figure 8, where the semi-transparent planes represent 0 dB in the magnitude–frequency analysis and ±180° in the phase–frequency analysis.

- When a_1 is treated as a variable, the trend of the Bode plot of the impedance ratio is shown in Figure 8a. As shown in the left part of Figure 8a, the global maximum value shows up when the frequency is twice as much as the initial bandwidth of the PLL (f_{bpll0}). When a_1 is near 0, the local maximum value in the low-frequency band shows up and a local minimum point shows up as the frequency increases. In this case, the overall system stability would not be affected. As shown in the right part of Figure 8a, if $a_1 < 0$ but near 0, the phase–frequency curve crosses −180° in the low-frequency band and maintains at −270°. Then, the curve crosses −180° again in the low-frequency band and finally can be stabilized at 360° as a_1 gradually increases.
- When a_0 is treated as a variable, the trend of the Bode plot of the impedance ratio is as is shown in Figure 8b. In this situation, the phase–frequency characteristic is hardly effected by a_0. For the magnitude–frequency characteristic, which is shown in the left part of Figure 8b, only the local maximum value in the low-frequency band is slightly increased by the increment in a_0, while the values in the high-frequency band are almost unaffected. Therefore, the effect of a_0 on the stability of the DESS is very limited.
- When b is treated as a variable, the Bode plot of the impedance ratio is as is shown in Figure 8c. In this situation, as shown in the left part of Figure 8c, if $b < 0$, the curve crosses 0 dB and shows a high peak in the low-frequency band before $0.5f_{bpll0}$. Meanwhile, as shown in the right part of Figure 8c, the curve of the phase–frequency characteristic crosses −180°, which causes the amplitude of the system impedance ratio to fall below 0. In this case, the system will step into an unstable state. If $b > 0$, as shown in the left part of Figure 8c, there will be no very high peak values in the low-frequency band; only a local maximum value shows up in the low-frequency band. Meanwhile, as shown in the right part of Figure 8c, the frequency, at which the phase–frequency curve crosses −180° in the low-frequency band, will be raised by the increment in b. However, if b is overlarge, the phase–frequency curve in the

low frequency will be reversed and crosses 0°, which is not conducive to the phase stability of the DESS.

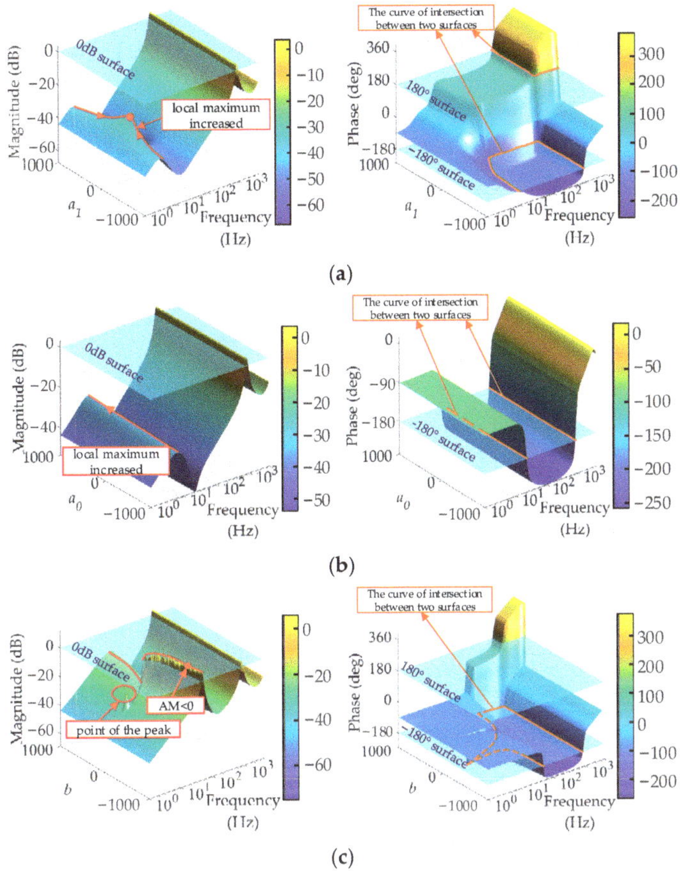

Figure 8. The effects of the parameters of H_c on the system impedance ratio: (**a**) effect of a_1 on the Bode plot of the impedance ratio; (**b**) effect of a_0 on the Bode plot of the impedance ratio; (**c**) effect of b on the Bode plot of the impedance ratio.

According to the analysis above, in order to take the dynamic change in the characteristics of the DESS into consideration, a_0 is defined as the ratio of the RMS value of the output voltage U_{bx} and the PWM carrier amplitude U_{tri}. With the generalized Nyquist impedance criterion, the following expression can be obtained.

$$\begin{cases} G_0 = 1 - [(\frac{-1}{G_f G_c G_{PWM}} + G_d)(I_{gd} G_c G_{PWM} + U_{gd}) G_{pll\_cl} G_{cl}] \\ \left| \frac{1+G_f G_c G_{PWM} G_d}{G_f G_0} \right| > |Z_g| \qquad \omega = \omega_x \\ \arg(\frac{1+G_f G_c G_{PWM} G_d}{G_f G_0}) + 90^\circ > 0 \qquad \omega = \omega_c \\ a_0 = U_{bx}/U_{tri} \end{cases} \tag{13}$$

By neglecting the high-order fractional components in the transfer function, the ranges of a_1 and b are given by (14) and (15)

$$\begin{cases} G_1 = -1 + I_{gd}K_{\mathrm{Pc}}G_{\mathrm{PWM}} + U_{gd} \\ G_2 = -K_{\mathrm{Ipll}}U_{gd}(L_f{}^2 + L_g{}^2G_1)\omega^4 \\ G_3 = I_{gd}K_{\mathrm{Ppll}}G_{\mathrm{PWM}}L_g{}^2U_{\mathrm{gd}}K_{\mathrm{Ipll}}\omega^4(K_{\mathrm{Pc}}\omega + K_{\mathrm{Ic}}) \\ G_4 = K_{\mathrm{Ipll}}(L_f{}^2 + L_g{}^2G_1) \\ G_5 = (K_{\mathrm{Ipll}}{}^2 + K_{\mathrm{Ppll}}{}^2\omega^2) \cdot [2K_{\mathrm{Ic}}G_{\mathrm{PWM}}L_f - (L_f{}^2 - L_g{}^2G_1{}^2)\omega^2] \\ G_6 = K_{\mathrm{Ic}}\cos(3T_s\omega/2) + K_{\mathrm{Pc}}\omega\sin(3T_s\omega/2) \\ G_7 = K_{\mathrm{Pc}}{}^2G_{\mathrm{PWM}}{}^2 + (L_f{}^2 - L_g{}^2)\omega^2 - 2G_{\mathrm{PWM}}L_fG_6 \\ G_8 = U_{\mathrm{gd}}\omega^2\sqrt{G_4{}^2\omega^4 + G_5G_7} \\ G_9 = U_{gd}{}^2G_{\mathrm{PWM}}K_{\mathrm{Ic}}(2L_f + I_{gd}{}^2K_{\mathrm{Ic}}G_{\mathrm{PWM}}L_g{}^2) - (L_f{}^2 - L_g{}^2G_1{}^2)\omega^2 \\ G_{10} = (G_2 - G_3 + G_8)/G_9 \\ G_{11} = K_{\mathrm{Pc}}{}^2G_{\mathrm{PWM}} + K_{\mathrm{Ic}}{}^2L_f{}^2\omega^2 + \sqrt{2G_4G_6} \end{cases} \tag{14}$$

$$\begin{cases} \frac{G_{11}}{L_g{}^2G_1{}^2\omega_c{}^2} \le b \le G_{10}\sqrt{(a_0 - \omega_x{}^2)^2 + (a_1\omega_x)^2} \\ 0 < a_1 < (a_0\omega_c{}^2G_{10} + G_1)/\left|U_{gd} - G_{PWM}\right| \end{cases} \tag{15}$$

where G_{10} and G_{11} are functions of ω_x and ω_c, respectively. However, in order to make the system stable after the introduction of the compensation link, values should be tuned within the range determined by (15). According to the previous analysis in Figure 8, ω_x is a variable near twice of ω_{bpll0}, and ω_c is less than $0.5\omega_{\mathrm{bpll0}}$. Therefore, the values of a_1 and b can be determined by

$$\begin{cases} a_1 = (a_0\omega_{\mathrm{bpll0}}{}^2G_{10} + G_1)/4\left|U_{gd} - G_{\mathrm{PWM}}\right| \\ b = G_{10}(\omega_{\mathrm{bpll0}})\sqrt{(a_0 - \omega_{\mathrm{bpll0}}{}^2)^2 + (a_1\omega_{\mathrm{bpll0}})^2} \end{cases} \tag{16}$$

According to the above equations, the parameters of the compensation link can be designed, before which both the grid impedance L_g and the parameters of the DESS should be obtained for the calculation process. In order to obtain the grid impedance L_g, a high-frequency small-signal current is injected into the grid-tied current loop shown in Figure 1. Then, the impedance of the distribution network can be obtained by

$$\begin{cases} |Z_g| = \frac{|u_{pcc}(f_i)|}{|i_L(f_i)|} \\ \angle Z_g = \angle u_{pcc}(f_i) - \angle i_L(f_i) \end{cases} \tag{17}$$

Since Z_g is considered to be an inductive component, only the amplitude needs to be calculated to observe the impedance of the distribution network L_{gt}(mH). To avoid frequency overlapping of the harmonic components in the network, the frequency of the injected small-signal current should avoid the integer multiple of the grid current, while the amplitude of the injected current should not be overlarge. On the other hand, in a practical system, u_b can be measured by existing voltage sensors in the inverter and the RMS value of u_b can thus be easily observed. Both obtained L_{gt} and U_{bx} should be updated to (12)~(16) in real time to achieve the adaptive compensation for the PLL.

3.3. Impedance Observation and Compensation Process

The flowchart of the implementation of the proposed compensation strategy for the PLL is given in Figure 9. As shown in the figure, total harmonic distortion (THD) is employed for the stability evaluation. Although the THD of the grid current controlled by the grid-tied inverter is generally limited below 5%, switching actions, a dead zone, grid distortion, and other factors might lead to the high current amplitude of low-order harmonics in a practical grid-tied system. Thus, it is necessary to correctly detect and comprehensively analyze harmonics in different orders, especially for low-order harmonic components. In this paper, harmonic detection is divided into two steps as shown in Figure 9.

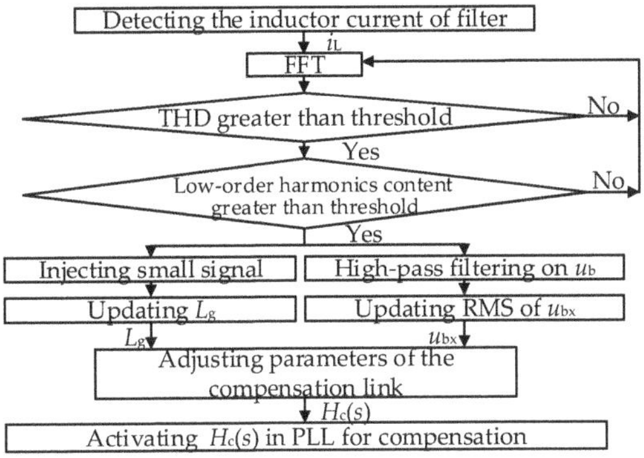

Figure 9. Flowchart of PLL compensation control strategy.

When the DESS is connected to the distribution network, the grid current i_L is detected and the THD value of the grid current is obtained by fast Fourier Transform (FFT). Then, by comparing the obtained THD with a threshold value, the controller decides whether to activate further comparison. If the THD is higher than the threshold, the fundamental component and the second–seventh harmonic components will be updated and further compared with the corresponding preset threshold. If the harmonic component is higher than the corresponding threshold, it means that there is an eigenvalue of low-level harmonic oscillation, which might lead to an unavoidable risk of destabilization. Thus, it is time to activate the compensation link. Thereafter, the detected and calculated L_g and u_{bx} are applied to adjust the parameters in the compensation link $H_C(s)$, which will be activated in the PLL and the current control loop for compensation.

4. Simulation and Experimental Results

In order to verify the correctness of the theoretical analysis and the effectiveness of the proposed PLL compensation strategy, a 3 MW grid-tied DESS model, which can be applied for user-side energy storage, is constructed on MATLAB/Simulink and the hardware-in-the-loop (HIL) experimental platform. Key parameters of the system are listed in Table 1.

Table 1. Main system parameters.

System and Control Parameters	Value
Filter inductor L_f	500 mH
RMS value of grid voltage u_g	10 kV
Grid frequency f_g	50 Hz
Switching frequency f_s	5 kHz
Reference amplitude of grid-tied current I_{gd}	141.4 A
Control parameters of current loop	$K_{Pc} = 0.1$, $K_{Ic} = 10$
Control parameters of PLL	$K_{Ppll} = 0.26$, $K_{Ipll} = 500$
Frequency of the injected small-signal current	$f_i = 670$ Hz
Amplitude of the injected small-signal current	$i_i = 0.7$ A

4.1. Simulation Results

To study the dynamic stability characteristics of the grid-tied DESS when the impedance varies, the time-varying grid impedance is set to simulate different operating conditions of the distribution network. The grid impedance is configured to 20 mH, 30 mH, and 40 mH in 0~0.2 s, 0.2~0.4 s, and 0.4~0.9 s, respectively. Meanwhile, in order to quickly show the effect of SOC decrement, the SOC of the battery unit in the DESS decreases from 1 to 0.1 in 0.9 s in the simulation.

Figures 10–12 show time-domain waveforms, related harmonic components, and the THD of the output current of the DESS without a compensation link, respectively. In Figure 10, regions a and b are defined based on the change in grid impedance. In region a, grid impedance changes from 20 mH to 30 mH at 0.2 s; in region b, grid impedance changes from 20 mH to 40 mH at 0.4 s. This progression reflects typical grid conditions that the proposed strategy is designed to address. As shown in Figure 10, during 0~0.2 s (the grid impedance is 20 mH), the DESS is discharged and the discharging current flows to the distribution network via a short transient stage; during 0.2~0.4 s (the grid impedance is increased to 30 mH), the THD of the grid current exceeds 5%, which is marked as the threshold. The RMS value of the lower-order harmonic component is also increased significantly. According to the previous theoretical analysis, it can be seen that the stability of the grid-tied DESS might have deteriorated. During 0.4~0.9 s, the grid impedance is increased to 40 mH, the THD of the output current is further increased, and the harmonic components of the current spectrum are also significantly enlarged. The phase of the output current is slightly shifted and the current amplitude undergoes unexpected oscillation, which reveals the instability of the grid-tied DESS caused by impedance fluctuations from the grid side.

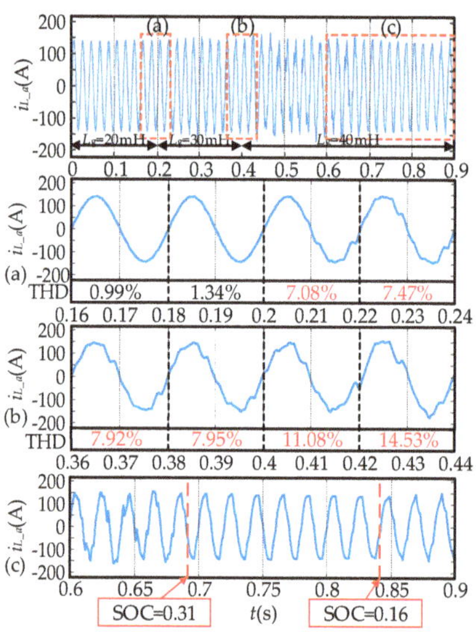

Figure 10. Output current waveform of the DESS without compensation.

In region c of Figure 10, the SOC decreases from 0.31 to 0.16. When the SOC of the DESS drops to 0.31 (at about 0.68 s), which is shown at the bottom of Figure 10, it can be seen that the system's performance is affected by the variation in the terminal voltage and internal impedance in the DESS. However, the amplitude of the output current is not decreased, and hence, the battery is still able to support the stable operation of the system. When the SOC drops to 0.16 at 0.84 s, which can also be found at the bottom of Figure 10, the SOC can no longer support the normal operation of the DESS.

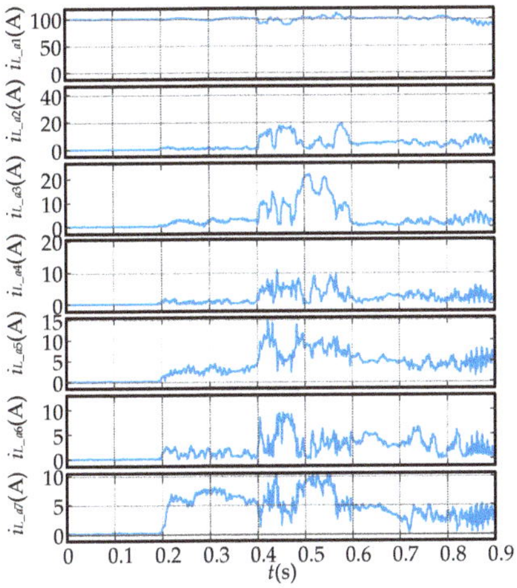

Figure 11. Harmonic component of output current of the DESS without compensation.

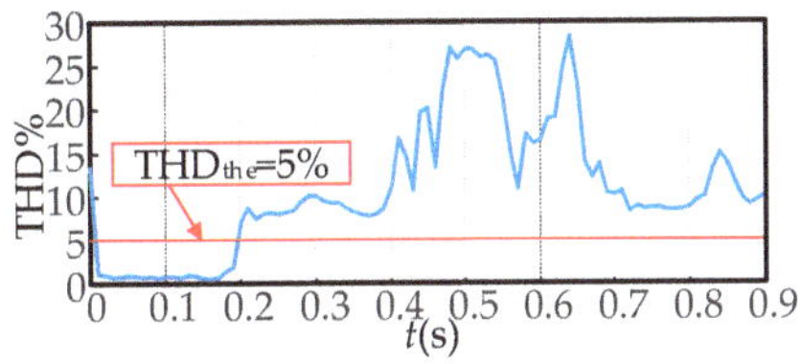

Figure 12. THD of the output current of the DESS without compensation.

In the proposed compensation strategy, the grid impedance is timely measured by the high-frequency signal injection method. The injected small-signal, with a frequency of 670 Hz, is used to detect grid impedance and evaluate the system's response to controlled disturbances. The real-time measured grid impedance is shown in Figure 13. It can be seen that the grid impedance can be correctly detected under different operating conditions of the distribution network.

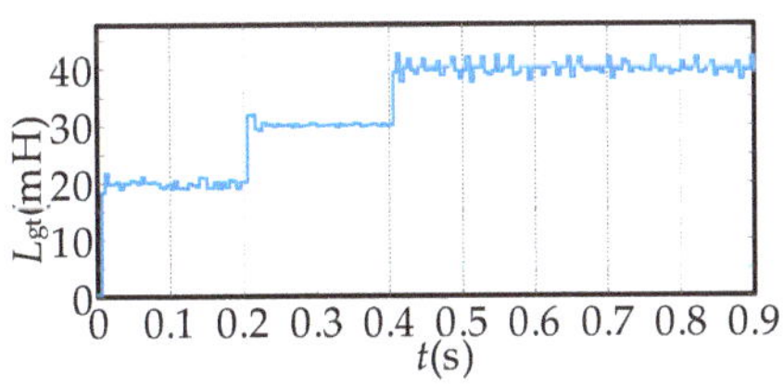

Figure 13. Real-time measured grid impedance by high-frequency injection method.

Meanwhile, the output voltage u_b of the DESS is sampled and processed to obtain the RMS value of its high-frequency ripple u_{bx}; the measured result of u_b and the obtained u_{bx} are shown in Figure 14. It can be observed that the RMS value of u_{bx} is increased by the decrement in the SOC in the DESS.

By sampling and calculating L_{gt} and u_{bx}, the proposed compensation link for the PLL can be constructed. Current waveforms and the measured low-order harmonic components with the proposed compensation link are shown in Figures 15a and 15b, respectively.

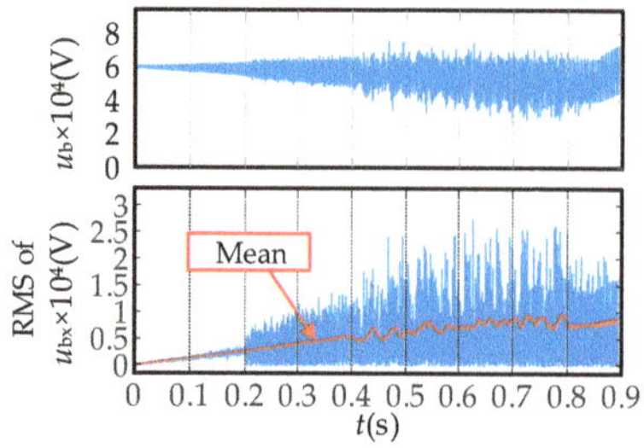

Figure 14. High-frequency ripple of output voltage high-frequency ripple.

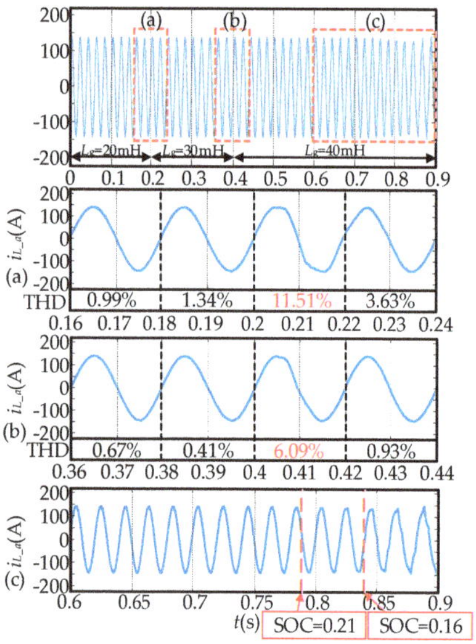

(**a**)

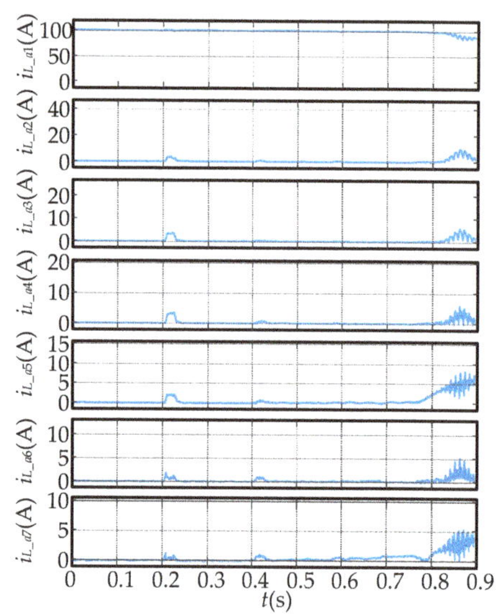

(**b**)

Figure 15. Effect of proposed compensation link in DESS: (**a**) the output current waveform; (**b**) the harmonic component of output current.

As shown in Figure 15, during 0~0.2 s, the DESS operates steadily, and the compensation link for the PLL is nonactivated. At 0.2 s, the grid impedance is increased to 30 mH, and the RMS value of the lower harmonic component is also significantly increased. These results imply that the energy storage-based power system is operated with unstable behaviors. The compensation link is thus activated for the PLL during the time interval 0.2~0.22 s and the process of compensation is shown in region a in Figure 15a. After 0.22 s, the detected THD and the harmonics of the discharging current satisfy the system stabilization criterion. However, the proposed compensation link is still active for the PLL, and the grid-tied DESS stays in a stable operation state. At 0.4 s, the energy storage-based power system is destabilized due to the further increase in grid impedance to 40 mH. Meanwhile, the proposed compensation link is reconstructed for this new condition automatically, and the grid-tied DESS can be enforced to stable operation after 0.42 s. The corresponding process of compensation is shown in region b in Figure 15a.

Referring to the SOC decrement, as shown in region c of Figure 15a, low-order harmonic components of the output current show up at 0.79 s. However, the whole system still operates steadily until the DESS is no longer able to support the system.

The aforementioned simulation results demonstrate that the proposed compensation strategy can be adaptively activated and reconstructed to improve the system's stability effectively under the condition of impedance fluctuation from both the distribution network and the DESS.

4.2. Experimental Results

To validate the effectiveness of the proposed PLL compensation strategy, a hardware-in-the-loop (HIL) experimental platform was established, the photograph of which is shown in Figure 16. The HIL system utilizes an RT-Box real-time simulator that is employed to emulate the dynamic behavior of a grid-tied DESS, and a TMS320F28335 digital signal processor (DSP) is employed to implement the proposed strategy. The RT-Box executes a high-fidelity DESS model designed in the PLECS environment, including grid impedance, inverter, and battery characteristics. The model operates with a simulation step size of 10 μs, ensuring precise real-time emulation of grid dynamics. The sampling period of the DSP is set to 50 kHz. A 10 kHz control signal, which includes a phase adjustment and modulation index, is generated by the DSP and transmitted to the RT-Box via a high-speed communication interface. This configuration enables accurate evaluation of the compensation strategy under various disturbance scenarios.

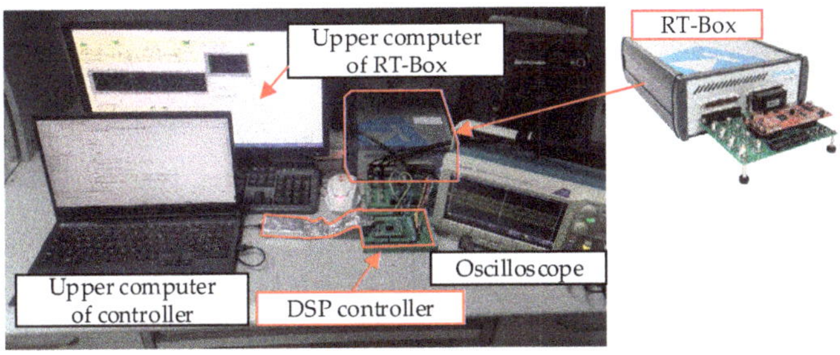

Figure 16. HIL experimental platform for grid-tied DESS systems.

The experiments focused on two key scenarios.

Grid Impedance Variation: The grid impedance was increased from 20 mH to 30 mH, simulating a weak grid condition. This setup tested the system's ability to maintain stability and suppress harmonic distortions under dynamically changing grid conditions.

SOC Decline: The SOC of the DESS was gradually reduced from 1.0 to 0.1 to examine the system's robustness as the battery approaches depletion.

This scenario evaluates the effectiveness of the compensation strategy in stabilizing the system under challenging conditions brought by batteries in the DESS.

Without the compensation link for the PLL, the waveforms of the grid-tied current i_L in consideration of the impedance fluctuation of both the grid and DESS sides are shown in Figure 17, in which the grid impedance is changed from 20 mH to 30 mH to emulate the grid impedance fluctuation.

As shown in Figure 17a, without the proposed compensation link, the low-order harmonics of current i_L with high amplitude can be immediately detected. Figure 17a highlights the distortion caused by odd harmonics, which leads to sharp peaks and valleys, and even harmonics, which induces waveform asymmetry. The current waveform is evidently distorted with a THD > 5%, and the system operates with instability. In Figure 17b, the impedance variation in the DESS is emulated by reducing the SOC. The impedance as well as the terminal voltage of the DESS are changed, which causes unstable variation in current i_L. As the SOC is decreased to 0.16, it can no longer support the grid-tied system, and the amplitude of i_L suffers attenuation and a small fluctuation.

With the proposed compensation link for the PLL, the measured waveforms of the grid-tied current i_L of the system with regards to the impedance fluctuation from both the grid and the DESS are shown in Figure 18.

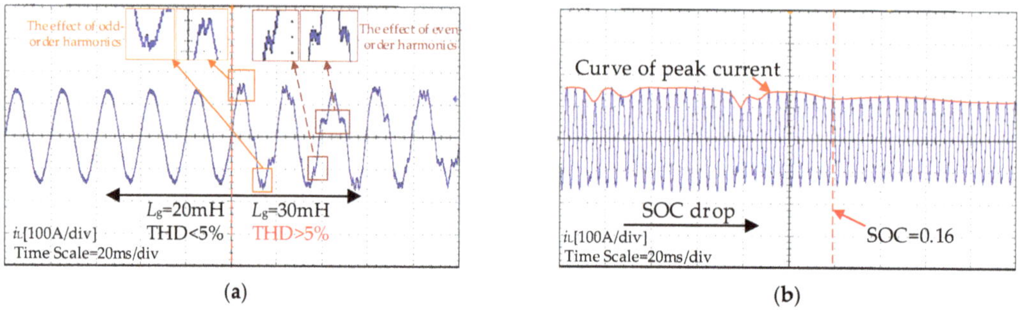

Figure 17. Grid-tied current i_L without proposed compensation strategy: (**a**) i_L with impedance fluctuation of grid; (**b**) i_L with impedance fluctuation of DESS.

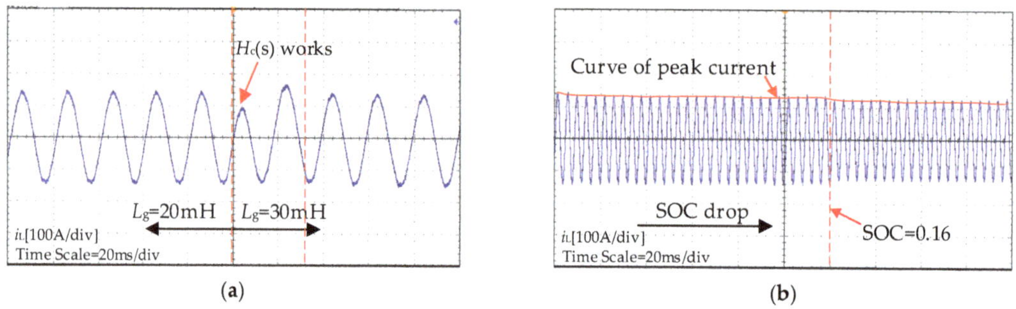

Figure 18. Grid-tied current i_L with the proposed compensation strategy: (**a**) i_L with impedance fluctuation of grid; (**b**) i_L with impedance fluctuation of DESS.

As the compensation control strategy is activated, as shown in Figure 18a, $H_c(s)$ is added to the PLL under grid impedance fluctuations. Compared with Figure 17a, the measured waveforms in Figure 18a demonstrate the effectiveness of the proposed compensation strategy in suppressing these distortions, achieving a smoother and more stable current waveform with a THD below 5%. The output voltage and current can be stabilized with very low harmonics. During the compensation process, the current i_L undergoes a short transition and can be returned to a steady operation state within two periods.

As shown in Figure 18b, it can be observed that when the SOC of the DESS is decreased, the grid-tied current i_L can be stabilized without many harmonics and distortions. When the SOC is dropped to 0.16, the SOC of the DESS can hardly maintain stable operation of the grid-tied system, and the amplitude of current i_L is reduced.

According to the above experimental results, it can be concluded that the proposed compensation strategy can adaptively suppress the instability of the grid-tied DESS caused by impedance fluctuation from both the input (DESS) and output (grid) sides.

5. Conclusions

In this paper, the instability problem of a grid-tied DESS is investigated by comprehensively considering the impedance fluctuation in a DESS and the impedance variation in a distribution network. The impedance model of the grid-tied DESS is established, and a second-order compensation link is introduced into the PLL controller. Different structures and compensation coefficients of this compensation link are analyzed, and an adaptive compensation strategy is thus proposed to deal with instability caused by impedance variations. Meanwhile, the system's dynamic characteristics, the grid impedance, and the DESS impedance are evaluated and analyzed. Then, a second-order compensation link is designed, which can be adaptively adjusted according to the detected value of grid impedance and the output voltage ripple of the DESS to suppress the system's instability. Compared with existing approaches, the proposed strategy effectively addresses the combined effects of dual-side impedance variations, which have often been treated separately in previous studies. Simulation and experimental results demonstrate that the proposed strategy can effectively address dual-side impedance variations, maintaining a THD below 5% under normal operation, significantly reducing low-order harmonics, and improving robustness under low SOC conditions. As a result, this study enhances the stability and adaptability of the system and provides a practical solution to mitigate the instability of modern distribution networks. Our future work will focus on extending the strategy to more complex grid conditions and practical applications.

Author Contributions: Conceptualization, Y.N. and Q.W.; methodology, X.W. and H.L.; software, Z.Y. and T.W.; validation, Z.Y. and Y.D.; formal analysis, X.W. and Q.W.; data curation, T.W. and Q.W.; writing—original draft preparation, Y.H., H.L., and T.W.; supervision, Y.N. and Q.W.; project administration, Y.N. and Q.W. All authors have read and agreed to the published version of the manuscript.

Funding: This research was funded by the National Natural Science Foundation of China (Grant No. 51967013) and the Natural Science Foundation of Jiangxi Province (Grant No. 20212BAB214061).

Data Availability Statement: Data are contained within the article.

Conflicts of Interest: Authors Y.N., H.L., X.W., Y.H. and Z.Y. are employed by Power China Jiangxi Hydropower Engineering Burea Co., Ltd. Authors Q.W. and Y.D. are employed by Nanchang University. Author T.W. is working towards his Master degree in Nanchang University. All authors declare that the research was conducted in the absence of any commercial or financial relationship that could construed as a potential conflict of interest.

Nomenclature

DESS, PLL	Distributed energy storage system, phase-locked loop
SOC	State of charge of battery unit in DESS
PWM	Pulse-width modulation
$G_{pll\_cl}(s)$, $G_{pll\_cl}{}'(s)$	Transfer function of traditional PLL, transfer function of PLL with proposed compensation link
u_b, L_f, i_L, u_{pcc}	DC-link voltage, inductance of filter, output current of DESS, voltage at grid connection point
i_q, i^*_q, i_d, i^*_d	Actual q-axis current, required q-axis current, actual d-axis current, required d-axis current
U_d, U_q	d axis voltage and q-axis obtained by PLL
$G_c(s)$, $G_d(s)$	Transfer function of PI controller, transfer function of controller delay
ω_0 and θ_{pll}	Required angular frequency and acquired phase of PLL
$\Delta\theta_{pll}$, Δv_{gt}	Small increment in phase position obtained by PLL, small increment in transient grid voltage
K_{Ppll}, K_{Ipll}	Proportional and integral coefficient in PLL loop
U_{gd}	d-axis voltage magnitude of distribution network
ζ, ω_n	Damping ratio coefficient and natural frequency for second-order function
f_{bpll}, ω_{bpll}	Bandwidth of PLL, natural angular frequency of PLL
$G_{cl}(s)$	Transfer function of q-axis current control loop
i_{gq}, i^*_{gq}	Actual q-axis current, required q-axis current
$G_{PWM}(s)$ and $G_f(s)$	Transfer function of DES and filter
U_{tri}	Amplitude of PWM carrier
u_b, U_{bx}	Transient voltage of DESS inverter, RMS value of output voltage ripple
u_{OC}, R_p, C_p, and R_0	Open-circuit battery voltage, equivalent resistance of polarization effect in battery, equivalent capacitance of polarization effect in battery, and internal resistance of battery
Z_g, Z_{GCI}	Impedance of distribution network, impedance of grid-connected inverter in DESS

L_g	Inductance of distribution network
$H_C(s)$	Proposed compensation link in PLL
a_1, a_0, and b	Primary coefficient of denominator, constant term of denominator, and constant term of numerator in $H_c(s)$
$\mid Z_g \mid$, $\angle Z_g$	Amplitude and phase of impedance of distribution network
$i_{L\_ak}(k = 1, 2\ldots, 7)$	The amplitude of kth-order harmonic component of output current of DESS. ($k = 1, 2\ldots, 7$)

References

1. Dunn, B.; Kamath, H.; Tarascon, J.-M. Electrical energy storage for the grid: A battery of choices. *Science* **2011**, *334*, 928–935. [CrossRef]
2. Zhong, W.; Xie, K.; Liu, Y.; Yang, C.; Xie, S.; Zhang, Y. Online control and near-optimal algorithm for distributed energy storage sharing in smart grid. *IEEE Trans. Smart Grid* **2020**, *11*, 2552–2562. [CrossRef]
3. Nunna, H.; Doolla, S. Energy management in microgrids using demand response and distributed storage-A multiagent approach. *IEEE Trans. Power Deliv.* **2013**, *28*, 939–947. [CrossRef]
4. Peng, P.; Li, Y.; Li, D.; Guan, Y.; Yang, P.; Hu, Z.; Zhao, Z.; Liu, D. Optimized economic operation strategy for distributed energy storage with multi-profit mode. *IEEE Access* **2020**, *9*, 8299–8311. [CrossRef]
5. Zhong, W.; Xie, K.; Liu, Y.; Yang, C.; Xie, S. Topology-aware vehicle-to-grid energy trading for active distribution systems. *IEEE Trans. Smart Grid* **2019**, *10*, 2137–2147. [CrossRef]
6. Gong, Q.; Wang, Y.; Fang, J.; Qiao, H.; Liu, D. Optimal configuration of the energy storage system in ADN considering energy storage operation strategy and dynamic characteristic. *IET Gener. Transm. Distrib.* **2020**, *14*, 1005–1011. [CrossRef]
7. Sistani, A.; Hosseini, S.A.; Sadeghi, V.S.; Taheri, B. Fault Detection in a Single-Bus DC Microgrid Connected to EV/PV Systems and Hybrid Energy Storage Using the DMD-IF Method. *Sustainability* **2023**, *15*, 16269. [CrossRef]
8. Liu, K.; Li, K.; Peng, Q.; Zhang, C. A brief review on key technologies in the battery management system of electric vehicles. *Front. Mech. Eng.* **2019**, *14*, 47–64. [CrossRef]
9. Tran, D.T.; Habibullah, A.F.; Kim, K.-H. Seamless Power Management for a Distributed DC Microgrid with Minimum Communication Links under Transmission Time Delays. *Sustainability* **2022**, *14*, 14739. [CrossRef]
10. Kim, T.; Qiao, W.; Qu, L. An enhanced hybrid battery model. *IEEE Trans. Energy Convers.* **2019**, *34*, 1848–1858. [CrossRef]
11. Ning, B.; Cao, B.; Wang, B.; Zou, Z. Adaptive sliding mode observers for lithium-ion battery state estimation based on parameters identified online. *Energy* **2018**, *153*, 732–742. [CrossRef]
12. Jiang, L.; Huang, Y.; Li, Y.; Yu, J.; Qiao, X.; Huang, C.; Cao, Y. Optimization of variable-current charging strategy based on SOC segmentation for Li-ion battery. *IEEE Trans. Intell. Transp. Syst.* **2021**, *22*, 622–662. [CrossRef]
13. Cao, W.; Liu, K.; Kang, H.; Wang, S.; Fan, D.; Zhao, J. Resonance Detection Strategy for Multi-Parallel Inverter-Based Grid-Connected Renewable Power System Using Cascaded SOGI-FLL. *Sustainability* **2019**, *11*, 4839. [CrossRef]
14. Lv, Z.; Kong, W.; Zhang, X.; Jiang, D.; Lv, H.B.; Lu, X.H. Intelligent security planning for regional distributed energy internet. *IEEE Trans. Ind. Inf.* **2020**, *16*, 3540–3547. [CrossRef]
15. Hu, S.; Liang, Z.; He, X. Research on the dynamic characteristics and regulation method of the energy stored quasi-z-source inverter system. *IEEE Trans. Ind. Electron.* **2020**, *67*, 4590–4599. [CrossRef]
16. Datta, U.; Kalam, A.; Shi, J. Battery energy storage system to stabilize transient voltage and frequency and enhance power export capability. *IEEE Trans. Power Syst.* **2019**, *34*, 1845–1857. [CrossRef]
17. Zhang, J.; Long, B.; He, Y.; Cheng, R.; Fan, L.; Li, B.; Gu, T.; Fu, X.; Li, H. Grouping consistency control strategy based on DMPC and energy storage unit constraints. *Int. J. Electr. Power Energy Syst.* **2023**, *148*, 108958. [CrossRef]
18. Liu, H.; Zhao, Y.; Gu, C.H.; Ge, S.Y.; Yang, Z. Adjustable capability of the distributed energy system: Definition, framework, and evaluation model. *Energy* **2021**, *222*, 119674. [CrossRef]
19. Du, W.; Dong, W.; Wang, Y.; Wang, H. Small-Disturbance stability of a wind farm with virtual synchronous generators under the condition of weak grid connection. *IEEE Trans. Power Syst.* **2021**, *36*, 5500–5511. [CrossRef]
20. Wang, Q.; Li, M.; Luo, P.; Wang, H.; Kang, P.; Sun, H.; Zhao, B. Analysis on the coupling characteristics of frequency stability and voltage stability in weak connected power grid with high ratio of clean energy. *Energy Rep.* **2022**, *8*, 1432–1437. [CrossRef]
21. Azizi, A.; Rahimi, M. Dynamic performance analysis, stability margin improvement and transfer power capability enhancement in DFIG based wind turbines at weak ac grid conditions. *Int. J. Electr. Power Energy Syst.* **2018**, *99*, 434–446. [CrossRef]
22. Li, C.Y.; Liu, W.; Liang, J.; Ding, X.; Cipcigan, L. Improved grid impedance compensation for phase-locked loop to stabilize the very-weak-grid connection of VSIs. *IEEE Trans. Power Deliv.* **2022**, *37*, 3863–3872. [CrossRef]
23. Eskandari, M.; Savkin, A.V.; Alhelou, H.H.; Blaabjerg, F. Explicit impedance modeling and shaping of grid-connected converters via an enhanced PLL for stabilizing the weak grid connection. *IEEE Access* **2022**, *10*, 128874–128889. [CrossRef]
24. Xu, Y.; Li, C.; Wang, X.; Zhang, H.; Yang, F.; Ma, L.; Wang, Y. Joint Estimation Method with Multi-Innovation Unscented Kalman Filter Based on Fractional-Order Model for State of Charge and State of Health Estimation. *Sustainability* **2022**, *14*, 15538. [CrossRef]
25. Miyatake, S.; Susuki, Y.; Hikihara, T.; Itoh, S.; Tanaka, K. Discharge characteristics of multicell lithium-ion battery with nonuniform cells. *J. Power Sources* **2013**, *241*, 736–774. [CrossRef]

26. Kim, T.; Qiao, W. A hybrid battery model capable of capturing dynamic circuit characteristics and nonlinear capacity effects. *IEEE Trans. Energy Convers.* **2011**, *26*, 1172–1180. [CrossRef]
27. Schneider, D.; Liebhart, B.; Endisch, C. Active state and parameter estimation as part of intelligent battery systems. *J. Energy Storage* **2021**, *39*, 102638. [CrossRef]
28. Liu, J. Impedance analysis of battery bidirectional DC-DC converter. *IEEE Access* **2020**, *8*, 30960–30968. [CrossRef]
29. Bazargan, D.; Filizadeh, S.; Gole, A. Stability analysis of converter-connected battery energy storage systems in the grid. *IEEE Trans. Sustain. Energy* **2014**, *5*, 1204–1212. [CrossRef]
30. Liu, F.; Hu, L.; Yuan, G.; Liu, B.; Bian, Y. Control loop stability criterion and interaction law analysis for grid-connected inverter in weak grid. *IEEE Access* **2023**, *11*, 12829–12842. [CrossRef]
31. Zhang, X.G.; Li, M.; Xu, D.G. PCC voltage perturbation path analysis and compensation for grid-connected voltage-source converter under weak grid. *IEEE Trans. Ind. Electron.* **2021**, *68*, 12331–12339. [CrossRef]
32. Jiang, X.; Han, X.; Sun, H.; Liu, L.; Pan, P.; Chen, G.; Si, Q.; Jiang, P. A Quantitative Stability Evaluation Method of VSC for Optimizing Control Parameters and Extending Stability Boundary to Avoid Oscillations. *Appl. Sci.* **2023**, *13*, 5663. [CrossRef]
33. Zhang, X.; Xia, D.; Fu, Z.; Wang, G.; Xu, D. An improved feedforward control method considering PLL dynamics to improve weak grid stability of grid-connected inverters. *IEEE Trans. Ind. Appl.* **2018**, *54*, 5143–5151. [CrossRef]
34. Han, Y.; Chen, H.; Li, Z.; Yang, P.; Xu, L. Stability analysis for the grid-connected single-phase asymmetrical cascaded multilevel inverter with SRF-PI current control under weak grid conditions. *IEEE Trans. Power Electron.* **2019**, *34*, 2052–2069. [CrossRef]
35. Chen, X.; Zhang, Y.; Wang, S.; Chen, J.; Gong, C. Impedance-phased dynamic control method for grid-connected inverters in a weak grid. *IEEE Trans. Power Electron.* **2017**, *32*, 274–283. [CrossRef]
36. Ding, X.; Zhang, D.; Cheng, J.; Wang, B.; Luk, P.C.K. An improved Thevenin model of lithium-ion battery with high accuracy for electric vehicles. *Appl. Energy* **2019**, *254*, 113615. [CrossRef]
37. Fang, J.; Li, X.; Li, H.; Tang, Y. Stability improvement for three-phase grid-connected converters through impedance reshaping in quadrature-axis. *IEEE Trans. Ind. Electron.* **2018**, *33*, 8365–8375. [CrossRef]
38. Sun, J. Impedance-based stability criterion for grid-connected inverters. *IEEE Trans. Power Electron.* **2011**, *26*, 3075–3078. [CrossRef]
39. Lennart, H.; Massimo, B.; Stefan, L. Input-admittance calculation and shaping for controlled voltage-source converters. *IEEE Trans. Ind. Electron.* **2007**, *54*, 3323–3334. [CrossRef]

electronics

Article

Short-Term Prediction Intervals for Photovoltaic Power via Multi-Level Analysis and Dual Dynamic Integration

Kaiyang Kuang [1], Jingshan Zhang [2], Qifan Chen [3], Yan Zhou [1,*], Yan Yan [4], Litao Dai [1] and Guanghu Wang [1]

1 School of Electronic Engineering, Jiangsu Ocean University, Lianyungang 222005, China; kkyjou@126.com (K.K.); juanuclear@163.com (L.D.); wgh980520@126.com (G.W.)
2 Solareast Holding Co., Ltd., Lianyungang 222243, China; zjs99107@solareast.com
3 State Grid Information & Telecommunication Co., Ltd., Beijing 100761, China; qifan-chen@sgcc.com.cn
4 State Grid Ningxia Electric Power Research Institute, Yinchuan 750011, China; yanyan503986706@163.com
* Correspondence: zhouyanxyzz@126.com

Abstract

There is an obvious correlation between the photovoltaic (PV) output of different physical levels; that is, the overall power change trend of large-scale regional (high-level) stations can provide a reference for the prediction of the output of sub-regional (low-level) stations. The current PV prediction methods have not deeply explored the multi-level PV power generation elements and have not considered the correlation between different levels, resulting in the inability to obtain potential information on PV power generation. Moreover, traditional probabilistic prediction models lack adaptability, which can lead to a decrease in prediction performance under different PV prediction scenarios. Therefore, a probabilistic prediction method for short-term PV power based on multi-level adaptive dynamic integration is proposed in this paper. Firstly, an analysis is conducted on the multi-level PV power stations together with the influence of the trend of high-level PV power generation on the forecast of low-level power generation. Then, the PV data are decomposed into multiple layers using the complete ensemble empirical mode decomposition with adaptive noise (CEEMDAN) and analyzed by combining fuzzy entropy (FE) and mutual information (MI). After that, a new multi-level model prediction method, namely, the improved dual dynamic adaptive stacked generalization (I-Stacking) ensemble learning model, is proposed to construct short-term PV power generation prediction models. Finally, an improved dynamic adaptive kernel density estimation (KDE) method for prediction errors is proposed, which optimizes the performance of the prediction intervals (PIs) through variable bandwidth. Through comparative experiments and analysis using traditional methods, the effectiveness of the proposed method is verified.

Keywords: PV power; multi-level; FE–MI; stacking ensemble learning; dynamic KDE; probabilistic prediction

1. Introduction

Energy is not only the foundation of economic development, but also plays an important role in addressing climate change and energy transition [1,2]. With the gradual depletion of fossil fuels and the growing call for environmental protection, the utilization of clean and renewable energy has become a global consensus [3]. Among these energy sources, solar PV power generation is regarded as one of the key technologies in the future energy field. The application of PV energy is becoming increasingly widespread, and accurately predicting PV power generation has become particularly important [4,5]. However,

PV power generation is susceptible to natural factors such as weather, season, and climate, and is characterized by volatility and uncertainty [6,7]. How to predict PV output efficiently has become a key issue in power grid dispatching. Therefore, accurate and reliable PV output prediction is of great value for the implementation of power grid dispatching and ensuring the stable and safe operation of PV power stations [8].

In recent years, with the advancement of prediction technology, the scale of PV power generation projects has been continuously expanding, demonstrating a promising development prospect [9]. At present, there have been relatively mature studies on PV power prediction methods, such as statistical methods [10], which have simple principles and are easy to implement. However, when the sample data are relatively complex, the prediction effect is average. Another type of method is machine learning algorithms [11], such as random forest (RF) [12], extreme learning machine (ELM) [13], and support vector machine (SVM) [14]. However, single-layer neural network prediction models are difficult to meet the prediction requirements. Deep learning methods can effectively solve these problems. Deep learning networks such as long short-term memory (LSTM) [15] and convolutional neural networks (CNNs) [16] have gradually been applied in PV prediction. The features extracted by CNNs are used as the input of a recurrent neural network to achieve the prediction of PV power generation [17]. More accurate PV power prediction is achieved by using LSTM autoencoders [18]. However, a single model has difficulty handling complex data distributions and is prone to overfitting or underfitting during training.

Ensemble learning captures more patterns and rules by integrating the prediction results of multiple base learners and has a stronger generalization ability. The PV production forecast is generated by using the basic models of four different LSTM architectures [19]. Two deep learning algorithms are utilized as the base learner, and the extreme gradient augmentation algorithm is used as the meta-learner to improve the accuracy of solar PV power generation prediction [20]. The three machine learning algorithms are integrated using the baseline linear regression model to achieve PV prediction. The results show that the integrated model is superior to the individual model [21]. A comparative analysis of the main features of various PV prediction technologies is shown in Table 1. However, the traditional stacking method adopts a fixed learner and lacks dynamic adjustment, which limits the model performance and prediction accuracy when dealing with complex and variable PV prediction scenarios. Therefore, it is necessary to explore a dynamic adaptive ensemble learning framework to improve the model's adaptability and generalization ability, which is of great significance for enhancing the performance of PV prediction models.

Table 1. Analysis of PV prediction technology.

Method	Model	Features	References
Statistics	Regression analysis, time series analysis	The principle of this method is simple, but it cannot handle complex data.	[10]
Machine learning	RF, SVM, ELM, CNN, LSTM	This method can handle complex data but is prone to overfitting or underfitting.	[11–18]
Ensemble learning	Stacking, XGBoost	This method has strong robustness and high generalization ability, but lacks adaptability.	[19–21]

In addition, PV prediction includes multiple meteorological characteristics, such as solar radiation, temperature, wind speed, humidity, etc. PV power generation is significantly affected by meteorological conditions, and meteorological factors need to be comprehensively considered [22]. However, meteorological characteristics have different influences on PV output. Considering all meteorological characteristics can affect the prediction accuracy of the model [23]. Meteorological factors with low correlation should be analyzed and filtered to remove redundant features, thereby improving the prediction efficiency of the model.

Moreover, for multi-level PV clusters, which contain multiple PV stations, due to the change of weather conditions, considering that the power mutation of sub-regional (low-level) station output is larger than that of large-scale regional (high-level) station output, high-level output can reflect the change trend of the power generation data of a period of time and a region as a whole, and establish the trend reference with high-level output to improve the prediction performance of low-level output. It can help low-level stations predict power generation more accurately. In traditional PV prediction methods, only single-level predictions are considered, and the correlations among multi-level PV outputs are not taken into account. Therefore, effective information mining of the relationship between multi-level PV data is a necessary condition for enhancing the precision of PV power forecasting.

The instability of PV power generation poses challenges to the dispatching decisions of energy storage systems [24,25]. Deterministic prediction cannot provide sufficient information to predict the actual fluctuations in power generation. Therefore, probabilistic prediction is needed to quantify the uncertainty of PV prediction and help energy managers make more flexible decisions [26,27]. The parameter estimation method combining the analytical method, the simulated annealing method, and the derived model is used to analyze the characteristics of solar PV [28]. Parametric probabilistic prediction usually assumes that the data follow a specific distribution, such as the normal distribution and Poisson distribution. If the assumption does not hold true, the prediction result will be biased. Nonparametric probabilistic prediction effectively solves the problem of unknown distribution. A dual-branch deep learning quantile regression (QR) PV power prediction method is proposed to improve the prediction performance of the model [29]. PV power prediction is carried out by using the least squares SVM and KDE, which realizes the combination of deterministic prediction and uncertain prediction [30]. However, in the traditional KDE, a fixed bandwidth is used, and the bandwidth parameter controls the width of the kernel function, which lacks the flexible processing of each data point estimation, thereby affecting the smoothness and accuracy of the estimation. Therefore, through an effective KDE strategy, a relatively reasonable PI can be determined to ensure the reliability of the interval.

As a consequence, in this paper, a new short-term probabilistic forecasting model of PV power generation that utilizes multi-level adaptive dynamic integration is proposed. Firstly, the correlation between the high-level output and the low-level output within a PV region is analyzed to explore the potential relationship between multi-level PV outputs and the corresponding analysis results are introduced into the low-level output prediction model. Then, the PV data are decomposed by CEEMDAN and reconstructed with FE and MI (FE–MI). Then, the improved dual dynamic ensemble learning method is used to construct short-term PV power generation prediction models, and the high-frequency parts are analyzed and corrected. Finally, the adaptive KDE prediction error method is used to make interval predictions under different confidence levels to improve the overall performance of PIs. Compared with traditional PV forecasting models, the main contributions of this paper are as follows:

1. The potential information on multi-level PV outputs is explored. The correlation between high-level PV output and low-level PV output is analyzed, and the prediction accuracy of low-level output is improved with the trend of high-level output;
2. The proposed FE–MI method is used to reconstruct and analyze the PV data to extract the potential information, and the improved dual dynamic stacking model is used to

make a deterministic prediction of PV power, thereby improving the accuracy of PV deterministic prediction;

3. The proposed adaptive KDE method is used to correct the point prediction error of the high-frequency time series and is used to predict the probability of PV power, thereby obtaining a more reliable PI and quantifying the uncertainty of PV power.

The rest of this paper is organized as follows: Section 2 establishes theoretical methods and model structure; Section 3 introduces the dataset and evaluation metrics; in Section 4, the prediction results are given and analyzed; Section 5 summarizes the paper.

2. Methodology

2.1. *Analysis of PV Data*

According to the characteristics of multi-level PV data, the PV data are deeply mined. The method consists of CEEMDAN and FE–MI. These methods can be used to deeply mine multi-level PV data, extract potential useful information, and provide reliable data for subsequent model prediction.

2.1.1. CEEMDAN Method for PV Power

The correlation between the PV output of high-level stations and the PV output of low-level stations within the region is explored. Based on the high-level output analysis, the corresponding analysis results are introduced into the sample analysis stage of the low-level output prediction model; that is, the PV power generation and meteorological data of the low-level stations and the historical output of the high-level stations are clustered. The high-level output reflects the overall regional PV output change trend and improves judgment of the future change of low-level output.

The PV power of different clusters is decomposed by CEEMDAN. By using Gaussian white noise with a specific standard deviation, the CEEMDAN algorithm can extract each component and change trend of time series in an adaptive way, thereby reducing the mode aliasing defect of the empirical mode decomposition (EMD) algorithm and ensemble empirical mode decomposition algorithm. The detailed steps of CEEMDAN are as follows:

Step 1: For the original PV data $F(t)$, n Gaussian white noise sequences with an average of 0 are added to construct the sequence $F_i(t)$ to be decomposed.

$$F_i(t) = F(t) + \lambda_0 \omega_i(t), i = 1, 2, 3, \cdots, n \tag{1}$$

where t is the moment; λ_0 is the signal-to-noise ratio; and $\omega_i(t)$ is the white noise sequence added for the i time.

Step 2: The EMD algorithm is applied to each $F_i(t)$, and the first modal component $IMF_1(t)$ and the first unique residual component $R_1(t)$ are obtained.

$$IMF_1(t) = \frac{1}{n}\sum_{i=1}^{n} E_1(F_i(t)) \tag{2}$$

$$R_1(t) = F(t) - IMF_1(t) \tag{3}$$

where $E_1(\cdot)$ is the first component obtained by EMD decomposition.

Step 3: Noise is added to the residual component obtained after decomposition, and EMD is continued for decomposition.

$$IMF_k(t) = \frac{1}{n}\sum_{i=1}^{n} E_1[R_{k-1}(t) + \lambda_{k-1}E_{k-1}(\omega_i(t))], k = 2, 3, \cdots, K \tag{4}$$

$$R_k(t) = R_{k-1}(t) - IMF_k(t) \tag{5}$$

where λ_{k-1} is the signal-to-noise ratio of $K-1$ white noise and K is the number of intrinsic mode functions (IMFs) after decomposition.

Step 4: When the residual time series is a monotone function, the calculation is terminated. The original time series is decomposed into n modal components and the residual term $R(t)$.

$$F(t) = \sum_{k=1}^{n} IMF_k(t) + R(t) \tag{6}$$

2.1.2. FE–MI

FE is one of the criteria by which to measure the complexity of time series. It is an optimization method of approximate entropy and sample entropy. Compared with the other two methods, FE uses fuzzy membership function as the threshold criterion of entropy to ensure that it is more suitable for classifying and reassembling the components after frequency domain decomposition. The larger the FE, the more complex the time series. The specific steps for FE are as follows:

Step 1: For the time series $\{x(i), i = 1, 2, \cdots, N\}$ of length N, dimension m is embeded to initialize the sequence and complete the phase space reconstruction, that is,

$$X(i) = \{x(i), x(i+1), \cdots, x(i+m-1)\} - u(i) \tag{7}$$

$$u(i) = \frac{1}{m}\sum_{k=0}^{m-1} x(i+k) \tag{8}$$

where $X(i)$ is the new sequence after reconstruction, $i = 1, 2, \cdots, N-m+1$, and $u(i)$ is the mean of m continuous variables $x(i)$.

Step 2: The distance between two vectors $X(i)$ and $X(j)$ is defined as the maximum absolute value of the difference between the elements of the two vectors, and its expression is as follows:

$$d_{ij}^{m} = \max\{|[x(i+k) - u(i)] - [x(j+k) - u(i)]|\} \tag{9}$$

where $1 \le i, j \le N-m+1$ and $i \ne j$.

Step 3: Fuzzy membership function is introduced to define the degree of similarity between vector $X(i)$ and $X(j)$, namely,

$$A_{ij}^{m} = \begin{cases} 1, d_{ij}^{m} = 0 \\ \exp\left[-\ln 2\left(\frac{d_{ij}^{m}}{r}\right)^{2}\right], d_{ij}^{m} > 0 \end{cases} \tag{10}$$

where r is the similar tolerance parameter, defined as R times the standard deviation of the original one-dimensional time series, that is, $r = R \times \delta$, and where δ is the standard deviation of the original one-dimensional time series.

Step 4: The function $C_i^m(r)$ is defined, as shown in Equation (11). On this basis, the relation dimension $\Phi^m(r)$ in m dimension is obtained by Equation (12).

$$C_i^m(r) = \frac{1}{N-m}\sum_{j=1, j\ne i}^{N-m+1} A_{ij}^{m} \tag{11}$$

$$\Phi^m(r) = \frac{1}{N-m}\sum_{i=1}^{N-m+1} C_i^m(r) \tag{12}$$

Step 5: The embedding dimension is increased by 1, and the above steps (1)~(4) are repeated for the $m+1$ dimensional vector to obtain the following relation:

$$\Phi^{m+1}(r) = \frac{1}{N-m}\sum_{i=1}^{N-m} C_i^{m+1}(r) \tag{13}$$

Step 6: Equation (14) is used to calculate FE, and the classification and recombination of components are completed according to the proximity of entropy.

$$FE(m,r,N) = \ln\Phi^m(r) - \ln\Phi^{m+1}(r) \tag{14}$$

where m is the embedding dimension parameter and N is the length of the original time series.

The results of CEEMDAN decomposition are grouped and reconstructed according to FE for high, medium, and low frequency, and the high-frequency time series is decomposed in the frequency domain. Because the fluctuation of PV power is easily affected by weather, light intensity, and other factors, high-frequency IMF contains instantaneous fluctuations and instability, which can easily interfere with the prediction model. Therefore, it is necessary to further analyze the high-frequency components to extract useful information, retain the true change trend, reveal the short-term fluctuation laws affecting PV power, and mine the time-varying and nonstationary characteristics of high-frequency data to provide more accurate time series information for the model and further improve the accuracy of the prediction model.

MI is a measure of the interdependence between two random variables, which describes the amount of information contained in one variable about the other variable. Through MI, the correlation, dependence and information flow between variables can be obtained. For discrete random variables X and Y, MI can be expressed as

$$I(X;Y) = \sum_{x\in X}\sum_{y\in Y} p(x,y)\log\frac{p(x,y)}{p(x)p(y)} \tag{15}$$

where $p(x,y)$ is the joint probability distribution function of X and Y, and $p(x)$ and $p(y)$ are the edge probability distribution functions of X and Y, respectively.

The FE can measure the complexity of the time series, and a higher FE usually means a higher complexity of the time series and a more chaotic structure. MI measures the correlation between two time series, and the greater the MI, the stronger the correlation. By selecting the time series with the smallest ratio of FE divided by MI, the mode containing both low complexity and low redundancy is found. Therefore, after frequency domain decomposition of high-frequency time data, the selection of the IMF with the smallest FE/MI ratio can provide cleaner, more accurate, and more representative data for subsequent forecasts. This not only helps to improve the prediction accuracy, training efficiency, and generalization ability of the model, but also reduces the negative impact of noise and redundant information on the prediction results.

2.2. Improved Dual Dynamic Stacked Ensemble Learning Model

Ensemble learning improves the overall prediction performance by combining the prediction results of multiple learning models, that is, combining multiple weak learners into one strong learner, thereby enhancing the accuracy, robustness, and generalization ability of the model. By taking advantage of the strengths of multiple models, the deviation and variance problems that occur in a single model can be effectively reduced, and the accuracy and stability of the final prediction can be improved. Ensemble learning has the following main forms: bootstrap aggregating (Bagging), Boosting, and Stacking. Bagging uses bootstrap sampling to improve stability. Boosting improves accuracy by adjusting the weights of training samples. However, the base learners of the above two are usually the same. Stacking combines multiple different types of base learners together, and each base learner generates a prediction result. Then, another model is used to conduct secondary learning on the prediction results of these base learners to generate the final prediction. Stacking is more flexible due to integrating multiple different types of models and is suitable for a wide variety of learners and datasets. It can capture more diverse features, thereby improving the prediction performance. Stacking usually adopts a double-layer structure for model fusion. The first layer is the base learner layer: Multiple different models are trained on the raw data to output predicted values. The second layer is the meta-learner layer: The prediction results of the base learner are taken as input features, and the meta-learning model is set for secondary training to generate the final prediction results. The Stacking model framework is shown in Figure 1.

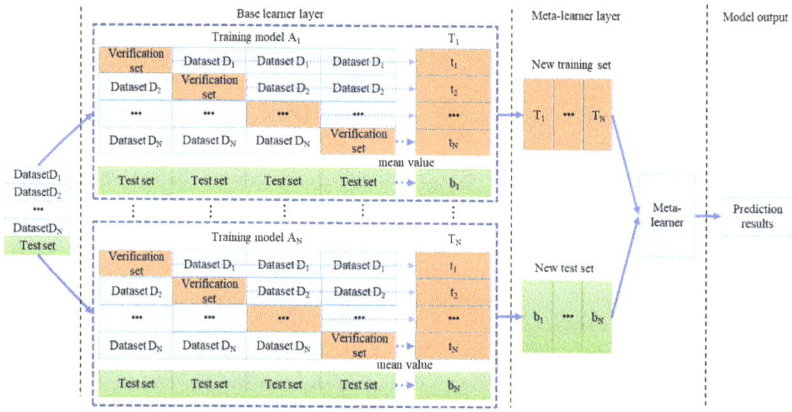

Figure 1. Stacking model framework.

Stacking combines different base learners to give full play to the different characteristics and advantages of the base model, thereby improving the performance of the overall model. However, in traditional Stacking, the base learner is usually fixed and remains unchanged throughout the training process, unable to cope with different data distributions and PV prediction scenarios, and lacks adaptability. Therefore, it is necessary to improve the Stacking model. A new dual dynamic adaptive Stacking model is proposed in this paper. For a single dataset, multiple training sets can be dynamically divided, and PV prediction can be adaptively carried out in the dual dynamic mode of dynamic datasets and dynamic models. Dynamically adjusted Stacking can dynamically select and adjust datasets and base learners according to the characteristics and changes of the data. This approach can adaptively adjust according to the current training environment and continuously optimize the selection and combination of base learners, thereby improving the overall model performance and having a stronger generalization ability. The improved dual dynamic stacked ensemble learning model is shown in Figure 2.

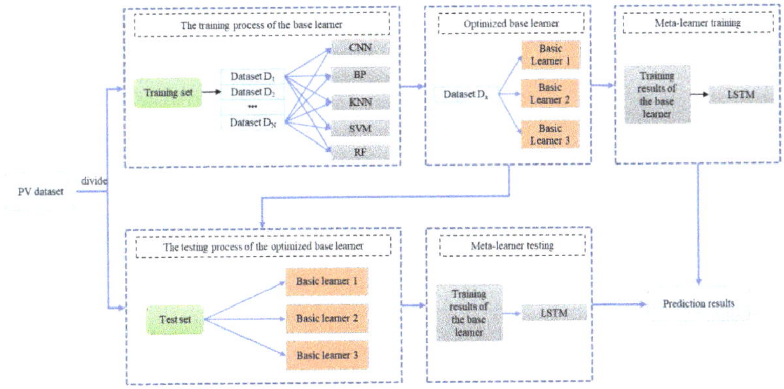

Figure 2. Improved dual dynamic stacked ensemble learning model.

A new dual dynamic stacked ensemble learning method for PV prediction based on the adaptive dual dynamic mode of data and models is proposed in this paper. The multiple datasets are dynamically divided, and the three optimal models are selected from the five pre-set basic models for model training. By continuously optimizing the selection and combination of data and base learners, the adaptability and generalization ability of the model are improved. In this paper, CNN, backpropagation (BP) neural networks, K-nearest neighbor (KNN), SVM, and RF are pre-set as the basic learners, and LSTM is set as the meta-learner. These base learners perform well on small datasets and have obvious advantages in computational efficiency and overfitting control, which can achieve better overall prediction results. LSTM is suitable for processing sequential data, capable of capturing the long-term dependencies in the prediction results of base learners, and can analyze the data distribution of each base learner with its unique memory mechanism.

2.3. Improved KDE

Nonparametric estimation is different from traditional parametric estimation in that it does not assume the distribution form of the data. Nonparametric estimation is used when the distribution of the data is not explicitly assumed, or the distribution of the data may be unknown or difficult to determine. KDE in nonparametric estimation is used to estimate the probability density function of a random variable, which approximates the true probability distribution by distributing data points around each point of the sample data to build a smooth curve. The principle is to specify a neighborhood range with a certain bandwidth for the target point, count the number of elements in the neighborhood, and calculate the average density. The KDE equation is as follows:

$$f(x) = \frac{1}{Nd}\sum_{i=1}^{N} K\left(\frac{x-\beta_i}{d}\right) \tag{16}$$

where N is the total number of sample points; x is the target sample point; β_i is the sample value of point i; d is the window width; and $K(\cdot)$ is a kernel function; common kernel functions include Gaussian kernel, uniform kernel, Epanechnikov kernel, etc. In this paper, the Gaussian kernel function is selected, and the expression of $f(x)$ becomes

$$f(x) = \frac{1}{\sqrt{2\pi}}\frac{1}{Nd}\sum_{i=1}^{N} \exp\left[-\frac{1}{2}\left(\frac{x-\beta_i}{d}\right)^2\right] \tag{17}$$

The choice of bandwidth in KDE is important because it directly affects the quality of the final estimated probability density function. The bandwidth determines the width of the kernel function, which in turn affects the smoothness of the estimation. Excessive bandwidth leads to excessive smoothing, ignoring detailed features in the data; on the other hand, too small a bandwidth easily leads to introducing noise, making the estimation unstable and not smooth.

For the characteristics of single-peak power and multi-peak power, the optimal bandwidth is determined by the standard deviation and standard quartile distance of the power prediction sample. The equation for calculating bandwidth d by the thumb method of the KDE model based on the Gaussian kernel function is

$$d = 1.06\min\left\{S, \frac{Q}{1.34}\right\}N^{-\frac{1}{5}} \tag{18}$$

where N is the number of sample points; S is the sample standard deviation; and Q is the standard quartile of the data.

The improved KDE method not only performs probabilistic prediction analysis on the deterministic prediction results of multi-level decomposition of PV power, but also corrects the error of deterministic prediction results of high-frequency time series after decomposition and reconstruction; that is, the point prediction results are corrected according to the PIs of probabilistic prediction. The correction equation is shown as follows:

$$corr = y_i + (\hat{y}_{i0} - y_{i0}) \cdot (q_{\overline{\alpha 0}} - q_{\underline{\alpha 0}}) \tag{19}$$

where y_i is the current predicted value; $\hat{y}_{i0}$ and y_{i0} are, respectively, the average true values and average predicted values at the same time over a known 5 days under similar weather conditions; and $q_{\overline{\alpha 0}}$ and $q_{\underline{\alpha 0}}$ are, respectively, the upper and lower boundaries of the mean division of the PIs at the same time over a known 5 days under similar weather conditions.

2.4. The Proposed Forecasting Model

The flow chart of the proposed model is shown in Figure 3. First, the correlation analysis of PV data and meteorological data is carried out to remove unnecessary meteorological characteristics. Then, the PV data for high-level and low-level stations are divided into clusters, and the output prediction of low-level stations is optimized according to the output trend of high-level stations. The output trend of high-level stations is the predicted value of the actual PV power, which is predicted by LSTM. The correlation between the high and low levels of output within the PV area is analyzed, and the corresponding analysis results are introduced into the low-level output prediction model. Then, the PV data are decomposed by CEEMDAN, and the high-frequency IMF, medium-frequency IMF, and low-frequency IMF are divided according to FE. Because the fluctuation of PV power is easily affected by weather, light intensity, and other factors, high-frequency IMF contains instantaneous fluctuations and instability, which can easily interfere with the prediction model. Therefore, further analysis of the high-frequency components is needed to extract useful information. The high-frequency time series is decomposed again. The high-frequency IMF is reconstructed and analyzed based on FE and MI entropy. The components with low complexity and high correlation to the original high-frequency time series are extracted as the basic analytical quantities of the high-frequency time series; that is, the decomposition quantities with low complexity and certain patterns are further mined from the high-frequency IMF for point prediction analysis. Medium-frequency IMF and low-frequency IMF indicate the long-term trend of PV power, including relatively stable and regular changes, which can be directly input into the prediction model to improve the prediction efficiency and accuracy of the model. Then, the decomposed multi-level data are input into the improved Stacking for training to obtain the deterministic prediction values, and the improved KDE method is used to optimize the prediction results of the filtered high-frequency time series (high-frequency reanalysis components). Finally, the uncertainty of PV prediction is quantified by using adaptive KDE to make probabilistic prediction of the overall forecast output through the deterministic predicted value and prediction error.

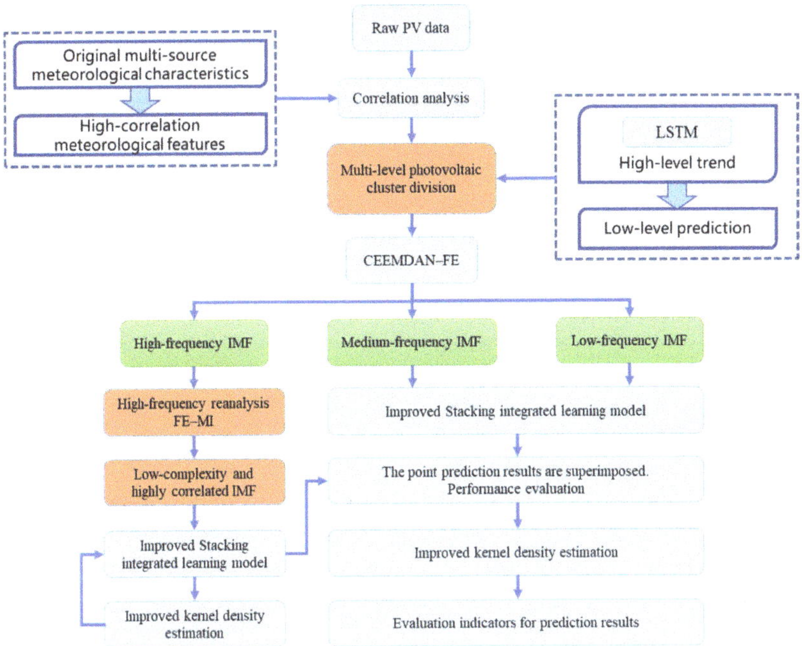

Figure 3. Flowchart of the hybrid forecasting model.

3. Materials and Metrics

3.1. Description of Dataset

Several PV power plants with an installed capacity of 30 MW in China are used for validation in this paper, among which the installed capacity of the high-level PV station is 6 MW, and the installed capacity of the low-level PV station is 300 kW, and the high-level PV power station includes multiple low-level PV power stations. It is a 24 h all-weather dataset with a data resolution of 15 min. The meteorological data consist of 24 weather variables, including air temperature, wind direction, wind speed, air pressure, cloud cover, heat flux, heat radiation, precipitation, humidity, and other meteorological forecast information. The time scale for short-term PV prediction is the next 1 to 3 days. The input of the prediction model is meteorological characteristics, and the output is PV power. In this paper, datasets from January 1 to June 30 are used to forecast the actual PV power for the next 3 days. Since PV power at night is almost 0, the prediction is of little significance, so the periods without power generation in the early morning and at night are excluded during model training. The dataset is divided into a training set and a test set, with a ratio of 8:2. The sample sizes of the training set and the test set are 3371 and 843, respectively.

3.2. Evaluation Metrics

The error index is a unified standard for measuring prediction accuracy. Five different evaluation metrics, namely, root mean square error (RMSE), mean absolute error (MAE), mean absolute percentage error (MAPE), coefficient of determination (R^2), and accuracy rate, are used in this paper to assess the discrepancy between forecasted and actual values and quantify the deterministic performance of the prediction model. The calculation equations for each metric are as follows:

$$RMSE = \sqrt{\frac{1}{N}\sum_{i=1}^{N}\left(\frac{\hat{y}_i - y_i}{c_i}\right)^2} \tag{20}$$

$$MAE = \frac{1}{N}\sum_{i=1}^{N}\left|\frac{\hat{y}_i - y_i}{c_i}\right| \tag{21}$$

$$MAPE = \frac{1}{N}\sum_{i=1}^{N}\left|\frac{\hat{y}_i - y_i}{y_i}\right| \times 100\% \tag{22}$$

$$R^2 = 1 - \sum_{i=1}^{N}(\hat{y}_i - y_i)^2 / \sum_{i=1}^{N}(\overline{y}_i - y_i)^2 \tag{23}$$

$$Accuracy = 1 - RMSE \tag{24}$$

where N is the number of predicted samples; y_i is the predicted value; $\hat{y}_i$ is the true value; $\overline{y}_i$ is the mean of the true values; and c_i is the installed capacity.

For probabilistic prediction, PI coverage probability (PICP), PI normalized average width (PINAW), and interval score (IS) are used as evaluation indicators in this paper. The evaluations include reliability and overall performance, which are given in Equations (25)–(30).

$$PICP = \frac{1}{T_p}\sum_{i=1}^{T_p}\eta_i \tag{25}$$

$$\eta_i = \begin{cases} 1 & t_i \in I_\alpha \\ 0 & t_i \notin I_\alpha \end{cases} \tag{26}$$

$$AW^\alpha = \sum_{i=1}^{T_p} W(x_i) \tag{27}$$

$$W(x_i) = q_{\overline{\alpha}}(x_i) - q_{\underline{\alpha}}(x_i) \tag{28}$$

$$S_\alpha(x_i) = \begin{cases} -2(1-\alpha)W(x_i) - 4(q_{\underline{\alpha}}(x_i) - t_i) & t_i < q_{\underline{\alpha}}(x_i) \\ -2(1-\alpha)W(x_i) & t_i \in I_\alpha(x_i) \\ -2(1-\alpha)W(x_i) - 4(t_i - q_{\overline{\alpha}}(x_i)) & t_i > q_{\overline{\alpha}}(x_i) \end{cases} \tag{29}$$

$$IS_\alpha = \frac{1}{T_p}\sum_{i=1}^{T_p} S_\alpha(x_i) \tag{30}$$

where T_p is the number of predicted samples; η_i is the Boolean quantity; t_i is the predicted value; I_a is the forecast interval; AW^α is the width of the PI; $W(x_i)$ is the width of each PI; $q_{\overline{\alpha}}$ is the upper boundary; $q_{\underline{\alpha}}$ is the lower boundary; $S_\alpha(x_i)$ is the bandwidth deviation of the PI of the i iteration; and α is the quantile.

4. Case Studies

4.1. Multi-Level PV Data Analysis Results

The validity of meteorological characteristics should be taken into account in the prediction of PV power generation capacity. Some meteorological characteristics are dispensable for the prediction effect. Therefore, it is necessary to analyze the meteorological characteristics, remove the features with low correlation, and avoid affecting the prediction performance. In this paper, the Spearman [31] correlation coefficient is used to calculate the correlation between PV power and various meteorological characteristics, and the meteorological characteristics with higher correlations are obtained for subsequent analysis. Figure 4 shows the calculation results of the correlation coefficient.

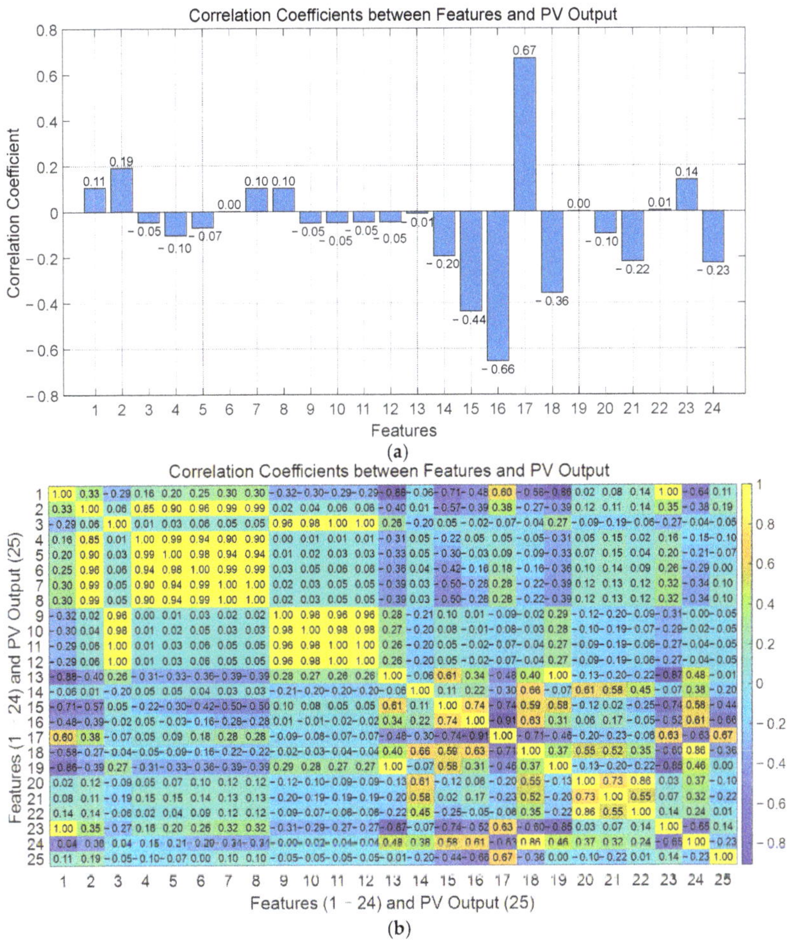

Figure 4. Spearman correlation analysis results: (**a**) coefficient chart; (**b**) heat map.

Figure 4a shows a bar chart of the correlation between 24 meteorological features and PV power. Figure 4b shows the heat map of the correlations between 24 meteorological features and PV power, respectively. According to Spearman's evaluation criterion, meteorological and geographical features with an absolute value of the correlation coefficient of 0.3 or above should be selected. Therefore, in the PV power prediction of this paper, sensible heat flux, latent heat flux, short-wave radiation, and long-wave radiation are taken as meteorological characteristics data.

By exploring the correlation between the high-level output and the low-level output within the region, the high-level output reflects the change trend of the overall PV output in the region so as to improve the judgment of the future change of the low-level output. Combining the historical data of low-level and high-level stations and the predicted output of the high-level stations, the training sample is optimized. In this paper, K-means [32] is used for clustering, the power output of the high-level stations is taken as one of the clustering input variables, and the PV output of the low-level stations is taken as the forecast object. The clustering results are shown in Figure 5.

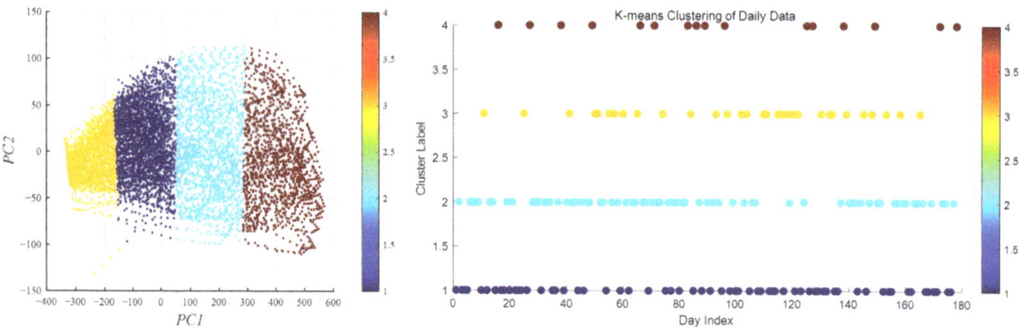

Figure 5. Clustering results of high-level and low-level PV: (**a**) by sampling points; (**b**) by day.

Since the PV data at night are almost 0, it is of little significance to predict it. The PV data from 6:00 to 18:00 are selected for clustering in this paper. The optimal number of clusters is selected by comparing the performance of model prediction under different numbers of clusters. As can be seen from Figure 5, the PV data of high-level and low-level stations are clustered into four categories. Figure 5a shows the clustering situation of each sampling point, and Figure 5b shows the clustering situation of 49 samples per day. By analyzing the correlation between high-level PV power and low-level PV power, the trend of high-level PV power is used to optimize the prediction of low-level PV power. LSTM is used to predict the high-level PV power, and the input of the low-level PV output prediction model, that is, meteorological characteristics, and the prediction results of high-level PV power are taken as analysis samples. The clustering of PV data is carried out to improve the accuracy of the samples used in the prediction model, achieve multi-level correlation analysis, and optimize the judgment of the prediction model on the overall output and the change trend of weather processes in the region.

In this paper, the CEEMDAN algorithm is adopted to decompose the PV data of each cluster after clustering to obtain multiple components, and the high-frequency, medium-frequency, and low-frequency IMF are obtained through EF. For the high-frequency IMF, variational mode decomposition [33] is adopted for further decomposition, and the high-frequency reanalysis components with low complexity and high correlation are selected through EF–MI. Figure 6 shows the decomposition results of CEEMDAN and the high-frequency, medium-frequency, and low-frequency IMF components obtained after the final optimization.

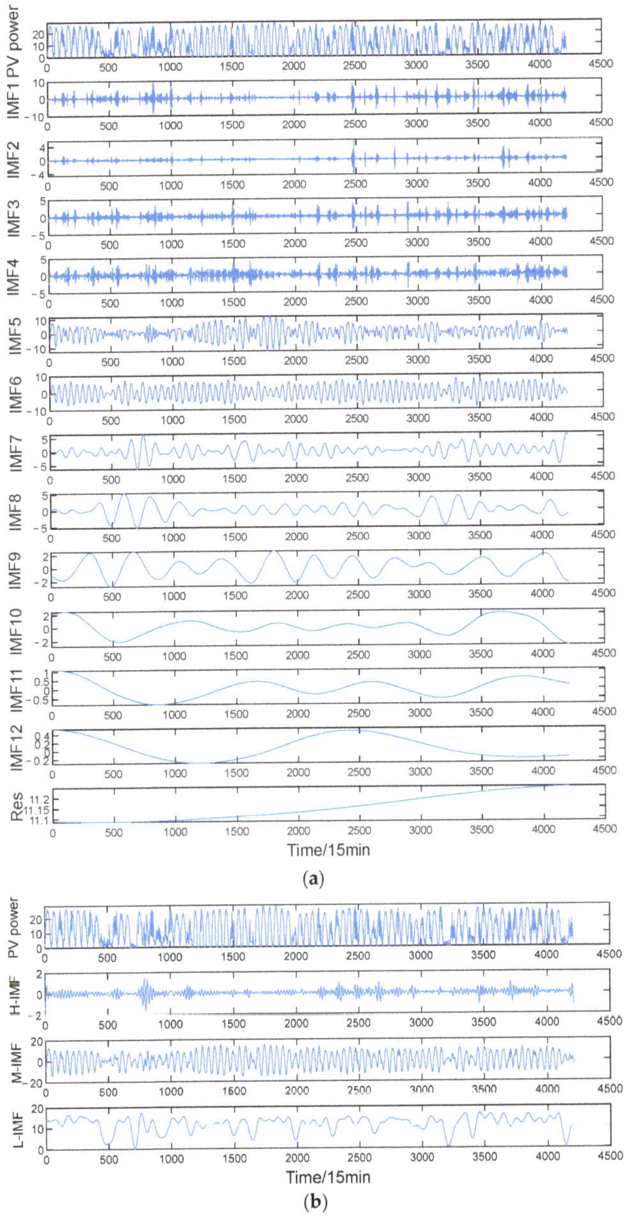

Figure 6. PV data decomposition results: (**a**) CEEMDAN; (**b**) CEEMDAN–FE–MI.

4.2. Analysis of Deterministic Prediction Results

The training set is dynamically divided into three training sets, and the proportion of each training set is half of the original training set, that is, to make each training set contain similar data as much as possible to avoid the contingency of model training. The number of samples in each of the three training sets of cluster I is 3371. The parameter configurations of each model in this paper are shown in Table 2.

Table 2. Configuration of model parameters.

Model	CNN	BP	KNN	SVM	RF	LSTM	ELM
Optimizer	Adam	Levenberg–Marquardt	-	SMO	Bagging	Adam	-
Convolutional kernels	10	-	-	-	-	-	-
Kernel	-	-	-	RBF	-	-	-
Trees	-	-	-	-	50	-	-
Neighbors	-	-	5	-	-	-	-
Hidden units	-	10	-	-	-	10	25
Max iterations	100	100	-	-	-	100	-
Learning rate	0.01	0.01	-	-	-	0.01	-

The pre-set CNN, BP, KNN, SVM, and RF are trained, respectively, on three training sets. Each model is independent and does not interfere with any other. Table 3 presents the performance indicators of the five models on the three training sets.

Table 3. Evaluation index of base learners in each training set.

Training Set	Prediction Model	RMSE/MW	MAE/MW	MAPE/%	Accuracy/%
1	CNN	0.2377	0.1686	1.70	76.23
	BP	0.2324	0.1687	1.67	76.76
	KNN	0.3044	0.2114	2.14	69.56
	SVM	0.1543	0.1082	0.80	84.57
	RF	0.2278	0.1595	1.62	77.22
2	CNN	0.2852	0.2235	1.52	71.48
	BP	0.2491	0.1979	1.27	75.09
	KNN	0.3655	0.2857	2.02	63.45
	SVM	0.1887	0.1342	0.78	81.13
	RF	0.2792	0.2178	1.48	72.08
3	CNN	0.2007	0.1570	0.88	79.93
	BP	0.2000	0.1584	0.79	80.00
	KNN	0.2715	0.1896	1.36	72.85
	SVM	0.1816	0.1248	0.76	81.84
	RF	0.1911	0.1475	0.83	80.89

It can be observed in Table 3 that the prediction performance of the five models is the best in the third training set. Therefore, the third training set is selected as the optimal training set. In the third training set, the prediction effects of CNN and KNN are poor. Therefore, CNN and KNN are abandoned, and BP, SVM, and RF are selected as the chosen base learners.

After determining the optimal training set and the Stacking base learner, the integrated learning model of high-frequency, medium-frequency, and low-frequency IMF is trained, respectively. Finally, the prediction results are superimposed, and the prediction errors are analyzed. For high-frequency IMF, adaptive KDE is used to make probabilistic predictions, and the prediction results are determined by ensemble learning of PI correction for high-frequency IMF. In order to prove the advantages of the Stacking ensemble learning approach for short-term PV power forecasting proposed in this paper, the proposed approach is evaluated against the single machine learning methods LSTM and ELM and compared with the traditional Stacking model, and the parameter settings in the comparison model are consistent with the model established in this paper. The traditional Stacking model that uses the data processing method in this paper is represented by ICEEMDAN–Stacking, and the proposed model is represented by ICEEMDAN–I-Stacking. Figure 7 shows the prediction results of cluster I dataset three days ahead, and the evaluation indicators are shown in Table 4.

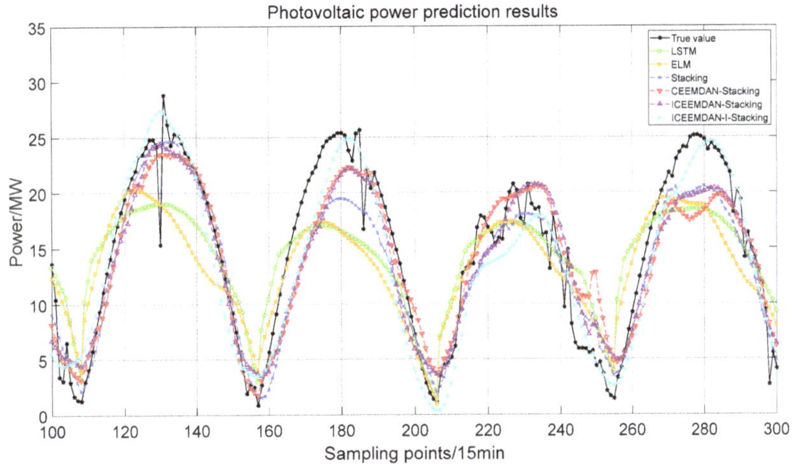

Figure 7. PV power prediction results of cluster I three days ahead.

Table 4. Evaluation indicators for forecast results of cluster I three days ahead.

Prediction Model	RMSE/MW	MAE/MW	MAPE/%	R^2	Accuracy/%	Training Time/s
LSTM	0.2052	0.1711	26.46	0.4152	79.48	35.56
ELM	0.1959	0.1626	26.07	0.4672	80.41	0.12
Stacking	0.1627	0.1216	20.86	0.6324	83.73	8.34
CEEMDAN–Stacking	0.1722	0.1317	22.18	0.5882	82.78	18.48
ICEEMDAN–Stacking	0.1553	0.1237	20.10	0.6651	84.47	19.99
ICEEMDAN–I-Stacking	0.1307	0.0999	16.26	0.7628	86.93	78.73

The predicted results of cluster II dataset three days ahead are shown in Figure 8, and Table 5 shows the evaluation indicators.

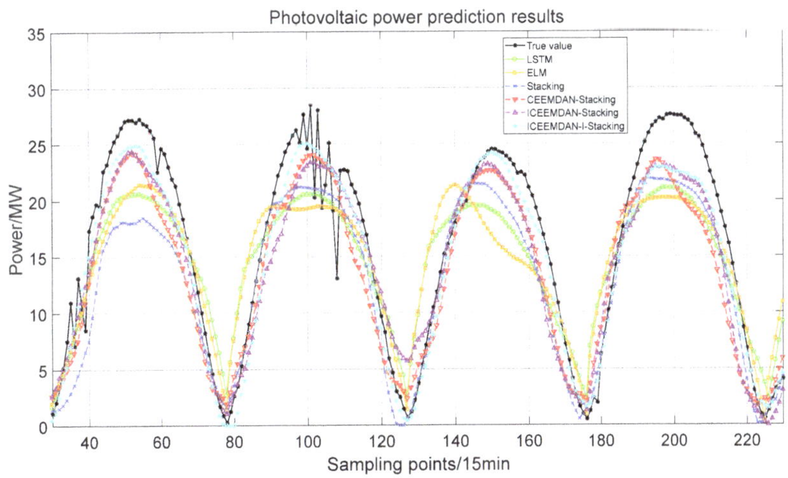

Figure 8. PV power prediction results of cluster II three days ahead.

Table 5. Evaluation indicators for forecast results of cluster II three days ahead.

Prediction Model	RMSE/MW	MAE/MW	MAPE/%	R^2	Accuracy/%	Training Time/s
LSTM	0.1731	0.1436	22.40	0.6116	82.69	39.83
ELM	0.1934	0.1596	25.18	0.5150	80.66	0.12
Stacking	0.1409	0.1041	18.95	0.7426	85.91	9.32
CEEMDAN–Stacking	0.1606	0.1201	21.41	0.6658	83.94	21.76
ICEEMDAN–Stacking	0.1396	0.1078	18.79	0.7473	86.04	25.24
ICEEMDAN–I-Stacking	0.1228	0.0915	15.97	0.8044	87.72	74.75

Figures 7 and 8, respectively, show the prediction result curves of each model in cluster I and cluster II three days ahead. It can be seen from the comparison of the three-day-ahead prediction errors for cluster I in Table 4 and for cluster II in Table 5 that the prediction accuracy of individual LSTM and ELM models is relatively low. The traditional Stacking model utilizes the advantages of multiple base learners to improve the prediction performance. Although PV data are analyzed in the CEEMDAN–Stacking model, the prediction accuracy is reduced due to insufficient mining. The ICEEMDAN–Stacking model further improves the prediction accuracy through sufficient data mining, but the prediction performance of this model is poor compared with the proposed ICEEMDAN–I-Stacking. The proposed I-Stacking can further optimize the integrated learning structure and obtain better prediction accuracy. The proposed model further optimizes the prediction accuracy by correcting the deterministic prediction results through the PIs, and is superior to other comparison models in each evaluation index. Compared to other models, the RMSE of the proposed method in the cluster I dataset decreased by 36.31%, 33.28%, 19.67%, 24.10%, and 15.84%, respectively, and the MAE decreased by 41.61%, 38.56%, 17.85%, 24.15%, and 19.24%, respectively. Compared to other models, the RMSE of the proposed method in the cluster II dataset decreased by 29.06%, 36.50%, 12.85%, 23.54%, and 12.03%, respectively, and the MAE decreased by 36.28%, 42.67%, 12.10%, 23.81%, and 15.12%, respectively. A personal computer configured with an Intel(R) Core(TM) i7-9750H CPU @ 2.60 GHz and 16.0 GB RAM is used for computing. The training time of the proposed model is less than 100 s, and the testing time is less than 1 s. The calculation efficiency can meet the practical application requirements of short-term PV power prediction. Among all the models, the predicted results have the highest fitting degree with the true values, and the accuracy is also the highest, indicating that the proposed method can effectively improve the accuracy of the three-day-ahead PV power prediction model.

To further verify the superiority of the proposed deterministic prediction model method, an example of day-ahead PV prediction is provided for illustration. Table 6 shows the performance indicators of each prediction model one day ahead.

Table 6. Evaluation indicators of the day-ahead PV power prediction result.

Prediction Model	RMSE/MW	MAE/MW	MAPE/%	R^2	Accuracy/%	Training Time/s
LSTM	0.2164	0.1838	31.15	0.4474	78.36	52.77
ELM	0.2246	0.1850	31.31	0.4049	77.54	0.19
Stacking	0.1601	0.1168	17.22	0.6976	83.99	15.66
CEEMDAN–Stacking	0.1548	0.1123	16.80	0.7173	84.52	19.71
ICEEMDAN–Stacking	0.1551	0.1172	17.69	0.7163	84.49	17.55
ICEEMDAN–I-Stacking	0.1276	0.0888	12.50	0.8081	87.24	66.59

It can be seen from the comparison of prediction errors in Table 6 that in the PV power prediction one day ahead, the prediction accuracy of both the LSTM and ELM models is relatively low. The traditional Stacking model and CEEMDAN–Stacking model can improve the prediction performance. The ICEEMDAN–Stacking model improves the prediction accuracy through sufficient data mining, but it is not as good as the proposed ICEEMDAN–I-Stacking model. The proposed method can further optimize the prediction accuracy and outperforms other comparison models in all evaluation indicators. Among all the models, the proposed method has the highest degree of fitting between the predicted results and the true values, as well as the highest accuracy. The training time of the proposed model is less than 100 s, and the testing time is less than 1 s. The calculation efficiency can meet the actual requirements. Therefore, the proposed method can effectively improve the accuracy of the day-ahead PV power generation prediction model.

4.3. Analysis of Probabilistic Prediction Results

The proposed deterministic prediction method improves the accuracy of short-term PV prediction, but it does not reflect the reliability and uncertainty of prediction. Probabilistic prediction can provide distribution information of predicted results, which helps decision makers to make more intelligent decisions and reduce risks. The probabilistic prediction results of the proposed improved KDE for the three-day-ahead PV power are shown in Figure 9 below.

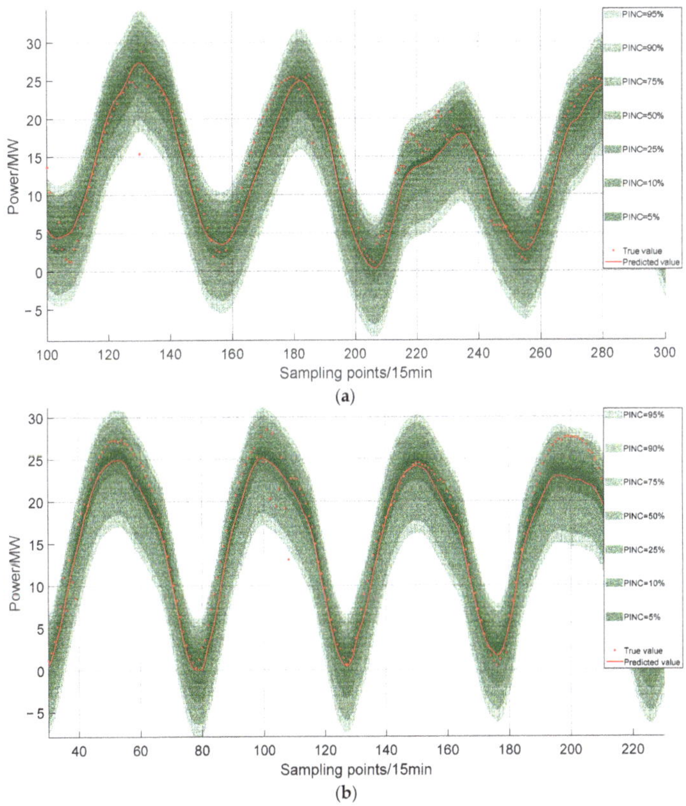

Figure 9. Probabilistic prediction results of PV power three days ahead: (**a**) cluster I; (**b**) cluster II.

Figure 9 shows that for cluster I and cluster II datasets, the three-day-ahead PV prediction has high reliability at both 90% PI normalized confidence (PINC) and 95% PINC, and the average width of PIs is narrow and the clarity is high. Therefore, the PI of the proposed approach has high reliability and good sharpness.

In order to verify the effectiveness of the probabilistic prediction method for improved KDE proposed in this paper, the proposed method is compared with traditional Gaussian, QR, Bootstrap, and KDE, and data with a PINC of 90% are selected for comparison. The proposed adaptive dynamic KDE method is represented by AKDE. The comparison results of error indicators three days ahead are shown in Table 7.

Table 7. Evaluation indicators for probabilistic prediction results three days ahead.

Dataset	Prediction Model	PICP/%	PINAW	IS
Cluster I	Gaussian	90.5101	11.6663	−3.3836
	QR	91.2218	12.1024	−3.3750
	Bootstrap	90.2728	13.1122	−3.8348
	KDE	90.9846	13.1025	−3.7638
	AKDE	91.9502	11.4966	−2.9702
Cluster II	Gaussian	90.6696	10.5198	−3.0555
	QR	91.6575	11.1686	−3.0618
	Bootstrap	90.5598	11.1644	−3.3623
	KDE	90.3403	11.2027	−3.4043
	AKDE	91.7311	10.1919	−2.8265

For probabilistic prediction, the higher the PICP and IS, and the smaller the PINAW, the better the prediction indicators. Therefore, as shown in the three-day-ahead probabilistic prediction results of Table 7, compared with Gaussian, QR, Bootstrap, and KDE methods, the overall performance of the proposed method is the best: PICP of cluster I is increased by 1.44%, 0.73%, 1.68%, and 0.97%, respectively, while PINAW is decreased by 1.45%, 5.01%, 12.32%, and 12.26%, respectively. PICP of cluster II is increased by 1.06%, 0.07%, 1.17%, and 1.39%, respectively, while PINAW is decreased by 3.12%, 8.75%, 8.71%, and 9.02%, respectively. The IS of the proposed AKDE method is also the highest in the two clusters. Therefore, the results show that under different datasets, compared with traditional probabilistic prediction methods, the proposed method overall has better reliability and prediction performance in three-day-ahead PV probabilistic prediction.

To further verify the superiority of the proposed probabilistic prediction model method, an example of day-ahead PV prediction is provided for illustration, and the data with a PINC of 90% are selected for comparison. Table 8 shows the evaluation indicators of each probabilistic prediction model one day ahead.

Table 8. Evaluation indicators for probabilistic prediction results one day ahead.

Prediction Model	PICP/%	PINAW	IS
Gaussian	90.6036	11.6845	−3.6911
QR	90.8994	11.8994	−3.6482
Bootstrap	89.7929	11.6996	−3.8475
KDE	90.5325	11.6355	−3.7656
AKDE	91.1467	10.9941	−3.4379

As shown in the day-ahead probabilistic prediction results of Table 8, compared with Gaussian, QR, Bootstrap, and KDE methods, the overall performance of the proposed method is the best. The PICP and IS of the proposed AKDE method are the highest, and the PINAW is the lowest. Therefore, the proposed method has better reliability and prediction performance in day-ahead PV probabilistic prediction.

5. Conclusions

In this paper, an improved PV short-term power prediction method based on multi-level adaptive dynamic integration is proposed to solve the problems of deterministic and probabilistic predictions of PV power generation time series. The theory that the trend of high-level output optimizes the prediction of low-level output has been applied, and the main environmental factors that constrain PV power generation have been fully considered in this method. The main conclusions are as follows:

1. Based on the relationship between multi-level PV data, considering that the power mutability of low-level output is larger than that of high-level output, the high-level output reflects the change trend of power generation and improves the prediction performance of low-level output;
2. Through the proposed FE–MI method, multi-layer reconstruction analysis of PV data is carried out to extract the potential information for PV power generation, improve the reliability of data input, and improve the accuracy of PV prediction;
3. A new dynamic adaptive Stacking integration method is proposed. Adaptive PV prediction is carried out through the dual dynamic mode of data and models to enhance the adaptability of the model and improve its generalization ability;
4. A new adaptive window width optimization strategy is proposed to establish PIs with high overall performance. The proposed method enhances the quality of prediction error analysis by comprehensively considering the reliability and overall performance of PIs.

Author Contributions: Conceptualization, K.K., J.Z. and Y.Z.; methodology, K.K., J.Z. and Y.Z.; software, K.K., J.Z. and Q.C.; validation, K.K., J.Z., Q.C. and Y.Z.; formal analysis, K.K., Q.C., Y.Y. and Y.Z.; investigation, K.K., Q.C. and Y.Z.; resources, J.Z. and Y.Z.; data curation, K.K., Q.C., Y.Z., Y.Y. and L.D.; writing—original draft preparation, K.K., J.Z., Y.Y. and G.W.; writing—review and editing, K.K., Q.C., Y.Z. and L.D.; visualization, K.K.; supervision, L.D. and G.W.; project administration, Y.Z., Y.Y. and J.Z.; funding acquisition, Y.Z., Y.Y. and J.Z. All authors have read and agreed to the published version of the manuscript.

Funding: This work is supported by the Lianyungang City Key Research and Development Program (Industrial Forward-Looking and Critical Core Technologies): Research on Ultra-Short-Term Probabilistic Forecasting of Photovoltaic Power Incorporating Micro-Meteorological Spatiotemporal Correlation (Project No. CG2315), and the Ningxia Natural Science Foundation Project under Grant 2023AAC03836.

Data Availability Statement: The data that support the findings of this study are available on request from the corresponding author. The data are not publicly available due to privacy or ethical restrictions.

Conflicts of Interest: Author Jingshan Zhang was employed by the company Solareast Holding Co., Ltd., and author Qifan Chen was employed by the company State Grid Information & Telecommunication Co., Ltd. The remaining authors declare that the research was conducted in the absence of any commercial or financial relationships that could be construed as a potential conflict of interest.

Abbreviations

The following abbreviations are used in this manuscript:

Bagging	Bootstrap aggregating
BP	Backpropagation
CEEMDAN	Complete ensemble empirical mode decomposition with adaptive noise
CNNs	Convolutional neural networks
ELM	Extreme learning machine
EMD	Empirical mode decomposition
FE	Fuzzy entropy

IMFs	Intrinsic mode functions
IS	Interval score
I-Stacking	Improved stacked generalization
KDE	Kernel density estimation
KNN	K-nearest neighbor
LSTM	Long short-term memory
MAE	Mean absolute error
MAPE	Mean absolute percentage error
MI	Mutual information
PICP	Coverage probability of the PI
PINAW	Normalized average width of the PI
PINC	Normalized confidence of the PI
PI	Prediction interval
PV	Photovoltaic
QR	Quantile regression
RMSE	Root mean square error
RF	Random forest
SVM	Support vector machine

References

1. Aghajari, H.A.; Niknam, T.; Shasadeghi, M.; Sharifhosseini, S.M.; Taabodi, M.H.; Sheybani, E.; Javidi, G.; Pourbehzadi, M. Analyzing complexities of integrating Renewable Energy Sources into Smart Grid: A comprehensive review. *Appl. Energy* **2025**, *383*, 125317. [CrossRef]
2. Wu, D.; Fu, Y.; Li, M.; Ouyang, L. A Combined Compensation Strategy for Steady-State Performance Improvement of Proportional-Proportional-Delay Controller. *IET Power Electron.* **2025**, *18*, e70022. [CrossRef]
3. Husein, S.M.; Gago, E.J.; Hasan, B.; Pegalajar, M.C. Towards energy efficiency: A comprehensive review of deep learning-based photovoltaic power forecasting strategies. *Heliyon* **2024**, *10*, e33419. [CrossRef] [PubMed]
4. Wu, D.; Zhi, Q.; Li, M.; Zhao, Y.; Tian, H. Control strategy for L-type grid-connected inverters under ultra-weak grid conditions. *J. Power Electron.* **2025**, 1–12. [CrossRef]
5. Zhou, Y.; Wei, F.; Kuang, K.; Mahfoud, R.J. Research on a deep ensemble learning model for the ultra-short-term probabilistic prediction of wind power. *Electronics* **2024**, *13*, 475. [CrossRef]
6. Yang, M.; Guo, Y.; Fan, F. Ultra-short-term prediction of wind farm cluster power based on embedded graph structure learning with spatiotemporal information gain. *IEEE Trans. Sustain. Energy* **2025**, *16*, 308–322. [CrossRef]
7. Yang, M.; Han, C.; Zhang, W.; Wang, B. A short-term power prediction method for wind farm cluster based on the fusion of multi-source spatiotemporal feature information. *Energy* **2024**, *294*, 130770. [CrossRef]
8. Li, M.; Tian, Y.; Zhang, H.; Zhang, N. The source-load-storage coordination and optimal dispatch from the high proportion of distributed photovoltaic connected to power grids. *J. Eng. Res.* **2024**, *12*, 421–432. [CrossRef]
9. Yang, M.; Shen, X.; Huang, D.; Su, X. Fluctuation classification and feature factor extraction to forecast very short-term photovoltaic output powers. *CSEE J. Power Energy Syst.* **2025**, *11*, 661–670.
10. Fu, X. Statistical machine learning model for capacitor planning considering uncertainties in photovoltaic power. *Prot. Control Mod. Power Syst.* **2022**, *7*, 5. [CrossRef]
11. Yang, M.; Jiang, Y.; Xu, C.; Wang, B.; Wang, Z.; Su, X. Day-ahead wind farm cluster power prediction based on trend categorization and spatial information integration model. *Appl. Energy* **2025**, *388*, 125580. [CrossRef]
12. Dudáš, A.; Udristioiu, M.T.; Alkharusi, T.; Yildizhan, H.; Sampath, S.K. Examining effects of air pollution on photovoltaic systems via interpretable random forest model. *Renew. Energy* **2024**, *232*, 121066. [CrossRef]
13. Liu, H.; Cai, C.; Li, P.; Tang, C.; Zhao, M.; Zheng, X.; Li, Y.; Zhao, Y.; Liu, C.; Liu, C. Hybrid prediction method for solar photovoltaic power generation using normal cloud parrot optimization algorithm integrated with extreme learning machine. *Sci. Rep.* **2025**, *15*, 6491. [CrossRef] [PubMed]
14. Zhu, R.; Li, T.; Tang, B. Research on short-term photovoltaic power generation forecasting model based on multi-strategy improved squirrel search algorithm and support vector machine. *Sci. Rep.* **2024**, *14*, 14348. [CrossRef] [PubMed]
15. He, Q.; Zhao, M.; Li, S.; Li, X.; Wang, Z. Machine Learning Prediction of Photovoltaic Hydrogen Production Capacity Using Long Short-Term Memory Model. *Energies* **2025**, *18*, 543. [CrossRef]
16. Srikanth, D.; Sukumar, G.D.; Sobhan, P.V. A convolutional neural network based energy management system for photovoltaic/battery systems in microgrid using enhanced coati optimization approach. *J. Energy Storage* **2025**, *119*, 116252. [CrossRef]

17. Babalhavaeji, A.; Radmanesh, M.; Jalili, M.; González, S.A. Photovoltaic generation forecasting using convolutional and recurrent neural networks. *Energy Rep.* **2023**, *9*, 119–123. [CrossRef]
18. Sabri, M.; El Hassouni, M. Photovoltaic power forecasting with a long short-term memory autoencoder networks. *Soft Comput.* **2023**, *27*, 10533–10553. [CrossRef]
19. Sarmas, E.; Spiliotis, E.; Stamatopoulos, E.; Marinakis, V.; Doukas, H. Short-term photovoltaic power forecasting using meta-learning and numerical weather prediction independent Long Short-Term Memory models. *Renew. Energy* **2023**, *216*, 118997. [CrossRef]
20. Khan, W.; Walker, S.; Zeiler, W. Improved solar photovoltaic energy generation forecast using deep learning-based ensemble stacking approach. *Energy* **2022**, *240*, 122812. [CrossRef]
21. Ait Abdelmoula, I.; Elhamaoui, S.; Elalani, O.; Ghennioui, A.; El Aroussi, M. A photovoltaic power prediction approach enhanced by feature engineering and stacked machine learning model. *Energy Rep.* **2022**, *8*, 1288–1300. [CrossRef]
22. Yang, M.; Peng, T.; Zhang, W.; Su, X.; Han, C.; Fan, F. Abnormal data identification and reconstruction based on wind speed characteristics. *CSEE J. Power Energy Syst.* **2025**, *11*, 612–622.
23. Zhi, Y.; Sun, T.; Yang, X. A physical model with meteorological forecasting for hourly rooftop photovoltaic power prediction. *J. Build. Eng.* **2023**, *75*, 106997. [CrossRef]
24. Zhou, Y.; Sun, Y.; Wang, S.; Mahfoud, R.J.; Hou, D.; Wang, J. Very short-term probabilistic prediction for regional wind power generation based on OPNPIs. *CSEE J. Power Energy Syst.* **2024**, 1–10. [CrossRef]
25. Zhou, Y.; Sun, Y.; Wang, S.; Bai, L.; Hou, D.; Mahfoud, R.J.; Wang, P. A very short-term probabilistic prediction method of wind speed based on ALASSO-nonlinear quantile regression and integrated criterion. *CSEE J. Power Energy Syst.* **2023**, *9*, 2121–2129.
26. Zhou, Y.; Sun, Y.; Wang, S.; Mahfoud, R.J.; Alhelou, H.H.; Hatziargyriou, N.; Siano, P. Performance improvement of very short-term prediction intervals for regional wind power based on composite conditional nonlinear quantile regression. *J. Mod. Power Syst. Clean Energy* **2021**, *10*, 60–70. [CrossRef]
27. Sun, Y.; Zhou, Y.; Wang, S.; Mahfoud, R.J.; Alhelou, H.H.; Sideratos, G.; Hatziargyriou, N.; Siano, P. Nonparametric probabilistic prediction of regional PV outputs based on granule-based clustering and direct optimization programming. *J. Mod. Power Syst. Clean Energy* **2023**, *11*, 1450–1461. [CrossRef]
28. Jadli, U.; Thakur, P.; Shukla, R.D. A new parameter estimation method of solar photovoltaic. *IEEE J. Photovolt.* **2017**, *8*, 239–247. [CrossRef]
29. Ren, X.; Liu, Y.; Zhang, F.; Li, L. A Deep Learning Quantile Regression Photovoltaic Power-Forecasting Method under a Priori Knowledge Injection. *Energies* **2024**, *17*, 4026. [CrossRef]
30. Li, G.; Wei, X.; Yang, H. A method for accurate prediction of photovoltaic power based on multi-objective optimization and data integration strategy. *Appl. Math. Model.* **2024**, *136*, 115643. [CrossRef]
31. Amiri, A.F.; Chouder, A.; Oudira, H.; Silvestre, S.; Kichou, S. Improving photovoltaic power prediction: Insights through computational modeling and feature selection. *Energies* **2024**, *17*, 3078. [CrossRef]
32. Pei, T.; Yu, C.; Zhong, Y.; Lian, L.; Xiang, X. A self-error corrector integrating K-means clustering with Markov model for marine craft maneuvering prediction with experimental verification. *Ocean Eng.* **2023**, *285*, 115420. [CrossRef]
33. Liu, Y.; Liu, Y.; Cai, H.; Zhang, J. An innovative short-term multihorizon photovoltaic power output forecasting method based on variational mode decomposition and a capsule convolutional neural network. *Appl. Energy* **2023**, *343*, 121139. [CrossRef]

MDPI AG
Grosspeteranlage 5
4052 Basel
Switzerland
Tel.: +41 61 683 77 34

Electronics Editorial Office
E-mail: electronics@mdpi.com
www.mdpi.com/journal/electronics

www.ingramcontent.com/pod-product-compliance
Lightning Source LLC
Chambersburg PA
CBHW061501210926
51850CB00008B/1789